Elasticsearch 8.x Cookbook

Fifth Edition

Over 180 recipes to perform fast, scalable, and reliable searches for your enterprise

Alberto Paro

BIRMINGHAM—MUMBAI

Elasticsearch 8.x Cookbook
Fifth Edition

Publishing Product Manager: Devika Battike
Senior Editor: Nathanya Dias
Content Development Editor: Sean Lobo
Technical Editor: Rahul Limbachiya
Copy Editor: Safis Editing
Project Coordinator: Aparna Ravikumar Nair
Proofreader: Safis Editing
Indexer: Manju Arasan
Production Designer: Ponraj Dhandapani
Marketing Coordinator: Priyanka Mhatre

First published: December 2013
Second edition: January 2015
Third edition: February 2017
Fourth edition: April 2019
Fifth edition: May 2022

Production reference: 1280422

Published by Packt Publishing Ltd.
Livery Place
35 Livery Street
Birmingham
B3 2PB, UK.

ISBN 978-1-80107-981-5

www.packt.com

Contributors

About the author

Alberto Paro is an engineer, manager, and software developer. He currently works as the technology architecture delivery associate director of the Accenture Cloud First data and AI team in Italy. He loves to study emerging solutions and applications, mainly related to cloud and big data processing, NoSQL, **Natural Language Processing** (**NLP**), software development, and machine learning. In 2000, he graduated in computer science engineering from Politecnico di Milano. Then, he worked with many companies, mainly using Scala/Java and Python on knowledge management solutions and advanced data mining products, using state-of-the-art big data software. A lot of his time is spent teaching others how to effectively use big data solutions, NoSQL data stores, and related technologies.

About the reviewers

Kyle Davis is the senior developer advocate with OpenSearch and Open Distro for Elasticsearch at **Amazon Web Services** (**AWS**). Kyle has a long history of working in software development, starting in the late 1990s. His experience runs the gamut from frontend development to microcontrollers, but his most passionate area of interest is NoSQL databases. He has blogged and presented extensively about technology and is the author of *Redis Microservices for Dummies*. Kyle is based out of Edmonton, Alberta, Canada.

Mahipalsinh Rana is currently **chief technology officer** (**CTO**) of Inexture Solutions LLP. At Inexture, he specializes in enterprise searching, Python, Java, and ML/AI. He has 15 years of experience. His stint with search technologies started in 2010 when he started working with Solr. He then started working with Elastic and has done various large-scale implementations and consultations. At the start of his career, he worked for Sun Microsystems, where he worked on **internationalization** (**i18n**). He likes exploring emerging technology trends such as NLP and intuitive searching for e-commerce. He plans to develop a search engine for people who are still in the early stages of technological advancement to provide them with information at ease. He has also worked on *Liferay Beginner's Guide* by Packt.

Arpit Dubey is a big data engineer with over 14 years of experience in building large-scale, data-intensive applications. He has experience in envisioning enterprise-wide data strategies, roadmaps, and architecture for large internet companies, with varied use cases. He specializes in building event-driven architectures and real-time analytical solutions, using distributed systems such as Kafka, Flink, Spark, the Hadoop stack, NoSQL databases, and graph databases. He has been an active public speaker on various technology topics and has spoken at Kafka Summit, Druid Summit, and several other technology meetups.

I would like to thank my entire family for always being my guiding light for every path I choose and every step I take.

Table of Contents

2
Managing Mappings

3

Basic Operations

5

Text and Numeric Queries

6

Relationships and Geo Queries

7

Aggregations

8
Scripting in Elasticsearch

9

Managing Clusters

10

Backups and Restoring Data

11

User Interfaces

12
Using the Ingest Module

13
Java Integration

14
Scala Integration

15
Python Integration

16

Plugin Development

17

Big Data Integration

18

X-Pack

Index

Other Books You May Enjoy

Preface

Welcome to the fifth edition of *Elasticsearch Cookbook* targeting Elasticsearch 8.x. It's a long journey (about 12 years) that I have been on with both Elasticsearch and readers of my books. Every version of Elasticsearch brings breaking changes and new functionalities, and the evolution of already present components is a continuous cycle of product and marketing evolution.

Elasticsearch, once a very niche product, is now one of the most used databases in the world (ranked seventh in April 2022 – source: `https://db-engines.com/en/ranking`), and both the on-premises (bare metal, Docker, or K8S) and multi-cloud markets provided by Elastic on Amazon, Azure, and Google will rank it as one of the next best solutions for cloud searching and storage.

The growth of Elasticsearch is mainly due to it being one of the best solutions for searching, storage, and providing analytics on unstructured content in petabyte-sized datasets, and these are the main pillars of modern data-centered companies.

In this book, you'll be guided through comprehensive recipes on Elasticsearch 8.x and see how you can create and run complex queries and analytics.

Packed with recipes on performing index mapping, aggregation, and scripting using Elasticsearch, this fifth edition of *Elasticsearch Cookbook* will get you acquainted with numerous solutions and quick techniques to perform both everyday and uncommon tasks, such as how to deploy Elasticsearch nodes, integrate other tools into Elasticsearch, and create different visualizations with Kibana. Finally, you will integrate your Java, Scala, Python, and big data applications, such as Apache Spark and Pig, and create efficient data applications powered by enhanced functionalities and custom plugins.

By the end of this book, you will have gained in-depth knowledge of implementing Elasticsearch architecture, and you'll be able to manage, search, and store data efficiently and effectively using Elasticsearch.

IMHO, this book is the last of a long series and, due to continuous refinements, technical/ stylistic improvements, and the suggestions of about 10 years of readers, it's probably one of the most complete and effective books on Elasticsearch.

Dear reader, thus, it is a technical book. I hope you'll enjoy it from the bottom of your heart!

Sincerely,

Alberto

Who this book is for

If you're a software engineer, big data infrastructure engineer, or Elasticsearch developer, you'll find this book useful. This Elasticsearch book will also help data professionals working in the e-commerce and FMCG industries who use Elasticsearch for metrics evaluation and search analytics to get deeper insights for better business decisions.

Prior experience with Elasticsearch will help you get the most out of this book in the latter chapters, which cover more advanced topics.

What this book covers

Chapter 1, *Getting Started*, covers the basic steps to start using Elasticsearch, from the simple installation to the cloud. We also cover several setup cases.

Chapter 2, *Managing Mappings*, covers the correct definition of the data fields to improve both indexing and searching quality.

Chapter 3, *Basic Operations*, introduces the most common actions that are required to ingest data in Elasticsearch and manage it.

Chapter 4, *Exploring Search Capabilities*, talks about executing searches, sorting, and related API calls. The APIs discussed in this chapter are the essential ones.

Chapter 5, *Text and Numeric Queries*, talks about the search DSL part of text and numeric fields – the core of the search functionalities of Elasticsearch.

Chapter 6, *Relationships and Geo Queries*, talks about queries that work on related documents (child/parent and nested) and geo-located fields.

Chapter 7, *Aggregations*, covers another capability of Elasticsearch, the possibility to execute analytics on search results to improve both the user experience and to drill down on the information contained in Elasticsearch.

Chapter 8, *Scripting in Elasticsearch*, shows how to customize Elasticsearch with scripting and how to use the scripting capabilities in different parts of Elasticsearch (search, aggregation, and ingestion) using different languages. The chapter is mainly focused on Painless, the new scripting language developed by the Elastic team.

Chapter 9, *Managing Clusters*, shows how to analyze the behavior of a cluster/node to understand common pitfalls.

Chapter 10, *Backups and Restoring Data*, covers one of the most important components in managing data: backing up. It shows how to manage a distributed backup and the restoration of snapshots.

Chapter 11, *User Interfaces*, describes two of the most common user interfaces for Elasticsearch: Cerebro, mainly used for admin activities, and Kibana, with X-Pack as a common UI extension for Elasticsearch.

Chapter 12, *Using the Ingest Module*, talks about the ingest functionality for importing data into Elasticsearch via an ingestion pipeline.

Chapter 13, *Java Integration*, describes how to integrate Elasticsearch in a Java application using both REST and native protocols.

Chapter 14, *Scala Integration*, describes how to integrate Elasticsearch in Scala using elastic4s – an advanced type-safe and feature-rich Scala library based on the native Java API.

Chapter 15, *Python Integration*, covers the usage of the official Elasticsearch Python client.

Chapter 16, *Plugin Development*, describes how to create native plugins to extend Elasticsearch functionalities. Some examples show the plugin skeletons, the setup process, and the building of them.

Chapter 17, *Big Data Integration*, covers how to integrate Elasticsearch in common big data tools, such as Apache Spark and Apache Pig.

Chapter 18, *X-Pack*, covers the extra functionalities provided by XPack, including security, machine learning, SQL, and reporting.

To get the most out of this book

Basic knowledge of Java, Scala, and Python would be beneficial.

If you are using the digital version of this book, we advise you to type the code yourself or access the code via the GitHub repository (link available in the next section). Doing so will help you avoid any potential errors related to the copying and pasting of code.

Download the example code files

You can download the example code files for this book from GitHub at `https://github.com/PacktPublishing/Elasticsearch-8.x-Cookbook`. In case there's an update to the code, it will be updated on the existing GitHub repository.

We also have other code bundles from our rich catalog of books and videos available at `https://github.com/PacktPublishing/`. Check them out!

Download the color images

We also provide a PDF file that has color images of the screenshots/diagrams used in this book. You can download it here: `https://static.packt-cdn.com/downloads/9781801079815_ColorImages.pdf`.

Conventions used

There are a number of text conventions used throughout this book.

`Code in text`: Indicates code words in the text, database table names, folder names, filenames, file extensions, pathnames, dummy URLs, user input, and Twitter handles. Here is an example: "Mount the downloaded `WebStorm-10*.dmg` disk image file as another disk in your system."

A block of code is set as follows:

```
html, body, #map {
  height: 100%;
  margin: 0;
  padding: 0
}
```

When we wish to draw your attention to a particular part of a code block, the relevant lines or items are set in bold:

```
[default]
exten => s,1,Dial(Zap/1|30)
exten => s,2,Voicemail(u100)
exten => s,102,Voicemail(b100)
exten => i,1,Voicemail(s0)
```

Any command-line input or output is written as follows:

```
$ mkdir css
$ cd css
```

Bold: Indicates a new term, an important word, or words that you see on screen. For example, words in menus or dialog boxes appear in the text like this. Here is an example: "Select **System info** from the **Administration** panel."

> **Tips or Important Notes**
> Appear like this.

Sections

In this book, you will find several headings that appear frequently (*Getting ready*, *How to do it...*, *How it works...*, *There's more...*, and *See also*).

To give clear instructions on how to complete a recipe, use these sections as follows:

Getting ready

This section tells you what to expect in the recipe and describes how to set up any software or any preliminary settings required for the recipe.

How to do it...

This section contains the steps required to follow the recipe.

How it works...

This section usually consists of a detailed explanation of what happened in the previous section.

There's more...

This section consists of additional information about the recipe in order to enhance your knowledge of it.

See also

This section provides helpful links to other useful information for the recipe.

Get in touch

Feedback from our readers is always welcome.

General feedback: If you have questions about any aspect of this book, mention the book title in the subject of your message and email us at `customercare@packtpub.com`.

Errata: Although we have taken every care to ensure the accuracy of our content, mistakes do happen. If you have found a mistake in this book, we would be grateful if you would report this to us. Please visit `www.packtpub.com/support/errata`, selecting your book, clicking on the Errata Submission Form link, and entering the details.

Piracy: If you come across any illegal copies of our works in any form on the internet, we would be grateful if you would provide us with the location address or website name. Please contact us at `copyright@packt.com` with a link to the material.

If you are interested in becoming an author: If there is a topic that you have expertise in and you are interested in either writing or contributing to a book, please visit `authors.packtpub.com`.

Share Your Thoughts

Once you've read *Elasticsearch 8.x Cookbook*, we'd love to hear your thoughts! Scan the QR code below to go straight to the Amazon review page for this book and share your feedback.

https://packt.link/r/1-801-07981-1

Your review is important to us and the tech community and will help us make sure we're delivering excellent quality content.

1
Getting Started

In this chapter, we will start using **Elasticsearch** by downloading the correct version for our operating system, configuring it to perform at its best, and extending it via plugins. By the end of the chapter, we will see how to set it up on Docker, and in a cluster using **Elastic Cloud Enterprise** (Docker/Kubernetes).

We will cover the following recipes:

- Downloading and installing Elasticsearch
- Setting up networking
- Setting up a node
- Setting up Linux systems
- Setting up different node roles
- Setting up a coordinating-only node
- Setting up an ingestion node
- Installing plugins in Elasticsearch
- Removing a plugin
- Changing logging settings
- Setting up a node via Docker
- Deploying on Elastic Cloud Enterprise

Technical requirements

Elasticsearch runs on Linux/macOS/Windows, and a browser to access Kibana.

All the examples and code in this book are available at `https://github.com/PacktPublishing/Elasticsearch-8.x-Cookbook`.

If you don't want to go into the details of installing and configuring your Elasticsearch instance, and instead want to quickly set up your environment for developing or fun purposes, you can skip and go straight to the *Setting up a node via Docker* recipe to fire it up via Docker Compose. This will quickly help you install an Elasticsearch instance with Kibana and other tools.

Downloading and installing Elasticsearch

Elasticsearch has an active community, and the release cycles are very fast; generally, new minor releases are available every 2 or 3 weeks.

Since Elasticsearch depends on many common Java libraries (Lucene, Guice, and Jackson are the most famous ones), the Elasticsearch community tries to keep them updated and fix bugs that are discovered in them and in the Elasticsearch core.

The large user base is also a source of new ideas and features for improving Elasticsearch use cases.

For these reasons, if possible, it's best to use the latest available release; this is usually the most stable, with plenty of rich features, and bug-free as well. At the time of writing this book, the version is 8.0.0.

Getting ready

To install Elasticsearch, you need a supported operating system (Linux/macOS X/Windows) and a web browser, which is required to download the Elasticsearch binary release. At least 1 GB of free disk space is required to install Elasticsearch.

How to do it...

The following steps will show how Elasticsearch can be downloaded and successfully installed:

1. We will start by downloading Elasticsearch from the web.

   Elasticsearch is distributed in two different versions: the commercial one with integrated X-Pack, whose latest version is always downloadable at `https://www.elastic.co/downloads/elasticsearch`.

   The versions that are available for different operating systems are as follows:

   - `elasticsearch-{version-number}-windows-x86_64.zip` and `elasticsearch-{version-number}.msi` are for the Windows operating systems.
   - `elasticsearch-{version-number}-darwin-x86_64.tar.gz` is for macOS X.
   - `elasticsearch-{version-number}-linux-x86_64.tar.gz` is for Linux.
   - `elasticsearch-{version-number}-x86_64.deb` is for Debian-based Linux distributions (this also covers the Ubuntu family); this is installable with Debian by using the `dpkg -i elasticsearch-*.deb` command.
   - `elasticsearch-{version-number}-x86_64.rpm` is for Red Hat-based Linux distributions (this also covers the Cent OS family). This is installable with the `rpm -i elasticsearch-*.rpm` command.

   The preceding packages contain everything to start Elasticsearch (the application and a bundled **Java Virtual Machine (JVM)** for running it). This book targets version 8.x or higher. The latest and most stable version of Elasticsearch is 8.0.0. To check out whether this is the latest version when you read this, visit `https://www.elastic.co/downloads/elasticsearch`.

2. Extract the binary content. After downloading the correct release for your platform, the installation involves expanding the archive in a working directory.

   Choose a working directory that is safe from charset problems and does not have a long path. This prevents problems when Elasticsearch creates its directories to store index data.

   For the Windows platform, a good directory in which to install Elasticsearch could be `c:\es`, on Unix, and `/opt/es` on macOS X.

3. Let's start Elasticsearch to check whether everything is working. To start your Elasticsearch server, just access the directory, and for Linux and macOS X execute the following command:

```
# bin/elasticsearch
```

Alternatively, you can type the following command line for Windows:

```
# bin\elasticserch.bat
```

Your server should now start up and show logs similar to the following (I commented out the most important part. *Pay attention to the credential part for accessing Elasticsearch/Kibana*):

```
[2022-02-13T11:18:17,230][INFO ][o.e.n.Node
] [iMacParo] version[8.0.0], pid[57579], build[default/
tar/1b6a7ece17463df5ff54a3e1302d825889aa1161/2022-
02-03T16:47:57.507843096Z], OS[Mac OS X/11.1/
x86_64], JVM[Eclipse Adoptium/OpenJDK 64-Bit Server
VM/17.0.1/17.0.1+12]
```

```
[2022-02-13T11:18:17,235][INFO ][o.e.n.Node
] [iMacParo] JVM home [/opt/elasticsearch-8.x-cookbook/
elasticsearch/jdk.app/Contents/Home], using bundled JDK
[true] …
```

Module and plugin loading:

```
[2022-02-13T11:18:20,382][INFO ][o.e.p.PluginsService
] [iMacParo] loaded module [aggs-matrix-stats] …
```

Setup node networking functionalities:

```
[2022-02-13T11:18:20,454][INFO ][o.e.e.NodeEnvironment
] [iMacParo] using [1] data paths, mounts [[/System/
Volumes/Data (/dev/disk1s1)]], net usable_space
[141.7gb], net total_space [931.6gb], types [apfs]
```

```
[2022-02-13T11:18:20,454][INFO ][o.e.e.NodeEnvironment
] [iMacParo] heap size [31gb], compressed ordinary object
pointers [true] …
```

Current license:

```
[2022-02-13T11:18:26,646][INFO ][o.e.x.s.a.Realms
] [iMacParo] license mode is [trial], currently licensed
security realms are [reserved/reserved,file/default_
file,native/default_native] …
```

Binding Transport Protocol Network address:

```
[2022-02-13T11:18:29,642][INFO ][o.e.t.TransportService
] [iMacParo] publish_address {127.0.0.1:9300}, bound_
addresses {[::1]:9300}, {127.0.0.1:9300} …
```

Binding HTTP Protocol Network address:

```
[2022-02-13T11:18:30,550][INFO ]
[o.e.h.AbstractHttpServerTransport] [iMacParo]
publish_address {192.168.1.31:9200}, bound_addresses
{[::1]:9200}, {127.0.0.1:9200}, {192.168.1.31:9200}
[2022-02-13T11:18:30,551][INFO ][o.e.n.Node
] [iMacParo] started …
```

Registering new index patterns:

```
[2022-02-13T11:18:30,972][INFO ]
[o.e.c.m.MetadataIndexTemplateService] [iMacParo]
adding template [.monitoring-kibana] for index patterns
[.monitoring-kibana-7-*]    …
```

Registering license check:

```
[2022-02-13T11:18:35,079][INFO ]
[o.e.x.i.a.TransportPutLifecycleAction] [iMacParo] adding
index lifecycle policy [.fleet-actions-results-ilm-
policy]
[2022-02-13T11:18:35,335][INFO ][o.e.l.LicenseService
] [iMacParo] license [880f6db9-75b6-4106-8e2e-
0c06cb0e8b30] mode [basic] - valid
[2022-02-13T11:18:35,336][INFO ][o.e.x.s.a.Realms
] [iMacParo] license mode is [basic], currently licensed
security realms are [reserved/reserved,file/default_
file,native/default_native]
[2022-02-13T11:18:36,244][INFO ]
[o.e.c.m.MetadataCreateIndexService] [iMacParo] [.geoip_
databases] creating index, cause [auto(bulk api)],
templates [], shards [1]/[0]
```

Generation of token to connect other nodes:

```
[2022-02-13T11:18:39,862][INFO ]
[o.e.x.s.e.InternalEnrollmentTokenGenerator] [iMacParo]
Will not generate node enrollment token because node is
only bound on localhost for transport and cannot connect
to nodes from other hosts
```

```
[2022-02-13T11:18:39,950][INFO ]
[o.e.c.m.MetadataCreateIndexService] [iMacParo]
[.security-7] creating index, cause [api], templates [],
shards [1]/[0]…
```

Credentials:

```
Elasticsearch security features have been automatically
configured!
```

```
Authentication is enabled and cluster connections are
encrypted.    … truncated…
```

```
i  Configure Kibana to use this cluster:
```

```
• Run Kibana and click the configuration link in the
terminal when Kibana starts.
```

```
• Copy the following enrollment token and paste it into
Kibana in your browser (valid for the next 30 minutes):
```

```
eyJ2ZXIiOiI4LjAuMCIsImFkciI6WyIxOTIuMTY4LjEuMzE6OTIwM
CJdLCJmZ3IiOiJjNDRkMTZmNWEzODljODhkMDhlY2MxNjNmZDEyM
GQyNGUzMzYwOTBlOTRmNTc3NjQ1MWVhNzU5MDY4MWE1MTAyIiwia2V
5IjoiREt1WDhuNEJRY19MRXFtN2Q5YkY6UnZzNVU1Wk1UY3llQm9SZ
HRtTG5DdyJ9 … truncated…
```

Download of `geoip` data:

```
[2022-02-13T11:18:41,922][INFO ][o.e.i.g.GeoIpDownloader
] [iMacParo] successfully downloaded geoip database
[GeoLite2-City.mmdb]
```

```
… truncated…
```

How it works…

The Elasticsearch package generally contains the following directories:

- `bin`: This contains the scripts to start and manage Elasticsearch.
- `elasticsearch.bat`: This is the main executable script to start Elasticsearch.
- `elasticsearch-plugin.bat`: This is a script to manage plugins.
- `config`: This contains the Elasticsearch configurations. The most important ones are as follows:
  - `elasticsearch.yml`: This is the main `config` file for Elasticsearch.
  - `log4j2.properties`: This is the logging `config` file.

- `data`: This stores all the ingested data in Elasticsearch.

- `jdk.app`: The name of this directory can change based on the operating system. It contains a bundled JVM 11 version to be used with Elasticsearch.

- `lib`: This contains all the libraries required to run Elasticsearch.

- `logs`: This directory is empty at installation time, but in the future, it will contain the application logs.

- `modules`: This contains the Elasticsearch default plugin modules.

- `plugins`: This directory is empty at installation time, but it's the place where custom plugins will be installed.

During Elasticsearch startup, the following events happen:

- A node name is taken from the hostname of the machine. The default installed modules are loaded. The most important ones are as follows:

  - `aggs-matrix-stats`: This provides support for aggregation matrix statistics.

  - `analysis-common`: This is a common analyzer that extends the language processing capabilities of Elasticsearch.

  - `ingest-common/ingest-geoip/ingest-user-agent`: These include common functionalities for the ingest module plus geo/user agent management.

  - `kibana`: This sets up special indices for Kibana functionalities, including `.kibana*`, `.reporting*`, and `.apm*`.

  - `lang-expression/lang-mustache/lang-painless`: These are the default supported scripting languages of Elasticsearch.

  - `mapper-extras/mapper-version`: These provide extra mapper types to be used, such as `token_count` and `scaled_float`.

  - `parent-join`: This provides an extra query, such as `has_children` and `has_parent`.

  - `percolator`: This provides percolator capabilities.

  - `rank-eval`: This provides support for the experimental rank evaluation **Application Programming Interface (APIs)**. These are used to evaluate hit scoring based on queries.

  - `reindex`: This provides support for `reindex` actions (`reindex/update` by query).

- `repository-*`: These modules allow the use of external cloud services as repository storage (Azure, Google Cloud Storage, and S3).

  - `x-pack-*`: All the xpack modules depend on a subscription for their activation.

- If there are plugins, they are loaded.

- If not configured, Elasticsearch binds the following two ports on the `127.0.0.1` localhost automatically:

  - `9300`: This port is used for internal intranode communication.

  - `9200`: This port is used for the HTTP REST API.

- After starting, if indices are available, they are restored and ready to be used.

There are more events that are fired during the Elasticsearch startup. We'll see them in detail in other recipes.

There's more...

During a node's startup, a lot of required services are automatically started. The most important ones are as follows:

- **Cluster services**: These help you manage the cluster state and intranode communication and synchronization.

- **Indexing service**: This helps you manage all the index operations, initializing all active indices and shards.

- **Mapping service**: This helps you manage the document types stored in the cluster (we'll discuss mapping in *Chapter 2, Managing Mapping*).

- **Network services**: These include services such as HTTP REST services (default on port `9200`), and the internal Elasticsearch protocol (port `9300`).

- **Plugin service**: This manages the loading of the plugins.

- **Aggregation services**: These provide advanced analytics on stored Elasticsearch documents, such as statistics, histograms, and document grouping.

- **Ingesting services**: These provide support for document preprocessing before ingestion, such as field enrichment, **Natural Language Processing** (**NLP**), type conversion, and automatic field population.

- **Language scripting services**: These allow you to add new language scripting support to Elasticsearch.

See also

The *Setting up networking* recipe we're going to cover next will help you with the initial network setup. Check the official Elasticsearch download page at `https://www.elastic.co/downloads/elasticsearch` to get the latest version.

Setting up networking

Correctly setting up networking is very important for your nodes and cluster.

There are a lot of different installation scenarios and networking issues. The first step for configuring the nodes in order to build a cluster is to correctly set the node discovery.

Getting ready

To change configuration files, you will need a working Elasticsearch installation and a simple text editor, as well as your current networking configuration (your IP address).

How to do it...

To set up the networking, use the following steps:

1. Use a standard Elasticsearch configuration `config/elasticsearch.yml` file; your node will be configured to bind on the localhost interface (by default) so that it can't be accessed by external machines or nodes.

2. To allow another machine to connect to our node, we need to set `network.host` to our IP address (for example, I have `192.168.1.164`).

3. To be able to discover other nodes, we need to list them in the `discovery.zen.ping.unicast.hosts` parameter. This means that it sends signals to the machine in a unicast list and waits for a response. If a node responds to it, it can join a cluster.

4. In general, since Elasticsearch version 7.x, the node versions are compatible. You must have the same cluster name (the `cluster.name` option in `elasticsearch.yml`) to let nodes join with each other.

   The best practice is to have all the nodes installed with the same Elasticsearch version (`major.minor.release`). This suggestion is also valid for third-party plugins.

5. To customize the network preferences, you need to change some parameters in the `elasticsearch.yml` file, as follows:

```
cluster.name: ESCookBook
node.name: "Node1"
network.host: 192.168.1.164
discovery.zen.ping.unicast.hosts:
["192.168.1.164","192.168.1.165[9300-9400]"]
```

6. This configuration sets the cluster name to `Elasticsearch`, the node name, and the network address, and it tries to bind the node to the address given in the discovery section by performing the following tasks:

    *   We can check the configuration during node loading.

    *   We can now start the server and check whether networking is configured, as follows:

```
[2020-12-06T17:42:16,386][INFO ][o.e.c.s.MasterService
] [Node1] zen-disco-elected-as-master ([0]
nodes joined)[, ], reason: new_master {Node1}
{fyBySLMcR3uqKiYC32P5Sg}{IX1wpA01QSKkruZeSRP1Fg}
{192.168.1.164}{192.168.1.164:9300}{ml.machine_
memory=17179869184, xpack.installed=true, ml.max_open_
jobs=20, ml.enabled=true} [2020-12-06T17:42:16,390]
[INFO ][o.e.c.s.ClusterApplierService] [Node1] new_master
{Node1}{fyBySLMcR3uqKiYC32P5Sg}{IX1wpA01QSKkruZeSRP1Fg}
{192.168.1.164}{192.168.1.164:9300}{ml.machine_
memory=17179869184, xpack.installed=true, ml.max_open_
jobs=20, ml.enabled=true}, reason: apply cluster state
(from master [master {Node1}{fyBySLMcR3uqKiYC32P5Sg}
{IX1wpA01QSKkruZeSRP1Fg}{192.168.1.164}
{192.168.1.164:9300}{ml.machine_memory=17179869184, xpack.
installed=true, ml.max_open_jobs=20, ml.enabled=true}
committed version [1] source [zen-disco-elected-as-master
([0] nodes joined)[, ]]]) [2020-12-06T17:42:16,403][INFO
][o.e.x.s.t.n.SecurityNetty4HttpServerTransport] [Node1]
publish_address {192.168.1.164:9200}, bound_addresses
{192.168.1.164:9200} [2020-12-06T17:42:16,403][INFO ]
[o.e.n.Node ] [Node1] started [2020-12-06T17:42:16,600]
[INFO ][o.e.l.LicenseService ] [Node1] license [b2754b17-
a4ec-47e4-9175-4b2e0d714a45] mode [basic] - valid
```

As you can see from my screen dump, the transport is bound to `192.168.1.164:9300`. The REST HTTP interface is bound to `192.168.1.164:9200`.

How it works...

The following are the main important configuration keys for networking management:

- `cluster.name`: This sets up the name of the cluster. Only nodes with the same name can join together.
- `node.name`: If not defined, this is automatically assigned by Elasticsearch.

`node.name` allows defining a name for the node. If you have a lot of nodes on different machines, it is useful to set their names to something meaningful in order to easily locate them. Setting a valid name is easier to remember than a generated name such as `fyBySLMcR3uqKiYC32P5Sg`.

You must always set up `node.name` if you need to monitor your server. Generally, a node name is the same as a host server name for easy maintenance.

`network.host` defines the IP address of your machine to be used to bind the node. If your server is on different LANs, or you want to limit the bind on only one LAN, you must set this value with your server IP address.

`discovery.zen.ping.unicast.hosts` allows you to define a list of hosts (with ports or a port range) to be used to discover other nodes to join the cluster. The preferred port is the transport one, usually `9300`.

The addresses of the host list can be a mix of the following:

- Hostname, that is, `myhost1`
- IP address, that is, `192.168.1.12`
- IP address or hostname with the port, that is, `myhost1:9300`, `192.168.168.1.2:9300`
- IP address or hostname with a range of ports, that is, `myhost1:[9300-9400]`, `192.168.168.1.2:[9300-9400]`

See also

For more details, refer to the *Setting up a node* recipe in this chapter.

Setting up a node

Elasticsearch allows the customization of several parameters in an installation. In this recipe, we'll look at the most used ones to define where to store our data and improve the overall performance.

Getting ready

As described in the *Downloading and installing Elasticsearch* recipe, you need a working Elasticsearch installation and a simple text editor to change configuration files.

How to do it...

The steps required for setting up a simple node are as follows:

1. Open the `config/elasticsearch.yml` file with an editor of your choice.

2. Set up the directories that store your server data, as follows:

    -   For Linux or macOS X, add the following path entries (using `/opt/data` as the base path):

    ```
    path.conf: /opt/data/es/conf
    path.data: /opt/data/es/data1,/opt2/data/data2
    path.work: /opt/data/work
    path.logs: /opt/data/logs
    path.plugins: /opt/data/plugins
    ```

    -   For Windows, add the following path entries (using `c:\Elasticsearch` as the base path):

    ```
    path.conf: c:\Elasticsearch\conf
    path.data: c:\Elasticsearch\data
    path.work: c:\Elasticsearch\work
    path.logs: c:\Elasticsearch\logs
    path.plugins: c:\Elasticsearch\plugins
    ```

3. Set up the parameters to control the standard index shard and replication at creation. These parameters are as follows:

    ```
    index.number_of_shards: 1
    index.number_of_replicas: 1
    ```

How it works...

The `path.conf` parameter defines the directory that contains your configurations, mainly, `elasticsearch.yml` and `logging.yml`. The default is `$ES_HOME/config`, with `ES_HOME` to install the directory of your Elasticsearch server.

It's important to set up the `config` directory outside your application directory so that it is not necessary to obtain the details of the configuration files whenever you update your Elasticsearch server.

The `path.data` parameter is the most important one. This allows you to define one or more directories (in a different disk) where you can store your index data. When you define more than one directory, they are managed similarly to `RAID 0` (their space is summed up), favoring locations where most of the free space is available.

The `path.work` parameter is a location in which Elasticsearch stores temporary files.

The `path.log` parameter is where log files are put. These control how a log is managed in `logging.yml`.

The `path.plugins` parameter allows you to override the plugins path (the default is `$ES_HOME/plugins`). It's useful to put system-wide plugins in a shared path usually using **Network File System** (**NFS**) in case you want a single place in which to store your plugins for all of the clusters.

The main parameters are used to control index and shards in `index.number_of_shards`, which controls the standard number of shards for a newly created index, and `index.number_of_replicas`, which controls the initial number of replicas.

See also

Refer to the following points to learn more about topics related to this recipe:

- The *Setting up Linux systems* recipe
- The official Elasticsearch documentation at `https://www.elastic.co/guide/en/elasticsearch/reference/master/setup.html`

Setting up Linux systems

If you are using a Linux system (generally in a production environment), you need to manage an extra setup to improve performance or to resolve production problems with many indices.

This recipe covers the following two common errors that happen in production:

- Too many open files that can corrupt your indices and your data
- Slow performance in search and indexing due to the garbage collector

Big problems arise when you run out of disk space. In this scenario, some files can become corrupted. To prevent your indices from corruption and possible data loss, it is best to monitor the storage spaces. Default settings prevent index writing and block the cluster if your storage is over 95% full.

Getting ready

As we described in the *Downloading and installing Elasticsearch* recipe in this chapter, you need a working Elasticsearch installation and a simple text editor to change configuration files.

How to do it...

To improve the performance of Linux systems, we will perform the following steps:

1. First, you need to change the current limit for the user that runs the Elasticsearch server. In these examples, we will call this `elasticsearch`.

2. To allow Elasticsearch to manage a large number of files, you need to increment the number of file descriptors (number of files) that a user can manage. To do so, you must edit your `/etc/security/limits.conf` file and add the following lines at the end:

```
elasticsearch - nofile 65536
elasticsearch - memlock unlimited
```

3. Then, a machine restart is required to be sure that the changes have been made.

4. The new version of Ubuntu (that is, version 16.04 or later) can skip the `/etc/security/limits.conf` file in the `init.d` scripts. In these cases, you need to edit `/etc/pam.d/` and remove the following comment line:

```
# session required pam_limits.so
```

5. To control memory swapping, you need to set up the following parameter in `elasticsearch.yml`:

```
bootstrap.memory_lock: true
```

6. To fix the memory usage size of the Elasticsearch server, we need to set up the same values for Xms and Xmx in `$ES_HOME/config/jvm.options` (that is, we set 1 GB of memory in this case), as follows:

```
-Xms1g -Xmx1g
```

How it works...

The standard limit of file descriptors (`https://www.bottomupcs.com/file_descriptors.xhtml`) (maximum number of open files for a user) is typically 1,024 or 8,096. When you store a lot of records in several indices, you run out of file descriptors very quickly, so your Elasticsearch server becomes unresponsive and your indices may become corrupted, causing you to lose your data.

Changing the limit to a very high number means that your Elasticsearch doesn't hit the maximum number of open files.

The other setting for memory prevents Elasticsearch from swapping memory and gives a performance boost in an environment. This setting is required because, during indexing and searching, Elasticsearch creates and destroys a lot of objects in memory. This large number of **Create/Destroy** actions fragments the memory and reduces performance. The memory then becomes full of holes (`https://en.wikipedia.org/wiki/Fragmentation_(computing)`) and, when the system needs to allocate more memory, it suffers an overhead to find compacted memory. If you don't set `bootstrap.memory_lock: true`, Elasticsearch dumps the whole process memory on disk and defragments it back in memory, freezing the system. With this setting, the defragmentation step is done all in memory, with a huge performance boost.

There's more...

Generally, developers' machines do not have a lot of disk space: this prevents Elasticsearch from starting in write mode. To change the quota of free disk space for Elasticsearch, the following configuration can be used:

```
# no production safe
cluster.routing.allocation.disk.threshold_enabled: false
cluster.routing.allocation.disk.watermark.high: 99%
cluster.routing.allocation.disk.watermark.flood_stage: 99%
```

Setting up different node roles

Elasticsearch is natively designed for the cloud, so when you need to release a production environment with a huge number of records and you need **high availability (HA)** and good performance, you need to aggregate more nodes in a cluster.

Elasticsearch allows you to associate different roles to nodes to balance and improve overall performance.

Getting ready

As described in the *Downloading and installing Elasticsearch* recipe, you need a working Elasticsearch installation and a simple text editor to change the configuration files.

How to do it...

For the advanced setup of a cluster, there are some parameters that must be configured to define different node types.

These parameters are in the `config/elasticsearch.yml`, file and they can be set with the following steps:

1. Set up whether the node can only be a master, as follows:

   ```
   node.roles: [ master ]
   ```

2. Set up whether a node can only contain data, as follows:

   ```
   node.roles: [ data ]
   ```

3. Set up whether a node can only work as an ingest node, as follows:

   ```
   node.roles: [ ingest ]
   ```

How it works...

The node.roles parameter establishes that roles are associated with the actual node.

The default value for this value is to enable all the possible roles for the actual node, which are as follows:

- master: This is an arbiter for the cloud; it makes decisions about shard management, keeps the cluster status, and is the main controller of every index action. If your master nodes are on overload, all the nodes in the clusters will have performance penalties. The master node is the node that distributes the search across all data nodes and aggregates/rescores the result to return it to the user. In big data terms, it's a **Redux** layer in the **Map/Redux** search in Elasticsearch.

  The number of master nodes must always be odd.

- data: This allows you to store data in the node. This node will be a worker that is responsible for indexing and searching data.

- data_content, data_hot, data_warm, data_cold, and data_frozen: These are roles that allow defining different scopes in how the data is managed. Hot data is generally in **Solid State Drives** (**SSD**) of faster solutions to provide high frequency ingested/searched data. data_warm and data_cold are nodes used for infrequent searches.

- ingest: This role enables the usage of Elasticsearch ingest capabilities. See *Chapter 12, Using the Ingest Module.*

- ml: This role enables machine learning capabilities. This node will be able to run machine learning jobs and answer machine learning API calls.

- remote_cluster_client: This role enables cross-cluster integration. The node can connect to other clusters and execute a search on them.

- transform: This enables the transform functionalities that are automatic presets to copy data between indices.

The more frequent usage is mixing master and data roles. This allows you to have different node types with different scopes, as shown in the following table:

master	data	Node description
X	X	This is the default node. It can be the master, which contains data.
	X	This node never becomes a master node; it only holds data. It can be defined as a workhorse for your cluster.
X		This node only serves as a master in order to avoid storing any data and to have free resources. This will be the coordinator of your cluster.
		This node acts as a search load balancer (fetching data from nodes and aggregating results). This kind of node is also called a coordinator or client node.

Table 1.1 – Different setups between master and data settings

The most frequently used configuration is the first one, but if you have a very big cluster or special needs (such as defining a large group of data nodes), you can change the scopes of your nodes to better serve searches and aggregations.

There's more...

Related to the number of master nodes, there are settings that require at least half of them plus one to be available to ensure that the cluster is in a safe state (in order to avoid the risk of split-brain (https://www.elastic.co/guide/en/elasticsearch/reference/master/modules-node.html#split-brain). This setting is cluster.initial_master_nodes, and it must be set to the following equation:

```
(master_eligible_nodes / 2) + 1
```

To have an HA cluster, you need at least three nodes that are masters with the value of minimum_master_nodes set to 2.

See also

Refer to the following point to learn more about topics related to this recipe:

- The official Elasticsearch documentation about node setup at `https://www.elastic.co/guide/en/elasticsearch/reference/current/modules-node.html`

Setting up a coordinating-only node

The master nodes that we have seen previously are most important for cluster stability because they control node `join/leave`, index creation and mapping changes, and the allocation of resources. To prevent the queries and aggregations from creating instability in your cluster, coordinator (or client/proxy) nodes can be used to provide safe communication with the cluster.

Getting ready

You need a working Elasticsearch installation, as we described in the *Downloading and installing Elasticsearch* recipe in this chapter, and a simple text editor to change configuration files.

How to do it...

For the advanced setup of a cluster, there are some parameters that must be configured to define different node roles.

The parameter is in the `config/elasticsearch.yml` file, and you need to set the `nodes.roles` property to `empty,` as follows:

```
node.roles: []
```

How it works...

The coordinator node is a special node that works as a proxy/pass thought for the cluster. Its main advantages are as follows:

- It can easily be killed or removed from the cluster without causing any problems. It's not a master, so it doesn't participate in cluster functionalities, and it doesn't contain data, so there are no data relocations/replications due to its failure.

- It prevents the instability of the cluster due to a developer's/user's bad queries. Sometimes, a user executes aggregations that are too large (that is, date histograms with a range of some years and intervals of 10 seconds). Here, the Elasticsearch node could *crash*. In its newest version, Elasticsearch has a structure called **circuit breaker** to prevent similar issues, but there are always borderline cases that can bring instability using scripting, for example. The coordinator node is not a master and its overload doesn't cause any problems for cluster stability.

- If the coordinator or client node is embedded in the application, there are fewer round trips for the data, resulting in the speeding up of the application.

- You can add them to balance the search and aggregation throughput without generating changes and data relocation in the cluster.

Setting up an ingestion node

The main goals of Elasticsearch are indexing, searching, and analytics, but it's often necessary to modify or enhance the documents before storing them in Elasticsearch.

The following are the most common scenarios in this case:

- Preprocessing the log string to extract meaningful data

- Enriching the content of textual fields with NLP tools

- Enriching the content using ML computed fields

- Adding data modification or transformation during ingestion, such as the following:

  - Converting IP in geolocalization

  - Adding `DateTime` fields at ingestion time

  - Building custom fields (via scripting) at ingestion time

Getting ready

You need a working Elasticsearch installation, as described in the *Downloading and installing Elasticsearch* recipe, as well as a simple text editor to change configuration files.

How to do it...

To set up an ingest node, you need to edit the `config/elasticsearch.yml` file and set up the `nodes.roles` property to `ingest`, as follows:

```
node.roles: [ ingest]
```

Every time you change your `elasticsearch.yml` file, a node restart is required.

How it works...

The default configuration for Elasticsearch is to set the node as an ingest node (refer to *Chapter 12, Using the Ingest Module*, for more information on the ingestion pipeline).

As the coordinator node, using the ingest node is a way to provide functionalities to Elasticsearch without suffering cluster safety.

It's a best practice to disable this in the master and data nodes in order to prevent ingestion error issues and to protect the cluster. The coordinator node is the best candidate to be an ingest one.

If you are using NLP, attachment extraction (via the attachment ingest plugin), or logs ingestion, the best practice is to have a pool of coordinator nodes (no master, no data) with ingestion active.

The attachment and NLP plugins in the previous version of Elasticsearch were available in the standard data node or master node. These give a lot of problems to Elasticsearch due to the following reasons:

- High CPU usage for NLP algorithms that saturates all CPU on the data node, giving bad indexing and searching performances.

- Instability due to the bad format of attachment and/or Apache Tika bugs (the library used for managing document extraction).

- NLP or ML algorithms require a lot of CPU or stress the Java garbage collector (due to the high number of objects that are created and destroyed in memory), thereby decreasing the performance of the node.

The best practice is to have a pool of coordinator nodes with ingestion enabled to provide the best safety for the cluster and ingestion pipeline.

There's more...

Having known about the four kinds of Elasticsearch nodes, you can easily understand that a waterproof architecture designed to work with Elasticsearch should be similar to this one:

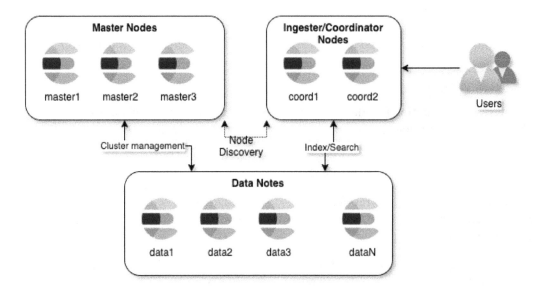

Figure 1.1 – Example of architecture with specialized nodes

Installing plugins in Elasticsearch

One of the main features of Elasticsearch is the possibility to extend it with **plugins**. Plugins extend Elasticsearch features and functionalities in several ways.

In Elasticsearch, these are native plugins. These are JAR files that contain application code, and are used for the following reasons:

- Script engines
- Custom analyzers, tokenizers, and scoring
- Custom mapping
- REST entry points
- Ingestion pipeline stages
- Supporting new storage (Hadoop and **Google Cloud Platform** (GCP) Cloud Storage)
- Extending X-Pack (that is, with a custom authorization provider)

Getting ready

You need a working Elasticsearch installation, as we described in the *Downloading and installing Elasticsearch* recipe, as well as a prompt/shell to execute commands in the Elasticsearch `install` directory.

How to do it...

Elasticsearch provides a script for automatic downloads and for the installation of plugins in `bin/directory` called `elasticsearch-plugin`.

The steps that are required to install a plugin are as follows:

1. Call the plugin and install the Elasticsearch command with the plugin name reference.

    For installing the ingested attachment plugin used to extract text from files, simply call and type the following command if you're using Linux:

    ```
    bin/elasticsearch-plugin install ingest-attachment
    ```

    For Windows, type the following command:

    ```
    elasticsearch-plugin.bat install ingest-attachment
    ```

2. If the plugin needs to change security permissions, a warning is prompted, and you need to accept this if you want to continue:

    ```
    -> Installing ingest-attachment
    -> Downloading ingest-attachment from elastic
    [=================================================] 100%
    @@@@@@@@@@@@@@@@@@@@@@@@@@@@@@@@@@@@@@@@@@@@@@@@@@@@@@@@@@@@@@
    @@
    @       WARNING: plugin requires additional permissions
    @
    @@@@@@@@@@@@@@@@@@@@@@@@@@@@@@@@@@@@@@@@@@@@@@@@@@@@@@@@@@@@@@
    @@
    * java.lang.RuntimePermission accessClassInPackage.sun.
    java2d.cmm.kcms
    * java.lang.RuntimePermission accessDeclaredMembers
    * java.lang.RuntimePermission getClassLoader
    * java.lang.reflect.ReflectPermission
    suppressAccessChecks
    ```

```
* java.security.SecurityPermission
createAccessControlContext

See https://docs.oracle.com/javase/8/docs/technotes/
guides/security/permissions.html

for descriptions of what these permissions allow and the
associated risks.


Continue with installation? [y/N]y

-> Installed ingest-attachment

-> Please restart Elasticsearch to activate any plugins
installed
```

3.   During the node's startup, check that the plugin is correctly loaded.

In the following screenshot, you can see the installation and the startup of the Elasticsearch server, along with the installed plugin:

```
[2022-02-13T11:54:59,211][INFO ][o.e.p.PluginsService    ] [iMacParo] loaded module [x-pack-stack]
[2022-02-13T11:54:59,211][INFO ][o.e.p.PluginsService    ] [iMacParo] loaded module [x-pack-text-structure]
[2022-02-13T11:54:59,211][INFO ][o.e.p.PluginsService    ] [iMacParo] loaded module [x-pack-voting-only-node]
[2022-02-13T11:54:59,212][INFO ][o.e.p.PluginsService    ] [iMacParo] loaded module [x-pack-watcher]
[2022-02-13T11:54:59,213][INFO ][o.e.p.PluginsService    ] [iMacParo] loaded plugin [ingest-attachment]
[2022-02-13T11:54:59,285][INFO ][o.e.e.NodeEnvironment   ] [iMacParo] using [1] data paths, mounts [[/System/Volumes/
[2022-02-13T11:54:59,286][INFO ][o.e.e.NodeEnvironment   ] [iMacParo] heap size [31gb], compressed ordinary object po
[2022-02-13T11:54:59,436][INFO ][o.e.n.Node              ] [iMacParo] node name [iMacParo], node ID [eNBAg76DT2-Ul3jN
sform, data_hot, ml, data_frozen, ingest]
```

Figure 1.2 – Server log line that shows the loaded ingest-attachment plugin

Remember that a plugin installation requires an Elasticsearch server restart.

How it works...

The elasticsearch-plugin.bat script is a wrapper for the Elasticsearch plugin manager. This can be used to install or remove a plugin (using the remove options).

There are several ways to install the plugin. Here are some examples:

- Passing the URL of the plugin (ZIP archive), as follows:

```
bin/elasticsearch-plugin install http://mywoderfulserve.
com/plugins/awesome-plugin.zip
```

- Passing the file path of the plugin (ZIP archive), as follows:

```
bin/elasticsearch-plugin install file:///tmp/awesome-
plugin.zip
```

- Using the `install` parameter with the GitHub repository of the plugin. The `install` parameter, which must be given, is formatted in the following way:

```
<username>/<repo>[/<version>]
```

During the installation process, the Elasticsearch plugin manager can do the following:

- Download the plugin
- Create a plugin directory in `ES_HOME/plugins`, if it's missing
- Optionally, ask whether the plugin wants special permission to be executed
- Unzip the `plugin` content in the `plugin` directory
- Remove temporary files

The installation process is completely automatic; no further actions are required. The user must only pay attention to the fact that the process ends with an `Installed` message to be sure that the install process has been completed correctly.

Restarting the server is always required to be sure that the plugin is correctly loaded by Elasticsearch.

> **Important Note**
> Elasticsearch needs the same plugins (binary/versions) *to be installed in all the nodes* of the cluster. If some nodes are missing some plugins, errors can arise during runtime operations.

There's more...

If your current Elasticsearch application depends on one or more plugins, a node can be configured to start up only if these plugins are installed and available. To achieve this behavior, you can provide the `plugin.mandatory` directive in the `elasticsearch.yml` configuration file.

For the previous example (`ingest-attachment`), the configuration line to be added is as follows:

```
plugin.mandatory:ingest-attachment
```

There are also some hints to remember while installing plugins; updating some plugins in a node environment can cause malfunctions due to different plugin versions in different nodes. If you have a big cluster for safety, it's better to check for updates in a separate environment to prevent problems (and remember to upgrade the plugin in all the nodes).

Updating the Elasticsearch version server can also break your custom binary plugins due to internal API changes, hence, the plugin needs to have the same version of the Elasticsearch server in its manifest.

Upgrading an Elasticsearch server version means upgrading all the installed plugins.

See also

On the Elasticsearch site, there is an updated list of available plugins: `https://www.elastic.co/guide/en/elasticsearch/plugins/current/index.html`.

The actual Elasticsearch documentation doesn't cover all available plugins. I suggest going to GitHub (`https://github.com`) and searching for them with these, or similar, queries: `elasticsearch plugin`, `elasticsearch lang`, and `elasticsearch ingest`.

Removing a plugin

You have installed some plugins, and now you need to remove a plugin because it's not required. Removing an Elasticsearch plugin is easy if everything goes right, otherwise, you will need to remove it manually.

This recipe covers both cases.

Getting ready

You need a working Elasticsearch installation, as described in the *Downloading and installing Elasticsearch* recipe, and a prompt or shell to execute commands in the Elasticsearch install directory. Before removing a plugin, it is safer to stop the Elasticsearch server to prevent errors due to the deletion of a plugin JAR file.

How to do it...

The steps to remove a plugin are as follows:

1. Stop your running node to prevent exceptions that are caused due to the removal of a file.

2. Use the Elasticsearch plugin manager, which comes with its script wrapper (`bin/elasticsearch-plugin`).

   On Linux and macOS X, type the following command:

   ```
   elasticsearch-plugin remove ingest-attachment
   ```

On Windows, type the following command:

```
elasticsearch-plugin.bat remove ingest-attachment
```

3. Restart the server.

How it works...

The plugin manager's `remove` command tries to detect the correct name of the plugin and remove the directory of the installed plugin.

If there are undeletable files in your plugin directory (or strange astronomical events that hit your server), the plugin script might fail to manually remove a plugin, so you need to follow these steps:

1. Go into the plugins directory.
2. Remove the directory with your plugin name.

Important Note

If the plugin is providing additional functionalities to your cluster, such as mappings, ingester processors, and storage capabilities, you cannot remove these plugins. To remove these plugins, you must be sure that all their functionalities are not used, otherwise, your Elasticsearch node will not start due to missing capability.

Changing logging settings

Standard logging settings work very well for general usage.

Changing the log level can be useful for checking for bugs or understanding malfunctions due to bad configuration or strange plugin behavior. A verbose log can be used from the Elasticsearch community to solve such problems.

If you need to debug your Elasticsearch server or change the logging, you need to change the `log4j2.properties` file.

Getting ready

You need a working Elasticsearch installation, as described in the *Downloading and installing Elasticsearch* recipe, and a simple text editor to change configuration files.

How to do it...

In the `config` directory in your Elasticsearch install directory, there is a `log4j2.properties` file that controls the working settings.

The steps that are required for changing the logging settings are as follows:

1. To emit every kind of logging that Elasticsearch could produce, you can change the current root level logging, which is as follows:

   ```
   rootLogger.level = info
   ```

2. This needs to be changed to the following:

   ```
   rootLogger.level = debug
   ```

3. Now, if you start Elasticsearch from the command line (with `bin/elasticsearch -f`), you should see a lot of information, which is not always useful (except to debug unexpected issues).

How it works...

The Elasticsearch logging system is based on the `log4j` library (`http://logging.apache.org/log4j/`).

Log4j is a powerful library that is used to manage logging. Covering all of its functionalities is outside the scope of this book; if a user needs advanced usage, there are a lot of books and articles on the internet about it.

Setting up a node via Docker

Docker (`https://www.docker.com/`) has become a common way to deploy application servers for testing or production.

Docker is a container system that makes it possible to easily deploy replicable installations of server applications. With Docker, you don't need to set up a host, configure it, download the Elasticsearch server, unzip it, or start the server—everything is done automatically by Docker.

Getting ready

You need a working Docker installation to be able to execute Docker commands (`https://www.docker.com/products/overview`).

How to do it...

To run Elasticsearch via Docker, you will need to execute the following steps:

1. If you want to start a *vanilla server*, execute the following command to fetch the Docker:

    ```
    docker pull docker.elastic.co/elasticsearch/
    elasticsearch:8.0.0
    ```

    An output similar to the following will be shown:

    ```
    8.0.0: Pulling from elasticsearch/elasticsearch
    4f96cc5f1d01: Pull complete
    b48e0e2deb60: Pull complete
    122b9d65e0b1: Pull complete
    b9ed64b36d59: Pull complete
    323db3a308c7: Pull complete
    Digest: sha256: 30dc26312256b810089b1e672f2ecd92ec6a400f-
    903d2198942e36c67a803611
    Status: Downloaded newer image for docker.elastic.co/
    elasticsearch/elasticsearch:8.0.0
    docker.elastic.co/elasticsearch/elasticsearch:8.0.0
    ```

2. After downloading the Elasticsearch image, we can start a `develop` instance that can be accessed outside from Docker:

    ```
    docker run -p 9200:9200 -p 9300:9300 -e "discovery.
    type=single-node" -e "http.host=0.0.0.0" -e "transport.
    host=0.0.0.0" docker.elastic.co/elasticsearch/
    elasticsearch:8.0.0
    ```

    You'll see the output of the Elasticsearch server starting.

3. In another window/terminal, to check whether the Elasticsearch server is running, execute the following command:

```
docker ps
```

The output will be similar to the following:

```
CONTAINER ID    IMAGE
COMMAND                     CREATED
STATUS          PORTS
NAMES
4c103c6c70a4    docker.elastic.co/elasticsearch/
elasticsearch:8.0.0    "/bin/tini -- /usr/l…"    26
seconds ago    Up 26 seconds    0.0.0.0:9200->9200/tcp,
0.0.0.0:9300->9300/tcp              admiring_mendeleev
```

The default exported ports are 9200 and 9300.

How it works...

The Docker container provides a Debian Linux installation with Elasticsearch installed.

The Elasticsearch Docker installation is easily repeatable and does not require a lot of editing and configuration.

The default installation can be tuned into in several ways, for example:

1. You can pass a parameter to Elasticsearch via the command line using the -e flag, as follows:

```
docker run -d docker.elastic.co/elasticsearch/
elasticsearch: 8.0.0 elasticsearch -e "node.
name=NodeName" -e "discovery.type=single-node" -e "http.
host=0.0.0.0" -e "transport.host=0.0.0.0"
```

2. You can customize the default settings of the environment that is providing custom Elasticsearch configuration by providing a volume mount point at /usr/share/elasticsearch/config, as follows:

```
docker run -d -v "$PWD/config":/usr/share/
elasticsearch/config docker.elastic.co/elasticsearch/
elasticsearch:8.0.0
```

3. You can persist the data between Docker reboots configuring a local data mount point to store index data. The path to be used as a mount point is `/usr/share/elasticsearch/config`, as follows:

```
docker run -d -v "$PWD/esdata":/usr/share/elasticsearch/
data docker.elastic.co/elasticsearch/elasticsearch:8.0.0
```

There's more...

The official Elasticsearch images are not only provided by Docker. There are also several customized images for custom purposes. Some of these are optimized for large cluster deployments or more complex Elasticsearch cluster topologies than the standard ones.

Docker is very handy for testing several versions of Elasticsearch in a clean way, without installing too much stuff on the host machine.

In the `ch01/docker/` code repository directory, there is a `docker-compose.yaml` file that provides a full environment that will set up the following elements:

- `elasticsearch`, which will be available at `http://localhost:9200`
- `kibana`, which will be available at `http://localhost:5601`
- `cerebro`, which will be available at `http://localhost:9000`

To install all the applications, you can simply execute `docker-compose up -d`. All the required binaries will be downloaded and installed in Docker, and they will then be ready to be used.

See also

- The official Elasticsearch Docker documentation at `https://www.elastic.co/guide/en/elasticsearch/reference/master/docker.html`
- The **Elasticsearch, Logstash, and Kibana** (**ELK**) Stack via Docker at `https://hub.docker.com/r/sebp/elk/`
- The Docker documentation at `https://docs.docker.com/`

Deploying on Elastic Cloud Enterprise

The Elasticsearch company provides **Elastic Cloud Enterprise (ECE)**, which is the same tool that's used in Elasticsearch Cloud (`https://www.elastic.co/cloud`) and is offered for free. This solution, which is available on **Platform as a Service (PaaS)** on AWS, Azure, or GCP, can be installed on-premises to provide an enterprise solution on top of Elasticsearch.

If you need to manage multiple elastic deployments across teams or geographies, you can leverage ECE to centralize deployment management for the following functions:

- Provisioning
- Monitoring
- Scaling
- Replication
- Upgrades
- Backup and restoring

Centralizing the management of deployments with ECE enforces uniform versioning, data governance, backup, and user policies. Increased hardware utilization through better management can also reduce the total cost.

Getting ready

As this solution targets large installations of many servers, the minimum testing requirement is an 8 GB RAM node. The ECE solution lives at the top of Docker and must be installed on the nodes.

ECE supports only some operating systems; the compatibility matrix is available online at `https://www.elastic.co/support/matrix`.

On other configurations, ECE could work, but it is not supported in case of issues.

How to do it...

Before installing ECE, the following prerequisites are to be checked:

1. Your user must be a Docker-enabled one. In the case of an error due to a non-Docker user, add your user with `sudo usermod -aG docker $USER`.
2. In the case of an error when you try to access `/mnt/data`, give your user permission to access this directory.

3. You need to add the following line to `/etc/sysctl.conf` (a reboot is required): `vm.max_map_count = 262144`.

4. To be able to use ECE, it must initially be installed on the first host, as follows:

```
bash <(curl -fsSL https://download.elastic.co/cloud/
elastic-cloud-enterprise.sh) install
```

5. The installation process should manage these steps automatically, as shown in the following screenshot:

```
70a92fc4537d: Pull complete
18947ce6778e: Pull complete
Digest: sha256:95ea22be8d5b1b3494af108070a2ebd317e0e693d5da15c397c7745c577318cf
Status: Downloaded newer image for docker.elastic.co/cloud-enterprise/elastic-cloud-enterprise:3.0.0
~~~~~~~~~~~~~~~~~~~~~~~~~~~~~~~~~~~~~~~~~~~~~~~~~~~~~~~~~~~~~~~~~~~~~~~~~~~~~~~~~~~~~~~~~~~~~~~~~~~~~~~~~~~~
Elastic Cloud Enterprise Installer

Start setting up a new Elastic Cloud Enterprise installation by installing the software on your first host.
This first host becomes the initial coordinator and provides access to the Cloud UI, where you can manage your installation.
To learn more about the options you can specify, see the documentation.

NOTE: If you want to add this host to an existing installation, please specify the --coordinator-host and --roles-token flags
~~~~~~~~~~~~~~~~~~~~~~~~~~~~~~~~~~~~~~~~~~~~~~~~~~~~~~~~~~~~~~~~~~~~~~~~~~~~~~~~~~~~~~~~~~~~~~~~~~~~~~~~~~~~

-- Verifying Prerequisites --
Checking runner container does not exist... PASSED
Checking host storage root volume path is not root... PASSED
Checking host storage path is accessible... PASSED
Checking host storage path contents matches whitelist... PASSED
Checking Docker version... PASSED
Checking Docker SELinux support... PASSED
Checking Docker file system... PASSED
 - The installation with extfs can proceed; however, we recommend XFS
Checking Docker network settings... PASSED
 - The installation can proceed with ICC enabled; however, we recommend turning ICC off (with the 'icc=false' flag) as a security best practice.
Checking Docker storage driver... PASSED
Checking whether 'setuser' works inside a Docker container... PASSED
Checking memory settings... PASSED
 - Option '--memory-settings' not used. Default memory settings might be insufficient for production use!
Checking runner ip connectivity... PASSED
Checking OS IPv4 IP forward setting... PASSED
Checking metadata endpoint protection... PASSED
Checking OS max map count setting... PASSED
Checking OS kernel version... PASSED
Checking minimum required memory... PASSED
Checking OS kernel cgroup.memory... PASSED
 - OS setting 'cgroup.memory' should be set to cgroup.memory=nokmem
Checking OS minimum ephemeral port... PASSED
Checking OS max open file descriptors per process... PASSED
Checking OS max open file descriptors system-wide... PASSED
Checking OS file system and Docker storage driver compatibility... PASSED
Checking OS file system storage driver permissions... PASSED
Checking OS AppArmor status... PASSED
Checking OS SELinux status... PASSED
-- Completed Verifying Prerequisites --

- Running Bootstrap container
- Monitoring bootstrap process
- Loaded bootstrap settings {}
- Core service started {}
- Storing default ACLs on [/default_acls] {}
- Starting local runner {}
- Started local runner {}
- Waiting for runner container node {}
- Runner container node detected {}
- Waiting for coordinator candidate {}
```

Figure 1.3 – Installation of an ECE cluster

In the end, the installer should provide your credentials so that you can access your cluster in a similar output, as follows:

```
~~~~~~~~~~~~~~~~~~~~~~~~~~~~~~~~~~~~~~~~~~~~~~~~~~~~~~~~~~~~~~~~~~
~~~~~~~~~~~~~~~~~~~~~~~~~~~~~~~~~~~~~~~~~~~~~~~~~~~~~~~~~~~~~~~~~~
~~~
Elastic Cloud Enterprise installation completed successfully

Ready to copy down some important information and keep it safe?

To access the Cloud UI:
http://192.168.1.28:12400
https://192.168.1.28:12443

Admin username: admin
Password: wTXiQ9DP7Os5w9qbNtIaYr7TVP6ST5JoySWggxxR5PH
Read-only username: readonly
Password: 8EK0ksEs9AvuyIanxwIAks3903E90bCMQjbtUBGGjqX

... truncated ...

System secrets have been generated and stored in "/mnt/data/
elastic/bootstrap-state/bootstrap-secrets.json".
Keep the information in the bootstrap-secrets.json file secure
by removing the file and placing it into secure storage, for
example.
```

6. In my case, I can access the installed interface at
    `http://192.168.1.28:12400`.

    After logging in to the `admin` interface, you will see your actual cloud state,
    as follows:

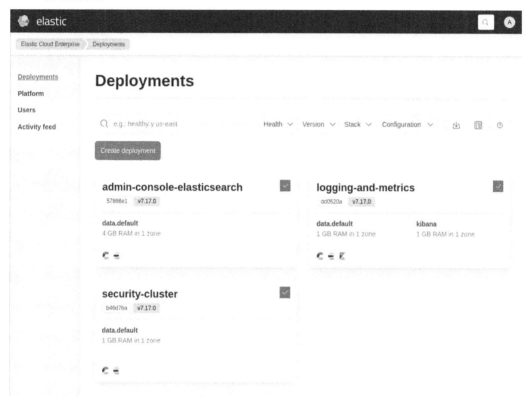

Figure 1.4 – ECE main view

7. You can now press on **Create deployment** to fire your first Elasticsearch cluster, as follows:

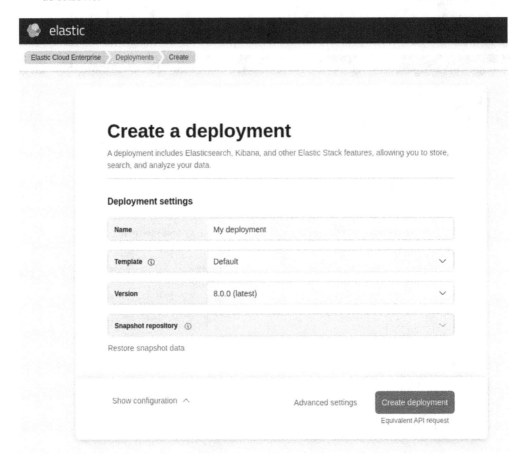

Figure 1.5 – ECE Create deployment

8.   You need to define a name (that is, a book cluster). Using standard options for this is okay. After pressing **Create deployment**, ECE will start to build your cluster, as follows:

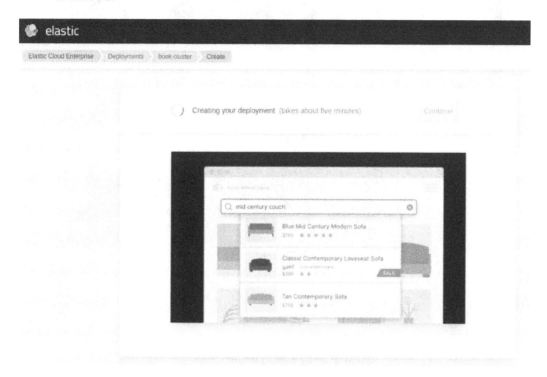

Figure 1.6 – ECE cluster building

9. After a few minutes, the cluster should be up and running, as follows:

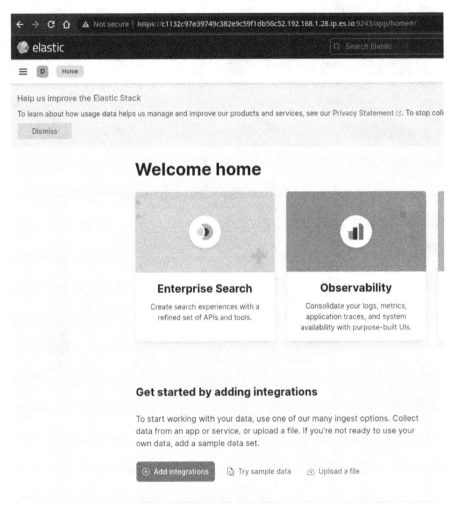

Figure 1.7 – Cluster up and running with URL pointing to local installation

How it works...

ECE allows you to manage a large Elasticsearch cloud service that can create an instance via deployments. By default, the standard deployment will fire an Elasticsearch node with 4 GB RAM, a 32 GB disk, and a Kibana instance.

You can define a lot of parameters during the deployments for ElasticSearch, such as the following:

- The RAM used for instances from 1 GB to 64 GB. The storage is proportional to the memory, so you can go from 1 GB RAM and 128 GB storage to 64 GB RAM and 2 TB storage.

- Whether the node requires ML.

- Master configurations if you have more than six data nodes.

- The plugins that are required to be installed.

For Kibana, you can only configure the memory (from 1 GB to 8 GB) and pass extra parameters (usually used for custom maps).

ECE does all the provisioning and, if you want a monitoring component and other X-Pack features, it is able to autoconfigure your cluster to manage all the required functionalities.

ECE is very useful if you need to manage several Elasticsearch/Kibana clusters because it eliminates all the infrastructure problems.

A benefit of using a deployed Elasticsearch cluster is that, during deployment, a proxy is installed. This is very handy for managing the debugging of Elasticsearch calls.

The URLs used to serve Kibana in your custom cluster via ECE are special ones (such as `https://c1132c97e39749c382e9c59f1db56c52.192.168.1.28.ip.es.io:9243/app/home#/`); they are composed of a hash (for example, `c1132c97e39749c382e9c59f1db56c52`) and a part that is the local redirect to your cluster IP address (for example, `192.168.1.28.ip.es.io`).

See also

You can refer to the following links for further information on the topics that were covered in this recipe:

- `https://www.elastic.co/cloud`, for a PAAS-managed cloud provider for ElasticSearch

- The complete documentation of Elastic Cloud Enterprise at `https://www.elastic.co/guide/en/cloud-enterprise/current/index.html`

- The ECE documentation for monitoring integrating Elastic Beats at `https://www.elastic.co/guide/en/cloud-enterprise/current/ece-cloud-id.html`

2
Managing Mappings

Mapping is a primary concept in Elasticsearch that defines how the search engine should process a document and its fields to be effectively used in search and aggregations.

Search engines perform the following two main operations:

- **Indexing**: This action is used to receive a document, process it, and store it in an index.

- **Searching**: This action is used to retrieve the data from the index based on a query.

These two operations are strictly connected; an error in the indexing step leads to unwanted or missing search results.

Elasticsearch, by default, has explicit mapping at the index level. When indexing, if a mapping is not provided, a default one is created and guesses the structure from the JSON data fields that the document is composed of. This new mapping is then automatically propagated to all the cluster nodes: it will begin part of the cluster's state.

The default type mapping has sensible default values, but when you want to change their behavior or customize several other aspects of indexing (object to special fields, storing, ignoring, completion, and so on), you need to provide a new mapping definition.

In this chapter, we'll look at all the possible mapping field types that document mappings are composed of.

In this chapter, we will cover the following recipes:

- Using explicit mapping creation
- Mapping base types
- Mapping arrays
- Mapping an object
- Mapping a document
- Using dynamic templates in document mapping
- Managing nested objects
- Managing a child document with a join field
- Adding a field with multiple mappings
- Mapping a GeoPoint field
- Mapping a GeoShape field
- Mapping an IP field
- Mapping an Alias field
- Mapping a Percolator field
- Mapping the Rank Feature and Feature Vector fields
- Mapping the Search as you type field
- Using the Range Field type
- Using the Flattened field type
- Using the Point and Shape field types
- Using the Dense Vector field type
- Using the Histogram field type
- Adding metadata to a mapping
- Specifying different analyzers
- Using index components and templates

Technical requirements

To follow and test the commands shown in this chapter, you must have a working Elasticsearch cluster installed on your system, as described in *Chapter 1, Getting Started*.

To simplify how you manage and execute these commands, I suggest that you install Kibana so that you have a more advanced environment to execute Elasticsearch queries.

Using explicit mapping creation

If we consider the index as a database in the SQL world, mapping is similar to the create table definition.

Elasticsearch can understand the structure of the document that you are indexing (reflection) and create the mapping definition automatically. This is called **explicit mapping creation**.

Getting ready

To execute the code in this recipe, you will need an up-and-running Elasticsearch installation, as described in the *Downloading and installing Elasticsearch* recipe of *Chapter 1, Getting Started*.

To execute these commands, you can use any HTTP client, such as curl (https://curl.haxx.se/), Postman (https://www.getpostman.com/), or similar platforms. I suggest using the Kibana console to provide code completion and better character escaping for Elasticsearch.

To understand the examples and code in this recipe, basic knowledge of JSON is required.

How to do it...

You can explicitly create a mapping by adding a new document to Elasticsearch. For this, perform the following steps:

1. Create an index, as shown in the following code:

```
PUT test
```

The output will be as follows:

```
{ "acknowledged" : true, "shards_acknowledged" : true,
  "index" : "test" }
```

2. Put a document in the index, as shown in the following code:

```
PUT test/_doc/1
{"name":"Paul", "age":35}
```

The output will be as follows:

```
{
    "_index" : "test", "_id" : "1", "_version" : 1,
    "result" : "created",
    "_shards" : {"total" : 2, "successful" : 1, "failed" :
0 },
    "_seq_no" : 0,   "_primary_term" : 1
}
```

3. Get the mapping with the following code:

```
GET test/_mapping
```

4. The mapping that's auto-created by Elasticsearch should look as follows:

```
{
  "test" : {
    "mappings" : {
      "properties" : {
        "age" : { "type" : "long" },
        "name" : {
          "type" : "text",
          "fields" : {
            "keyword" : {"type" : "keyword", "ignore_
above" : 256 }
} } } } }
```

5. To delete the index, you can use the following command:

```
DELETE test
```

The output will be as follows:

```
{ "acknowledged" : true }
```

How it works...

The first command line (*Step 1*) creates an index where we can configure the mappings in the future, if required, and store documents in it.

The second command (*Step 2*) inserts a document in the index (we'll learn how to create the index in the *Creating an index* recipe of *Chapter 3, Basic Operations*, and record indexing in the *Indexing a document* recipe of *Chapter 3, Basic Operations*).

Elasticsearch reads all the default properties for the field of the mapping and starts to process them as follows:

- If the field is already present in the mapping and the value of the field is valid (it matches the correct type), Elasticsearch does not need to change the current mappings.

- If the field is already present in the mapping but the value of the field is of a different type, it tries to upgrade the field type (that is, from integer to long). If the types are not compatible, it throws an exception, and the indexing process fails.

- If the field is not present, it tries to auto-detect the type of field. It updates the mappings with a new field mapping. (In the case of a null value, it skips the mapping update until it encounters a concrete type.)

There's more...

In Elasticsearch, every document has a unique identifier, called an ID for a single index, which is stored in the special _id field of the document.

The _id field can be provided at index time or can be assigned automatically by Elasticsearch if it is missing.

When a mapping type is created or changed, Elasticsearch automatically propagates mapping changes to all the nodes in the cluster so that all the shards are aligned to process that particular type.

In Elasticsearch 7.x, there was a default type (_doc): it was removed in Elasticsearch 8.x and above.

See also

Please refer to the following recipes in *Chapter 3*, *Basic Operations*:

- The *Creating an index* recipe, which is about putting new mappings in an index while it's being created
- The *Putting a mapping in an index* recipe, which is about extending a mapping in an index

Mapping base types

Using explicit mapping makes it possible to start to quickly ingest the data using a schemaless approach without being concerned about field types. Thus, to achieve better results and performance in indexing, it's required to manually define a mapping.

Fine-tuning mapping brings some advantages, such as the following:

- Reducing the index size on the disk (disabling functionalities for custom fields)
- Indexing only interesting fields (general speed up)
- Precooking data for fast search or real-time analytics (such as aggregations)
- Correctly defining whether a field must be analyzed in multiple tokens or considered as a single token
- Defining mapping types such as *geo point*, *suggester*, *vectors*, and so on

Elasticsearch allows you to use base fields with a wide range of configurations.

Getting ready

You will need an up-and-running Elasticsearch installation, as we described in the *Downloading and installing Elasticsearch* recipe of *Chapter 1*, *Getting Started*.

To execute the commands in this recipe, you can use any HTTP client, such as curl (`https://curl.haxx.se/`), Postman (`https://www.getpostman.com/`), or similar. I suggest using the Kibana console, which provides code completion and better character escaping for Elasticsearch.

To execute this recipe's examples, you will need to create an index with a `test` name, where you can put mappings, as explained in the *Using explicit mapping creation* recipe.

How to do it...

Let's use a semi real-world example of a shop order for our eBay-like shop:

1. First, we must define an order:

Name	Type	Description
id	identifier	Order identifier
date	date(time)	Date of order
customer_id	id reference	Customer ID reference
name	string	Name of the item
quantity	integer	How many items?
price	double	The price of the item
vat	double	VAT for item
sent	boolean	The order was sent

Figure 2.1 – Example of an order

2. Our `order` record must be converted into an Elasticsearch mapping definition, as follows:

```
PUT test/_mapping
{  "properties" : {
      "id" : {"type" : "keyword"},
      "date" : {"type" : "date"},
      "customer_id" : {"type" : "keyword"},
      "sent" : {"type" : "boolean"},
      "name" : {"type" : "keyword"},
      "quantity" : {"type" : "integer"},
      "price" : {"type" : "double"},
      "vat" : {"type" : "double", "index": false}
} }
```

Now, the mapping is ready to be put in the index. We will learn how to do this in the *Putting a mapping in an index* recipe of *Chapter 3, Basic Operations*.

How it works...

Field types must be mapped to one of the Elasticsearch base types, and options on how the field must be indexed need to be added.

The following table is a reference for the mapping types:

Type	ES-Type	Description
String, VarChar	keyword	This is a text field that is not tokenizable: CODE001.
String, VarChar, Text	text	This is a text field to be tokenized: a nice text.
Short	short	This is a short integer (16-bit): from -32768 to 32767.
Integer	integer	This is an Integer (32-bit): 1, 2, 3, or 4.
long	long	This is a long value (64-bit).
Float	float	This is a single precision floating-point number (32-bit): 1.2 or 32787324.5.
	half_float	This is a half precision floating-point number (16-bit): 1.2, or 4.5.
	scaled_float	This is a float stored in a long with a fixed scale factor.
Double	double	This is a floating-point number (64 bit).
Boolean	boolean	This is a Boolean value: true or false.
Date/Datetime	date	This is a date or datetime value: 2013-12-25, 2013-12-25T22:21:20.
Datetime with nanos	date_nanos	This is a datetime with a nano-precision second value: 2020-12-01T12:10:30.123456789Z.
Bytes/Binary	binary	This includes some bytes that are used for binary data, such as files or streams of bytes.

Figure 2.2 – Base type mapping

Depending on the data type, it's possible to give explicit directives to Elasticsearch when you're processing the field for better management. The most used options are as follows:

- `store` (default `false`): This marks the field to be stored in a separate index fragment for fast retrieval. Storing a field consumes disk space but reduces computation if you need to extract it from a document (that is, in scripting and aggregations). The possible values for this option are `true` and `false`. They are always retuned as an array of values for consistency.

  The stored fields are faster than others in aggregations.

- `index`: This defines whether or not the field should be indexed. The possible values for this parameter are `true` and `false`. Index fields are not searchable (the default is `true`).

- `null_value`: This defines a default value if the field is null.

- `boost`: This is used to change the importance of a field (the default is `1.0`).

  `boost` works on a term level only, so it's mainly used in term, terms, and match queries.

- `search_analyzer`: This defines an analyzer to be used during the search. If it's not defined, the analyzer of the parent object is used (the default is `null`).

- `analyzer`: This sets the default analyzer to be used (the default is `null`).

- `norms`: This controls the Lucene norms. This parameter is used to score queries better. If the field is only used for filtering, it's a best practice to disable it to reduce resource usage (`true` for analyzed fields and `false` for `not_analyzed` ones).

- `copy_to`: This allows you to copy the content of a field to another one to achieve functionalities, similar to the `_all` field.

- `ignore_above`: This allows you to skip the indexing string if it's bigger than its value. This is useful for processing fields for exact filtering, aggregations, and sorting. It also prevents a single term token from becoming too big and prevents errors due to the Lucene term's byte-length limit of 32,766. The maximum suggested value is `8191` (`https://www.elastic.co/guide/en/elasticsearch/reference/current/ignore-above.html`).

There's more...

From Elasticsearch version 6.x onward, as shown in the *Using explicit mapping creation* recipe, the explicit inferred type for a string is a multifield mapping:

- The default processing is `text`. This mapping allows textual queries (that is, term, match, and span queries). In the example provided in the *Using explicit mapping creation* recipe, this was `name`.

- The `keyword` subfield is used for `keyword` mapping. This field can be used for exact term matching and aggregation and sorting. In the example provided in the *Using explicit mapping creation* recipe, the referred field was `name.keyword`.

Another important parameter, available only for `text` mapping, is `term_vector` (the vector of terms that compose a string). Please refer to the Lucene documentation for further details at `https://lucene.apache.org/core/8_7_0/core/org/apache/lucene/index/Terms.html`.

`term_vector` can accept the following values:

- `no`: This is the default value; that is, skip term vector.
- `yes`: This is the store term vector.
- `with_offsets`: This is the store term vector with a token offset (start, end position in a block of characters).
- `with_positions`: This is used to store the position of the token in the term vector.
- `with_positions_offsets`: This stores all the term vector data.
- `with_positions_payloads`: This is used to store the position and payloads of the token in the term vector.
- `with_positions_offsets_payloads`: This stores all the term vector data with payloads.

Term vectors allow fast highlighting but consume disk space due to storing additional text information. It's a best practice to only activate it in fields that require highlighting, such as title or document content.

See also

You can refer to the following sources for further details on the concepts of this chapter:

- The online documentation on Elasticsearch provides a full description of all the properties for the different mapping fields at `https://www.elastic.co/guide/en/elasticsearch/reference/master/mapping-params.html`.
- The *Specifying a different analyzer* recipe at the end of this chapter shows alternative analyzers to the standard one.
- For newcomers who want to explore the concepts of tokenization, I would suggest reading the official Elasticsearch documentation at `https://www.elastic.co/guide/en/elasticsearch/reference/current/analysis-tokenizers.html`.

Mapping arrays

Array or multi-value fields are very common in data models (such as multiple phone numbers, addresses, names, aliases, and so on), but they're not natively supported in traditional SQL solutions.

In SQL, multi-value fields require you to create accessory tables that must be joined to gather all the values, leading to poor performance when the cardinality of the records is huge.

Elasticsearch, which works natively in JSON, provides support for multi-value fields transparently.

Getting ready

You will need an up-and-running Elasticsearch installation, as we described in the *Downloading and installing Elasticsearch* recipe of *Chapter 1, Getting Started*.

To execute the commands in this recipe, you can use any HTTP client, such as curl (`https://curl.haxx.se/`), Postman (`https://www.getpostman.com/`), or similar. I suggest using the Kibana console, which provides code completion and better character escaping for Elasticsearch.

How to do it...

To use an `Array` type in our mapping, perform the following steps:

1. Every field is automatically managed as an array. For example, to store tags for a document, the mapping would be as follows:

```
{  "properties" : {
      "name" : {"type" : "keyword"},
      "tag" : {"type" : "keyword", "store" : true},
      ...
   }
}
```

2. This mapping is valid for indexing both documents. The following is the code for `document1`:

```
{"name": "document1", "tag": "awesome"}
```

3. The following is the code for `document2`:

```
{"name": "document2", "tag": ["cool", "awesome",
"amazing"] }
```

How it works...

Elasticsearch transparently manages the array: there is no difference if you declare a single value or a multi-value due to its Lucene core nature.

Multi-values for fields are managed in Lucene, so you can add them to a document with the same field name. For people with a SQL background, this behavior may be quite strange, but this is a key point in the NoSQL world as it reduces the need for a join query and creates different tables to manage multi-values. An array of embedded objects has the same behavior as simple fields.

Mapping an object

The object type is one of the most common field aggregation structures in documental databases.

An object is a base structure (analogous to a record in SQL): in JSON types, they are defined as key/value pairs inside the { } symbols.

Elasticsearch extends the traditional use of objects (which are flat in DBMS), thus allowing for recursive embedded objects.

Getting ready

You will need an up-and-running Elasticsearch installation, as we described in the *Downloading and installing Elasticsearch* recipe of *Chapter 1, Getting Started*.

To execute the commands in this recipe, you can use any HTTP client, such as curl (https://curl.haxx.se/), Postman (https://www.getpostman.com/), or similar. Again, I suggest using the Kibana console, which provides code completion and better character escaping for Elasticsearch.

How to do it...

We can rewrite the mapping code from the previous recipe using an array of items:

```
PUT test/_doc/_mapping
{ "properties" : {
    "id" : {"type" : "keyword"},
    "date" : {"type" : "date"},
    "customer_id" : {"type" : "keyword", "store" : true},
    "sent" : {"type" : "boolean"},
    "item" : {
```

```
        "type" : "object",
      "properties" : {
        "name" : {"type" : "text"},
        "quantity" : {"type" : "integer"},
        "price" : {"type" : "double"},
        "vat" : {"type" : "double"}
} } } }
```

How it works...

Elasticsearch speaks native   ex JSON structure can be mapped in it.

When Elasticsearch is parsii   it tries to extract fields and processes them
as its defined mapping. If no   structure of the object using reflection.

The most important attribute   ct are as follows:

- properties: This is a   tion of fields or objects (we can consider them as
  columns in the SQL wor
- enabled: This establishes whether or not the object should be processed. If it's set
  to false, the data contained in the object is not indexed and it cannot be searched
  (the default is true).
- dynamic: This allows Elasticsearch to add new field names to the object using
  a reflection on the values of the inserted data. If it's set to false, when you try
  to index an object containing a new field type, it'll be rejected silently. If it's set to
  strict, when a new field type is present in the object, an error will be raised,
  skipping the indexing process. The dynamic parameter allows you to be safe about
  making changes to the document's structure (the default is true).

The most used attribute is properties, which allows you to map the fields of the object
in Elasticsearch fields.

Disabling the indexing part of the document reduces the index size; however, the data
cannot be searched. In other words, you end up with a smaller file on disk, but there is a
cost in terms of functionality.

See also

Some special objects are described in the following recipes:

- The *Mapping a document* recipe
- The *Managing a child document with a join field* recipe
- The *Mapping nested objects* recipe

Mapping a document

The **document mapping** is also referred to as the **root object**. This has special parameters that control its behavior, and they are mainly used internally to do special processing, such as routing or time-to-live of documents.

In this recipe, we'll look at these special fields and learn how to use them.

Getting ready

You will need an up-and-running Elasticsearch installation, as we described in the *Downloading and installing Elasticsearch* recipe of *Chapter 1, Getting Started*.

To execute the commands in this recipe, you can use any HTTP client, such as curl (`https://curl.haxx.se/`), Postman (`https://www.getpostman.com/`), or similar. I suggest using the Kibana console, which provides code completion and better character escaping for Elasticsearch.

How to do it...

We can extend the preceding order example by adding some of the special fields, like so:

```
PUT test/_mapping
{ "_source": { "store": true },
    "_routing": { "required": true },
    "_index": { "enabled": true },
    "properties": {} }
```

How it works...

Every special field has parameters and value options, such as the following:

- `_id`: This allows you to index only the ID part of the document. All the ID queries will speed up using the ID value (by default, this is not indexed and not stored).

- `_index`: This controls whether or not the index must be stored as part of the document. It can be enabled by setting the `"enabled":` `true` parameter (`enabled=false` is the default).

- `_source`: This controls how the document's source is stored. Storing the source is very useful, but it's a storage overhead, so it is not required. Consequently, it's better to turn it off (`enabled=true` is the default).

- `_routing`: This defines the shard that will store the document. It supports additional parameters, such as `required` (`true/false`). This is used to force the presence of the routing value, raising an exception if it's not provided.

Controlling how to index and process a document is very important and allows you to resolve issues related to complex data types.

Every special field has parameters to set particular configurations, and some of their behaviors could change in different releases of Elasticsearch.

See also

Please refer to the *Using dynamic templates in document mapping* recipe in this chapter and the *Putting a mapping in an index* recipe of *Chapter 3*, *Basic Operations*, to learn more.

Using dynamic templates in document mapping

In the *Using explicit mapping creation* recipe, we saw how Elasticsearch can guess the field type using reflection. In this recipe, we'll see how we can help it improve its guessing capabilities via dynamic templates.

The dynamic template feature is very useful. For example, it may be useful in situations where you need to create several indices with similar types because it allows you to move the need to define mappings from coded initial routines to automatic index-document creation. Typical usage is to define types for Logstash log indices.

Getting ready

You will need an up-and-running Elasticsearch installation, as we described in the *Downloading and installing Elasticsearch* recipe of *Chapter 1, Getting Started.*

To execute the commands in this recipe, you can use any HTTP client, such as curl (`https://curl.haxx.se/`), Postman (`https://www.getpostman.com/`), or similar. I suggest using the Kibana console, which provides code completion and better character escaping for Elasticsearch.

How to do it...

We can extend the previous mapping by adding document-related settings, as follows:

```
PUT test/_mapping
{
    "dynamic_date_formats":["yyyy-MM-dd", "dd-MM-yyyy"],\
    "date_detection": true,
    "numeric_detection": true,
    "dynamic_templates":[
      {"template1":{
        "match":"*",
        "match_mapping_type": "long",
        "mapping": {"type":" {dynamic_type}", "store": true}
      }}     ],
    "properties" : {...}
}
```

How it works...

The root object (document) controls the behavior of its fields and all its children object fields. In document mapping, we can define the following:

- `date_detection`: This allows you to extract a date from a string (`true` is the default).
- `dynamic_date_formats`: This is a list of valid date formats. This is used if `date_detection` is active.
- `numeric_detection`: This enables you to convert strings into numbers, if possible (`false` is the default).

- `dynamic_templates`: This is a list of templates that are used to change the explicit mapping inference. If one of these templates is matched, the rules that have been defined in it are used to build the final mapping.

A dynamic template is composed of two parts: the matcher and the mapping.

To match a field to activate the template, you can use several types of matchers, such as the following:

- `match`: This allows you to define a match on the field name. The expression is a standard GLOB pattern (`http://en.wikipedia.org/wiki/Glob_ (programming)`).

- `unmatch`: This allows you to define the expression to be used to exclude matches (optional).

- `match_mapping_type`: This controls the types of the matched fields; for example, string, integer, and so on (optional).

- `path_match`: This allows you to match the dynamic template against the full dot notation of the field; for example, `obj1.*.value` (optional).

- `path_unmatch`: This will do the opposite of `path_match`, excluding the matched fields (optional).

- `match_pattern`: This allows you to switch the matchers to `regex` (regular expression); otherwise, the glob pattern match is used (optional).

The dynamic template mapping part is a standard one but can use special placeholders, such as the following:

- `{name}`: This will be replaced with the actual dynamic field name.
- `{dynamic_type}`: This will be replaced with the type of the matched field.

The order of the dynamic templates is very important; only the first one that is matched is executed. It is good practice to order the ones with more strict rules first, and then the others.

There's more...

Dynamic templates are very handy when you need to set a mapping configuration to all the fields. This can be done by adding a dynamic template, similar to this one:

```
"dynamic_templates" : [
  { "store_generic" : {
      "match" : "*", "mapping" : { "store" : true }
} } ]
```

In this example, all the new fields, which will be added with explicit mapping, will be stored.

See also

- You can find the default Elasticsearch behavior for creating a mapping in the *Using explicit mapping creation* recipe and the base way of defining a mapping in the *Mapping a document* recipe.

- The glob pattern is available at `http://en.wikipedia.org/wiki/Glob_pattern`.

Managing nested objects

There is a special type of embedded object called a nested object. This resolves a problem related to Lucene's indexing architecture, in which all the fields of embedded objects are viewed as a single object (technically speaking, they are flattened). During the search, in Lucene, it is not possible to distinguish between values and different embedded objects in the same multi-valued array.

If we consider the previous order example, it's not possible to distinguish an item's name and its quantity with the same query since Lucene puts them in the same Lucene document object. We need to index them in different documents and then join them. This entire trip is managed by nested objects and nested queries.

Getting ready

You will need an up-and-running Elasticsearch installation, as we described in the *Downloading and installing Elasticsearch* recipe of *Chapter 1, Getting Started*.

To execute the commands in this recipe, you can use any HTTP client, such as `curl` (`https://curl.haxx.se/`), Postman (`https://www.getpostman.com/`), or similar. I suggest using the Kibana console, which provides code completion and better character escaping for Elasticsearch.

How to do it...

A nested object is defined as a standard object with the nested type.

Regarding the example in the *Mapping an object* recipe, we can change the type from `object` to `nested`, as follows:

```
PUT test/_mapping
{ "properties" : {
        "id" : {"type" : "keyword"},
        "date" : {"type" : "date"},
        "customer_id" : {"type" : "keyword"},
        "sent" : {"type" : "boolean"},
        "item" : {"type" : "nested",
            "properties" : {
                "name" : {"type" : "keyword"},
                "quantity" : {"type" : "long"},
                "price" : {"type" : "double"},
                "vat" : {"type" : "double"}
} } } }
```

How it works...

When a document is indexed, if an embedded object has been marked as `nested`, it's extracted by the original document before being indexed in a new external document and saved in a special index position near the parent document.

In the preceding example, we reused the mapping from the *Mapping an object* recipe, but we changed the type of the item from `object` to `nested`. No other action must be taken to convert an embedded object into a nested one.

The nested objects are special Lucene documents that are saved in the same block of data as its parent – this approach allows for fast joining with the parent document.

Nested objects are not searchable with standard queries, only with nested ones. They are not shown in standard query results.

The lives of nested objects are related to their parents: deleting/updating a parent automatically deletes/updates all the nested children. Changing the parent means Elasticsearch will do the following:

- Mark old documents as deleted.
- Mark all nested documents as deleted.
- Index the new document version.
- Index all nested documents.

There's more...

Sometimes, you must propagate information about the nested objects to their parent or root objects. This is mainly to build simpler queries about the parents (such as terms queries without using nested ones). To achieve this, two special properties of nested objects must be used:

- `include_in_parent`: This makes it possible to automatically add the nested fields to the immediate parent.
- `include_in_root`: This adds the nested object fields to the root object.

These settings add data redundancy, but they reduce the complexity of some queries, thus improving performance.

See also

- Nested objects require a special query to search for them – this will be discussed in the *Using nested queries* recipe of *Chapter 6, Relationships and Geo Queries*.
- The *Managing a child document with a join field* recipe shows another way to manage child/parent relationships between documents.

Managing a child document with a join field

In the previous recipe, we saw how it's possible to manage relationships between objects with the nested object type. The disadvantage of nested objects is their dependence on their parents. If you need to change the value of a nested object, you need to reindex the parent (this causes a potential performance overhead if the nested objects change too quickly). To solve this problem, Elasticsearch allows you to define child documents.

Getting ready

You will need an up-and-running Elasticsearch installation, as we described in the *Downloading and installing Elasticsearch* recipe of *Chapter 1, Getting Started*.

To execute the commands in this recipe, you can use any HTTP client, such as curl (`https://curl.haxx.se/`), Postman (`https://www.getpostman.com/`), or similar. I suggest using the Kibana console, which provides code completion and better character escaping for Elasticsearch.

How to do it...

In the following example, we have two related objects: an **Order** and an **Item**.

Their UML representation is as follows:

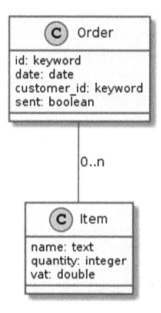

Figure 2.3 – UML example of an Order/Item relationship

The final mapping should merge the field definitions of both `Order` and `Item`, as well as use a special field (`join_field`, in this example) that takes the parent/child relationship.

To use `join_field`, follow these steps:

1. First, we must define the mapping, as follows:

```
PUT test1/_mapping
{ "properties": {
    "join_field": {
      "type": "join", "relations": { "order": "item" }
    },
    "id": { "type": "keyword" },
    "date": { "type": "date" },
    "customer_id": { "type": "keyword" },
    "sent": { "type": "boolean" },
    "name": { "type": "text" },
    "quantity": { "type": "integer" },
    "vat": { "type": "double" }
} }
```

The preceding mapping is very similar to the one in the previous recipe.

2. If we want to store the joined records, we will need to save the parent first and then the children, like so:

```
PUT test/_doc/1?refresh
{ "id": "1", "date": "2018-11-16T20:07:45Z", "customer_
id": "100", "sent": true, "join_field": "order" }
PUT test/_doc/c1?routing=1&refresh
  { "name": "tshirt", "quantity": 10, "price": 4.3, "vat":
8.5,
    "join_field": { "name": "item", "parent": "1" } }
```

The child item requires special management because we need to add `routing` with the parent (1 in the preceding example). Furthermore, we need to specify the parent name and its ID in the object.

How it works...

Mapping, in the case of multiple item relationships in the same index, needs to be computed as the sum of all the other mapping fields.

The relationship between objects must be defined in `join_field`.

There must only be a single `join_field` for mapping; if you need to provide a lot of relationships, you can provide them in the `relations` object.

The child document must be indexed in the same shard as the parent; so, when indexed, an extra parameter must be passed, which is `routing` (we'll learn how to do this in the *Indexing a document* recipe in *Chapter 3, Basic Operations*).

A child document doesn't need to reindex the parent document when we want to change its values. Consequently, it's fast in terms of indexing, reindexing (updating), and deleting.

There's more...

In Elasticsearch, we have different ways to manage relationships between objects, as follows:

- Embedding with `type=object`: This is implicitly managed by Elasticsearch and it considers the embedding as part of the main document. It's fast, but you need to reindex the main document to change the value of the embedded object.

- Nesting with `type=nested`: This allows you to accurately search and filter the parent by using nested queries on children. Everything works for the embedded object except for the query (you must use a nested query to search for them).

- External children documents: Here, the children are the external document, with a `join_field` property to bind them to the parent. They must be indexed in the same shard as the parent. The join with the parent is a bit slower than the nested one. This is because the nested objects are in the same data block as the parent in the Lucene index and they are loaded with the parent; otherwise, the child document requires more read operations.

Choosing how to model the relationship between objects depends on your application scenario.

> **Tip**
> There is also another approach that can be used, but on big data documents, it creates poor performance – decoupling a join relationship. You do the join query in two steps: first, collect the ID of the children/other documents and then search for them in a field of their parent.

See also

Please refer to the *Using the has_child query*, *Using the top_children query*, and *Using the has_parent query* recipes of *Chapter 6*, *Relationships and Geo Queries*, for more details on child/parent queries.

Adding a field with multiple mappings

Often, a field must be processed with several core types or in different ways. For example, a string field must be processed as `tokenized` for search and `not-tokenized` for sorting. To do this, we need to define a `fields` multifield special property.

The `fields` property is a very powerful feature of mappings because it allows you to use the same field in different ways.

Getting ready

You will need an up-and-running Elasticsearch installation, as we described in the *Downloading and installing Elasticsearch* recipe of *Chapter 1*, *Getting Started*.

To execute the commands in this recipe, you can use any HTTP client, such as curl (`https://curl.haxx.se/`), Postman (`https://www.getpostman.com/`), or similar. I suggest using the Kibana console, which provides code completion and better character escaping for Elasticsearch.

How to do it...

To define a multifield property, we need to define a dictionary containing the `fields` subfield. The subfield with the same name as a parent field is the default one.

If we consider the item from our `order` example, we can index the name like so:

```
{ "name": {
    "type": "keyword",
    "fields": {
      "name": {"type": "keyword"},
      "tk": {"type": "text"},
      "code": {"type": "text","analyzer": "code_analyzer"}
  } },
```

If we already have a mapping stored in Elasticsearch and we want to migrate the fields in a multi-field property, it's enough to save a new mapping with a different type, and Elasticsearch provides the merge automatically. New subfields in the `fields` property can be added without problems at any moment, but the new subfields will only be available while you're searching/aggregating newly indexed documents.

When you add a new subfield to already indexed data, you need to reindex your record to ensure you have it correctly indexed for all your records.

How it works...

During indexing, when Elasticsearch processes a `fields` property of the `multifield` type, it reprocesses the same field for every subfield defined in the mapping.

To access the subfields of a multifield, we must build a new path on the base field, plus use the subfield's name. In the preceding example, we have the following:

- `name`: This points to the default multifield subfield-field (the keyword one).
- `name.tk`: This points to the standard analyzed (tokenized) text field.
- `name.code`: This points to a field that was analyzed with a code extractor analyzer.

As you may have noticed in the preceding example, we changed the analyzer to introduce a code extractor analyzer that allows you to extract the item code from a string.

By using the multifield, if we index a string such as `Good Item to buy - ABC1234`, we'll have the following:

- `name` = `Good Item to buy - ABC1234` (useful for sorting)
- `name.tk`= `["good", "item", "to", "buy", "abc1234"]` (useful for searching)
- `name.code` = `["ABC1234"]` (useful for searching and aggregations)

In the case of the code analyzer, if the code is not found in the string, no tokens are generated. This makes it possible to develop solutions that carry out information retrieval tasks at index time and uses these at search time.

There's more...

The `fields` property is very useful in data processing because it allows you to define several ways to process field data.

For example, if we are working on documental content (such as articles, word documents, and so on), we can define fields as subfield analyzers to extract names, places, date/time, geolocation, and so on.

The subfields of a multifield are standard core type fields – we can perform every process we want on them, such as search, filter, aggregation, and scripting.

See also

To find out more about what Elasticsearch analyzers you can use, please refer to the *Specifying different analyzers* recipe.

Mapping a GeoPoint field

Elasticsearch natively supports the use of geolocation types – special types that allow you to localize your document in geographic coordinates (latitude and longitude) around the world.

Two main types are used in the geographic world: the point and the shape. In this recipe, we'll look at **GeoPoint** – the base element of geolocation.

Getting ready

You will need an up-and-running Elasticsearch installation, as we described in the *Downloading and installing Elasticsearch* recipe of *Chapter 1, Getting Started*.

To execute the commands in this recipe, you can use any HTTP client, such as curl (`https://curl.haxx.se/`), Postman (`https://www.getpostman.com/`), or similar. I suggest using the Kibana console, which provides code completion and better character escaping for Elasticsearch.

How to do it...

The type of the field must be set to `geo_point` to define a GeoPoint.

We can extend the order example by adding a new field that stores the location of a customer. This will result in the following output:

```
PUT test/_mapping
{ "properties": {
    "id": {"type": "keyword",},
    "date": {"type": "date"},
    "customer_id": {"type": "keyword"},
    "customer_ip": {"type": "ip"},
    "customer_location": {"type": "geo_point"},
    "sent": {"type": "boolean"}
} }
```

How it works...

When Elasticsearch indexes a document with a GeoPoint field (`lat_lon`), it processes the latitude and longitude coordinates and creates special accessory field data to provide faster query capabilities on these coordinates. This is because a special data structure is created to internally manage latitude and longitude.

Depending on the properties, given the latitude and longitude, it's possible to compute the geohash value (for details, I suggest reading `https://www.pubnub.com/learn/glossary/what-is-geohashing/`). The indexing process also optimizes these values for special computation, such as distance, ranges, and shape match.

GeoPoint has special parameters that allow you to store additional geographic data:

- `lat_lon` (the default is `false`): This allows you to store the latitude and longitude as the `.lat` and `.lon` fields. Storing these values improves the performance of many memory algorithms that are used in distance and shape calculus.

  It makes sense to set `lat_lon` to `true` so that you store them if there is a single point value for a field. This speeds up searches and reduces memory usage during computation.

- `geohash` (the default is `false`): This allows you to store the computed geohash value.

- `geohash_precision` (the default is `12`): This defines the precision to be used in geohash calculus.

For example, given a geo point value, `[45.61752, 9.08363]`, it can be stored using one of the following syntaxes:

- `customer_location = [45.61752, 9.08363]`
- `customer_location.lat = 45.61752`
- `customer_location.lon = 9.08363`
- `customer_location.geohash = u0n7w8qmrfj`

There's more...

GeoPoint is a special type and can accept several formats as input:

- `lat` and `lon` as properties, as shown here:

```
{ "customer_location": { "lat": 45.61752, "lon": 9.08363
},
```

- `lan` and `lon` as strings, as follows:

```
"customer_location": "45.61752,9.08363",
```

- `geohash` as a string, as shown here:

```
"customer_location": "u0n7w8qmrfj",
```

- As a `GeoJSON` array (note that here, `lat` and `lon` are reversed), as shown in the following code snippet:

```
"customer_location": [9.08363, 45.61752]
```

Mapping a GeoShape field

An extension of the concept of a point is its shape. Elasticsearch provides a type that allows you to manage arbitrary polygons in GeoShape.

Getting ready

You will need an up-and-running Elasticsearch installation, as we described in the *Downloading and installing Elasticsearch* recipe of *Chapter 1, Getting Started.*

To be able to use advanced shape management, Elasticsearch requires two JAR libraries in its `classpath` (usually the `lib` directory), as follows:

- Spatial4J (v0.3)
- JTS (v1.13)

How to do it...

To map a `geo_shape` type, a user must explicitly provide some parameters:

- `tree` (the default is `geohash`): This is the name of the `PrefixTree` implementation – `GeohashPrefixTree` and `quadtree` for `QuadPrefixTree`.

- `precision`: This is used instead of `tree_levels` to provide a more human value to be used in the tree level. The precision number can be followed by the unit; that is, 10 m, 10 km, 10 miles, and so on.

- `tree_levels`: This is the maximum number of layers to be used in the prefix tree.

- `distance_error_pct`: This sets the maximum errors that are allowed in a prefix tree (`0,025%` - `max 0,5%` by default).

The `customer_location` mapping, which we saw in the previous recipe using `geo_shape`, will be as follows:

```
"customer_location": {
   "type": "geo_shape",
   "tree": "quadtree",
   "precision": "1m" },
```

How it works...

When a shape is indexed or searched internally, a path tree is created and used.

A path tree is a list of terms that contain geographic information and are computed to improve performance in evaluating geo calculus.

The path tree also depends on the shape's type: point, linestring, polygon, multipoint, or multipolygon.

See also

To understand the logic behind the GeoShape, some good resources are the Elasticsearch page, which tells you about GeoShape, and the sites of the libraries that are used for geographic calculus (`https://github.com/spatial4j/spatial4j` and `http://central.maven.org/maven2/com/vividsolutions/jts/1.13/`, respectively).

Mapping an IP field

Elasticsearch is used in a lot of systems to collect and search logs, such as Kibana (`https://www.elastic.co/products/kibana`) and LogStash (`https://www.elastic.co/products/logstash`). To improve search when using IP addresses, Elasticsearch provides the IPv4 and IPv6 types, which can be used to store IP addresses in an optimized way.

Getting ready

You will need an up-and-running Elasticsearch installation, as we described in the *Downloading and installing Elasticsearch* recipe of *Chapter 1, Getting Started*.

How to do it...

You need to define the type of field that contains an IP address as `ip`.

Regarding the preceding order example, we can extend it by adding the customer IP, like so:

```
"customer_ip": { "type": "ip" }
```

The IP must be in the standard point notation form, as follows:

```
"customer_ip":"19.18.200.201"
```

How it works...

When Elasticsearch is processing a document and if a field is an IP one, it tries to convert its value into a numerical form and generates tokens for fast value searching.

The IP has special properties:

- `index` (the default is `true`): This defines whether the field must be indexed. If not, `false` must be used.

- `doc_values` (the default is `true`): This defines whether the field values should be stored in a column-stride fashion to speed up sorting and aggregations.

The other properties (`store`, `boost`, `null_value`, and `include_in_all`) work as other base types.

The advantage of using IP fields over strings is more speed in every range and filter and lower resource usage (disk and memory).

Mapping an Alias field

It is very common to have a lot of different types in several indices. Because Elasticsearch makes it possible to search in many indices, you should filter for common fields at the same time.

In the real world, these fields are not always called in the same way in all mappings (generally because they are derived from different entities); it's very common to have a mix of the `added_date`, `timestamp`, `@timestamp`, and `date_add` fields, all of which are referring to the same date concept.

The `alias` fields allow you to define an alias name to be resolved, as well as a query time to simplify the call for all the fields with the same meaning.

Getting ready

You will need an up-and-running Elasticsearch installation, as we described in the *Downloading and installing Elasticsearch* recipe of *Chapter 1, Getting Started*.

To execute the commands in this recipe, you can use any HTTP client, such as curl (`https://curl.haxx.se/`), Postman (`https://www.getpostman.com/`), or similar. I suggest using the Kibana console, which provides code completion and better character escaping for Elasticsearch.

How to do it...

If we take the order example that we saw in the previous recipes, we can add an alias for the price value to cost in the item subfield.

This can be achieved by following these steps:

1. To add this alias, we need to have a mapping that's similar to the following:

```
PUT test/_mapping
{ "properties": {
    "id": {"type": "keyword"},
    "date": {"type": "date"},
    "customer_id": {"type": "keyword"},
    "sent": {"type": "boolean"},
    "item": {
      "type": "object",
      "properties": {
        "name": {"type": "keyword"},
        "quantity": {"type": "long"},
        "price": {"type": "double"},
        "vat": {"type": "double"}
} } } }
```

2. Now, we can index a record, as follows:

```
PUT test/_doc/1?refresh
{ "id": "1", "date": "2018-11-16T20:07:45Z",
  "customer_id": "100", "sent": true,
  "item": [ { "name": "tshirt", "quantity": 10, "price":
4.3, "vat": 8.5 } ] }
```

3. We can search it using the cost alias, like so:

```
GET test/_search
{ "query": { "term": { "item.cost": 4.3 } } }
```

The result will be the saved document.

How it works...

The alias is a convenient way to use the same name for your search field without the need to change the data structure of your fields. An alias field doesn't need to change a document's structure, thus allowing more flexibility for your data models.

The alias is resolved when the search indices in the query are expanded and have no performance penalties due to its usage.

If you try to index a document with a value in an `alias` field, an exception will be thrown.

The `path` value of the `alias` field must contain the full resolution of the target field, which must be concrete and must be known when the alias is defined.

In the case of an alias in a nested object, it must be in the same nested scope as the target.

Mapping a Percolator field

The Percolator is a special type of field that makes it possible to store an Elasticsearch query inside the field and use it in a `percolator` query.

The Percolator can be used to detect all the queries that match a document.

Getting ready

You will need an up-and-running Elasticsearch installation, as we described in the *Downloading and installing Elasticsearch* recipe of *Chapter 1, Getting Started*.

To execute the commands in this recipe, you can use any HTTP client, such as curl (`https://curl.haxx.se/`), Postman (`https://www.getpostman.com/`), or similar. I suggest using the Kibana console, which provides code completion and better character escaping for Elasticsearch.

How to do it...

To map a `percolator` field, follow these steps:

1. We want to create a Percolator that matches some text in a `body` field. We can define the mapping like so:

    ```
    PUT test-percolator
    { "mappings": {
        "properties": {
          "query": { "type": "percolator"  },
          "body": { "type": "text" }
    } } }
    ```

2. Now, we can store a document with a `percolator` query inside it, as follows:

    ```
    PUT test-percolator/_doc/1?refresh
    { "query": { "match": { "body": "quick brown fox"  }}}
    ```

3. Now, let's execute a search on it, as shown in the following code:

    ```
    GET test-percolator/_search
    { "query": {
        "percolate": {
          "field": "query",
          "document": { "body": "fox jumps over the lazy dog"
    } } } }
    ```

4. This will result in us retrieving the hits of the stored document, as follows:

    ```
    {
        ... truncated...
        "hits" : [
          {
            "_index" : "test-percolator", "_id" : "1",
            "_score" : 0.13076457,
            "_source" : {
                "query" : {
                    "match" : { "body" : "quick brown fox" }
    ```

```
            }
        },
        "fields" : { "_percolator_document_slot" : [0]
    } } ] } }
```

How it works...

The `percolator` field stores an Elasticsearch query inside it.

Because all the Percolators are cached and are always active for performances, all the fields that are required in the query must be defined in the mapping of the document.

Since all the queries in all the Percolator documents will be executed against every document, for the best performance, the query inside the Percolator must be optimized so that they're executed quickly inside the `percolator` query.

Mapping the Rank Feature and Feature Vector fields

It's common to want to score a document dynamically, depending on the context. For example, if you need to score more documents that are inside a category, the classic scenario is to boost (increase low-scored) documents that are based on a value, such as page rank, hits, or categories.

Elasticsearch provides two new ways to boost your scores based on values. One is the Rank Feature field, while the other is its extension, which is to use a vector of values.

Getting ready

You will need an up-and-running Elasticsearch installation, as we described in the *Downloading and installing Elasticsearch* recipe of *Chapter 1, Getting Started*.

To execute the commands in this recipe, you can use any HTTP client, such as curl (`https://curl.haxx.se/`), Postman (`https://www.getpostman.com/`), or similar. I suggest using the Kibana console, which provides code completion and better character escaping for Elasticsearch.

How to do it...

We want to use the `rank_feature` type to implement a common PageRank scenario where documents are scored based on the same characteristics. To achieve this, follow these steps:

1. To be able to score based on a `pagerank` value and an inverse `url` length, we can use the following mapping:

```
PUT test-rank
{  "mappings": {
    "properties": {
      "pagerank": { "type": "rank_feature" },
      "url_length": {
        "type": "rank_feature",
        "positive_score_impact": false
} } } }
```

2. Now, we can store a document, as shown here:

```
PUT test-rank/_doc/1
{ "pagerank": 5, "url_length": 20 }
```

3. Now, we can execute a feature query on the `pagerank` value to return our record with a similar query, like so:

```
GET test-rank/_search
{ "query": { "rank_feature": { "field":"pagerank" }}}
```

> **Important Note**
>
> To query the special rank/`rank_features` types, we need to use the special `rank_feature` query type, which is only used for this special case.

The evolution of the previous feature's functionality is to define a vector of values using the `rank_features` type; usually, it can be used to score by topics, categories, or similar discerning facets. We can implement this functionality by following these steps:

1. First, we must define the mapping for the `categories` field:

```
PUT test-ranks
{ "mappings": {
    "properties": {
      "categories": { "type": "rank_features"  } } } }
```

2. Now, we can store some documents in the index by using the following commands:

```
PUT test-ranks/_doc/1
{ "categories": { "sport": 14.2, "economic": 24.3 } }
PUT test-ranks/_doc/2
{ "categories": { "sport": 19.2, "economic": 23.1 } }
```

3. Now, we can search based on the saved feature values, as shown here:

```
GET test-ranks/_search
{ "query": { "feature": { "field": "categories.sport"    }
} }
```

How it works...

rank_feature and rank_features are special type fields that are used for storing values and are mainly used to score the results.

> **Important Note**
> The values that are stored in these fields can only be queried using the feature query. This cannot be used in standard queries and aggregations.

The value numbers in rank_feature and rank_features can only be single positive values (multi-values are not allowed).

In the case of rank_features, the values must be a hash, composed of a string and a positive numeric value.

There is a flag that changes the behavior of scoring – positive_score_impact. This value is true by default, but if you want the value of the feature to decrease the score, you can set it to false. In the pagerank example, the length of url reduces the score of the document because the longer url is, the less relevant it becomes.

Mapping the Search as you type field

One of the most common scenarios is to provide the `Search as you type` functionality, which is typical of the Google search engine.

This capability is common in many use cases:

- Completing titles in media websites
- Completing product names in e-commerce websites
- Completing document names or authors in document management systems
- Suggesting best-associated terms to search on based on the actual knowledge base (collection of documents)

This type provides facilities to achieve this functionality.

Getting ready

You will need an up-and-running Elasticsearch installation, as we described in the *Downloading and installing Elasticsearch* recipe of *Chapter 1, Getting Started*.

To execute the commands in this recipe, you can use any HTTP client, such as curl (`https://curl.haxx.se/`), Postman (`https://www.getpostman.com/`), or similar. I suggest using the Kibana console, which provides code completion for Elasticsearch.

How to do it...

We want to use the `search_as_you_type` type to implement a completer (a widget that completes names/values) for titles for our media film streaming platform. To achieve this, follow these steps:

1. To be able to prove `"search as you type"` on a title field, we will use the following mapping:

```
PUT test-sayt
{ "mappings": {
    "properties": {
      "title": { "type": "search_as_you_type"  }
} } }
```

2. Now, we can store some documents, as shown here:

```
PUT test-sayt/_doc/1
{ "title": "Ice Age" }


PUT test-sayt/_doc/2
{ "title": "The Polar Express" }


PUT test-sayt/_doc/3
{ "title": "The Godfather" }
```

3. Now, we can execute a match query on the `title` value to return our records:

```
GET test-sayt/_search
{
  "query": {
    "multi_match": {
      "query": "the p", "type": "bool_prefix",
      "fields": [ "title", "title._2gram", "title._3gram"
    ]
} } }
```

The result will be something similar to the following:

```
{
  ...truncated...
    "hits" : [
      {
        "_index" : "test-sayt", "_id" : "2", "_score" :
2.4208174,
        "_source" : { "title" : "The Polar Express" }
      },
    ...truncated...
}
```

As you can see, more relevant results (that contain more code related to the search) score better!

How it works...

Due to the high demand for the *Search as you type* feature, this special mapping type was created.

This special mapping type is a helper that simplifies the process of creating a field with multiple subfields that can map the indexing requirements and provide an efficient *Search as you type* capability.

For example, for my `title` field, the following field and subfields are created:

- `title`: This contains the text to be used. It's processed as a standard text field and accepts the standard `text` parameters, as we saw regarding the `text` field in the *Mapping base types* recipe of this chapter.

- `title._2gram`: This contains the text with the applied shingle token filter (`https://www.elastic.co/guide/en/elasticsearch/reference/current/analysis-shingle-tokenfilter.html`) with a size of 2. This aggregates two contiguous terms.

- `title._3gram`: This is the same as `title._2gram` but uses a size of 3 to aggregate three contiguous terms.

- `title._index_prefix`: This wraps the maximum size gram (`_3gram`, in our case) with an Edge N-Gram Token Filter (`https://www.elastic.co/guide/en/elasticsearch/reference/current/analysis-edgengram-tokenfilter.html`) to be able to provide initial completion.

The `"search_as_you_type"` field can be customized using the `max_shingle_size` parameter (the default is 3). This parameter allows you to define the maximum size of the gram to be created.

The number of `ngram` subfields is given by the `max_shingle_size -1` value, but usually, the best values are 3 or 4. In the case of large values, it only increases the size of the index, but it doesn't generally provide query quality benefits.

See also

Please refer to the *Using a match query* recipe in *Chapter 5, Text and Numeric Queries*, to learn more about match queries.

Using the Range Fields type

Sometimes, we have values that represent a continuous range of values between an upper and lower bound. Some of the common scenarios of this are as follows:

- Price range (that is, from $4 to $10)
- Date interval (that is, from 8 A.M. to 8 P.M., December 2020, Summer 2021, Q3 2020, and so on)

In this case, for most queries, pointing to a value in the middle of them is not easy in Elasticsearch; for example, the worst case is to convert continuous values into discrete ones by extracting all the values using a prefixed interval. This kind of situation will largely increase the size of the index and reduce performance (queries).

Range mappings were created to provide continuous value support in Elasticsearch. For this reason, when it is not possible to store the exact value, but we have a range, we need to use range types.

Getting ready

You will need an up-and-running Elasticsearch installation, as we described in the *Downloading and installing Elasticsearch* recipe of *Chapter 1, Getting Started*.

To execute the commands in this recipe, you can use any HTTP client, such as curl (https://curl.haxx.se/), Postman (https://www.getpostman.com/), or similar. I suggest using the Kibana console, which provides code completion and better character escaping for Elasticsearch.

How to do it...

We want to use range types to implement stock mark values that are defined by low and high price values and the timeframe of the transaction. To achieve this, follow these steps:

1. To populate our stock, we need to create an index with range fields. Let's use the following mapping:

```
PUT test-range
{ "mappings": {
    "properties": {
        "price": { "type": "float_range" },
        "timeframe": { "type": "date_range" }
} } }
```

2. Now, we can store some documents, as shown here:

```
PUT test-range/_bulk
{"index":{"_index":"test-range","_id":"1"}}
{"price":{"gte":1.5,"lt":3.2},"timeframe":{"gte":"2022-
01-01T12:00:00","lt":"2022-01-01T12:00:01"}}
{"index":{"_index":"test-range","_id":"2"}}
{"price":{"gte":1.7,"lt":3.7},"timeframe":{"gte":"2022-
01-01T12:00:01","lt":"2022-01-01T12:00:02"}}
{"index":{"_index":"test-range","_id":"3"}}
{"price":{"gte":1.3,"lt":3.3},"timeframe":{"gte":"2022-
01-01T12:00:02","lt":"2022-01-01T12:00:03"}}
```

3. Now, we can execute a query for filtering on price and timeframe values to check the correct indexing of the data:

```
GET test-range/_search
{ "query": {
    "bool": {
      "filter": [
          { "term": { "price": { "value": 2.4 } } },
            { "term": { "timeframe": { "value": "2022-01-
01T12:00:02" } } }
] } } }
```

The result will be something similar to the following:

```
{
  ...truncated...
    "hits" : [
        { "_index" : "test-range", "_id" : "3", "_score" :
0.0,
          "_source" : {
            "price" : { "gte" : 1.3, "lt" : 3.3 },
            "timeframe" : {
              "gte" : "2022-01-01T12:00:02",
              "lt" : "2022-01-01T12:00:03"
        ...truncated...
  }
```

How it works...

Not all the base types that support ranges can be used in ranges. The possible range types that are supported by Elasticsearch are as follows:

- `integer_range`: This is used to store signed 32-bit integer values.
- `float_range`: This is used to store signed 32-bit floating-point values.
- `long_range`: This is used to store signed 64-bit integer values.
- `double_range`: This is used to store signed 64-bit floating-point values.
- `date_range`: This is used to store date values as 64-bit integers.
- `ip_range`: This is used to store IPv4 and IPv6 values.

These range types are very useful for all cases where the values are not exact.

When you're storing a document in Elasticsearch, the field can be composed using the following parameters:

- `gt` or `gte` for the lower bound of the range
- `lt` or `lte` for the upper bound of the range

> **Note**
> Range types can be used for querying values, but they have limited support for aggregation: they only support histogram and cardinality aggregations.

See also

- The *Using the range query* recipe in *Chapter 5, Text and Numeric Queries*, for range queries
- The *Executing histogram aggregations* recipe in *Chapter 7, Aggregation*

Using the Flattened field type

In many applications, it is possible to define custom metadata or configuration composed of key-value pairs. This use case is not optimal for Elasticsearch. Creating a new mapping for every key will not be easy to manage as they evolve into large mappings.

X-Pack provides a type (free for use) to solve this problem: the `flattened` field type.

As the name suggests, it takes all the key-value pairs (also nested ones) and indices them in a flat way, thus solving the problem of the mapping *explosion*.

Getting ready

You will need an up-and-running Elasticsearch installation, as we described in the *Downloading and installing Elasticsearch* recipe of *Chapter 1, Getting Started.*

To execute the commands in this recipe, you can use any HTTP client, such as curl (https://curl.haxx.se/), Postman (https://www.getpostman.com/), or similar. I suggest using the Kibana console, which provides code completion and better character escaping for Elasticsearch.

How to do it...

We want to use Elasticsearch to store configurations with a varying number of fields. To achieve this, follow these steps:

1. To create our configuration index with a `flattened` field, we will use the following mapping:

    ```
    PUT test-flattened
    { "mappings": {
        "properties": {
          "name": { "type": "keyword" },
          "configs": { "type": "flattened" } } } }
    ```

2. Now, we can store some documents that contain our configuration data:

    ```
    PUT test-flattened/_bulk
    {"index":{"_index":"test-flattened","_id":"1"}}
    {"name":"config1","configs":{"key1":"value1","
    key3":"2022-01-01T12:00:01"}}
    {"index":{"_index":"test-flattened","_id":"2"}}
    {"name":"config2","configs":{"key1":true,"key2":30}}
    {"index":{"_index":"test-flattened","_id":"3"}}
    {"name":"config3","configs":{"key4":"test","key2":30.3}}
    ```

3. Now, we can execute a query that's searching for the text in all the configurations:

    ```
    POST test-flattened/_search
    { "query": { "term": { "configs": "test" } } }
    ```

    Alternatively, we can search for a particular key in the `configs` object, like so:

    ```
    POST test-flattened/_search
    { "query": { "term": { "configs.key4": "test" } } }
    ```

The result for both queries will be as follows:

```
{ ...truncated...
    "hits" : [
        {
        "_index" : "test-flattened",
        "_id" : "3",  "_score" : 1.2330425,
        "_source" : {
          "name" : "config3",
          "configs" : { "key4" : "test", "key2" : 30.3
}
    ...truncated...
```

How it works...

This special field type can take a JSON object that's been passed in a document and flatten key/value pairs that can be searched without defining a mapping for fields in the JSON content.

This helps since the mapping can explode due to the JSON containing a large number of different fields.

During the indexing process, tokens are created for each leaf value of the JSON object using a `keyword` analyzer. Due to this, the number, date, IP, and other formats are converted into text and the only queries that can be executed are the ones that are supported by keyword tokenization. This includes `term`, `terms`, `terms_set`, `prefix`, `range` (this is based on text), `match`, `multi_match`, `query_string`, `simple_query_string`, and `exists`.

See also

See *Chapter 5*, *Text and Numeric Queries*, for more references on the cited query types.

Using the Point and Shape field types

The power of geoprocessing Elasticsearch is used to provide capabilities to a large number of applications. However, it has one limitation: it only works for world coordinates.

Using Point and Shape types, X-Pack extends the geo capabilities to every two-dimensional planar coordinate system.

Common scenarios for this use case include mapping and documenting building coordinates and checking if *documents* are inside a shape.

Getting ready

You will need an up-and-running Elasticsearch installation, as we described in the *Downloading and installing Elasticsearch* recipe of *Chapter 1, Getting Started*.

To execute the commands in this recipe, you can use any HTTP client, such as curl (`https://curl.haxx.se/`), Postman (`https://www.getpostman.com/`), or similar. I suggest using the Kibana console, which provides code completion and better character escaping for Elasticsearch.

How to do it...

We want to use Elasticsearch to map a device's coordinates in our shop. To achieve this, follow these steps:

1. To create our index for storing devices and their location, we will use the following mapping:

```
PUT test-point
{ "mappings": {
    "properties": {
      "device": { "type": "keyword" },
      "location": { "type": "point" } } } }
```

2. Now, we can store some documents that contain our device's data:

```
PUT test-point/_bulk
{"index":{"_index":"test-point","_id":"1"}}
{"device":"device1","location":{"x":10,"y":10}}
{"index":{"_index":"test-point","_id":"2"}}
{"device":"device2","location":{"x":10,"y":15}}
{"index":{"_index":"test-point","_id":"3"}}
{"device":"device3","location":{"x":15,"y":10}}
```

At this point, we want to create shapes in our shop so that we can divide it into parts and check if the people/devices are inside the defined shape. To do this, follow these steps:

1. First, let's create an index to store our shapes:

```
PUT test-shape
{ "mappings": {
    "properties": {
      "room": { "type": "keyword" },
      "geometry": { "type": "shape" } } } }
```

2. Now, we can store a document to test the mapping:

```
POST test-shape/_doc/1
{ "room":"hall",
  "geometry" : {
    "type" : "polygon",
    "coordinates" : [
        [ [8.0, 8.0], [8.0, 12.0], [12.0, 12.0], [12.0,
8.0], [8.0, 8.0]] ] } }
```

3. Now, let's search our devices in our stored shape:

```
POST test-point/_search
{ "query": {
    "shape": {
      "location": {
        "indexed_shape": { "index": "test-shape", "id":
"1", "path": "geometry" } } } } }
```

The result for both queries will be as follows:

```
{ ...truncated...
    "hits" : [ {
  "_index" : "test-point",   "_id" : "1", "_score" : 0.0,
        "_source" : {
          "device" : "device1",
          "location" : { "x" : 10, "y" : 10 }
    ...truncated...
```

How it works...

The `point` and `shape` types are used to manage every type of two-dimensional planar coordinate system inside documents. Their usage is similar to `geo_point` and `geo_shape`.

The advantage of storing shapes in Elasticsearch is that you can simplify how you match constraints between coordinates and shapes. This was shown in our query example, where we loaded the shape's geometry from the `test-shape` index and the search from the `test-point` index.

Managing coordinate systems and shapes is a very large topic that requires knowledge of shape types and geo models since they are strongly bound to data models.

See also

- The official documentation for Point types can be found at `https://www.elastic.co/guide/en/elasticsearch/reference/current/point.html`, while the official documentation for Shape types can be found at `https://www.elastic.co/guide/en/elasticsearch/reference/current/shape.html`.

- The official documentation about Shape Query can be found at `https://www.elastic.co/guide/en/elasticsearch/reference/current/query-dsl-shape-query.html`.

Using the Dense Vector field type

Elasticsearch is often used to store machine learning data for training algorithms. X-Pack provides the Dense Vector field to store vectors that have up to 2,048 dimension values.

Getting ready

You will need an up-and-running Elasticsearch installation, as we described in the *Downloading and installing Elasticsearch* recipe of *Chapter 1*, *Getting Started*.

To execute the commands in this recipe, you can use any HTTP client, such as curl (`https://curl.haxx.se/`), Postman (`https://www.getpostman.com/`), or similar. I suggest using the Kibana console, which provides code completion and better character escaping for Elasticsearch.

How to do it...

We want to use Elasticsearch to store a vector of values for our machine learning models. To achieve this, follow these steps:

1. To create an index to store a vector of values, we will use the following mapping:

```
PUT test-dvector
{ "mappings": {
    "properties": {
        "vector": { "type": "dense_vector", "dims": 4 },
        "model": { "type": "keyword" } } } }
```

2. Now, we can store a document to test the mapping:

```
POST test-dvector/_doc/1
{ "model":"pipe_flood", "vector" : [8.1, 8.3, 12.1, 7.32]
}
```

How it works...

The Dense Vector field is a helper field for storing vectors in Elasticsearch.

The ingested data for the field must be a list of floating-point values with the exact dimension of the value provided by the dims property of the mapping (4, in our example).

If the dimension of the vector field is incorrect, an exception is raised, and the document is not indexed.

For example, let's see what happens when we try to index a similar document with the wrong feature dimension:

```
POST test-dvector/_doc/1
{ "model":"pipe_flood", "vector" : [8.1, 8.3, 12.1] }
```

We will see a similar exception that enforces the right dimension size. Here, the document will not be stored:

```
{
  "error" : {
    "root_cause" : [
      {
        "type" : "mapper_parsing_exception",
```

```
            "reason" : "failed to parse"
      }
  ],
    "type" : "mapper_parsing_exception",
    "reason" : "failed to parse",
    "caused_by" : {
      "type" : "illegal_argument_exception",
      "reason" : "Field [vector] of type [dense_vector] of
doc [1] has number of dimensions [3] less than defined in the
mapping [4]"
    }
  },
  "status" : 400
}
```

Using the Histogram field type

Histograms are a common data type for analytics and machine learning analysis. We can store Histograms in the form of values and counts; they are not indexed, but they can be used in aggregations.

The `histogram` field type is a special mapping that's available in X-Pack that is commonly used to store the results of Histogram aggregations in Elasticsearch for further processing, such as to compare the aggregation results at different times.

Getting ready

You will need an up-and-running Elasticsearch installation, as described in the *Downloading and installing Elasticsearch* recipe of *Chapter 1, Getting Started*.

To execute the commands in this recipe, you can use any HTTP client, such as curl (`https://curl.haxx.se/`), Postman (`https://www.getpostman.com/`), or similar. I suggest using the Kibana console, which provides code completion and better character escaping for Elasticsearch.

How to do it...

In this recipe, we will simulate a common use case of Histogram data that is stored in Elasticsearch. Here, we will use a Histogram that specifies the millimeters of rain divided by year for our advanced analytics solution. To achieve this, follow these steps:

1. First, let's create an index for the Histogram by using the following mapping:

```
PUT test-histo
{ "mappings": {
    "properties": {
      "histogram": { "type": "histogram" },
      "model": { "type": "keyword" } } } }
```

2. Now, we can store a document to test the mapping:

```
POST test-histo/_doc/1
{ "model":"show_level", "histogram" : { "values" : [2016,
2017, 2018, 2019, 2020, 2021],  "counts" : [283, 337,
323, 312, 236, 232] } }
```

How it works...

The `histogram` field type specializes in storing Histogram data. I must be provided as a JSON object composed of the `values` and `counts` fields with the same cardinality of items. The only supported aggregations are the following ones. We will look at these in more detail in *Chapter 7, Aggregations*:

-   Metric aggregations such as `min`, `max`, `sum`, `value_count`, and `avg`
-   The `percentiles` and `percentile_ranks` aggregations
-   The `boxplot` aggregation
-   The `histogram` aggregation

The data is not indexed, but you can also check the existence of a document by populating this field with the `exist` query.

See also

- Aggregations will be discussed in more detail in *Chapter 7, Aggregations*
- The *Using the exist query* recipe in *Chapter 5, Text and Numeric Queries*

Adding metadata to a mapping

Sometimes, when we are working with our mapping, we may need to store some additional data to be used for display purposes, ORM facilities, permissions, or simply to track them in the mapping.

Elasticsearch allows you to store every kind of JSON data you want in the mapping with the special _meta field.

Getting ready

You will need an up-and-running Elasticsearch installation, as we described in the *Downloading and installing Elasticsearch* recipe of *Chapter 1, Getting Started*.

How to do it...

The _meta mapping field can be populated with any data we want in JSON format, like so:

```
{ "_meta": {
    "attr1": ["value1", "value2"],
    "attr2": { "attr3": "value3" }
  } }
```

How it works...

When Elasticsearch processes a new mapping and finds a _meta field, it stores it as-is in the global mapping status and propagates the information to all the cluster nodes. The content of the _meta files is only checked to ensure it's a valid JSON format. Its content is not taken into consideration by Elasticsearch. You can populate it with everything you need to be in JSON format.

_meta is only used for storing purposes; it's not indexed and searchable. It can be used to enrich your mapping with custom information that can be used by your applications.

It can be used for the following reasons:

- Storing type metadata:

```
{"name": "Address", "description": "This entity store
address information"}
```

- Storing **object relational mapping** (**ORM**)-related information (such as mapping class and mapping transformations):

```
{"class": "com.company.package.AwesomeClass",
"properties" : { "address":{"class": "com.company.
package.Address"}} }
```

- Storing type permission information:

```
{"read":["user1", "user2"], "write":["user1"]}
```

- Storing extra type information (that is, the `icon` filename, which is used to display the type):

```
{"icon":"fa fa-alert" }
```

- Storing template parts for rendering web interfaces:

```
{"fragment":"<div><h1>$name</h1><p>$description</p></
div>" }
```

Specifying different analyzers

In the previous recipes, we learned how to map different fields and objects in Elasticsearch, and we described how easy it is to change the standard analyzer with the `analyzer` and `search_analyzer` properties.

In this recipe, we will look at several analyzers and learn how to use them to improve indexing and searching quality.

Getting ready

You will need an up-and-running Elasticsearch installation, as we described in the *Downloading and installing Elasticsearch* recipe of *Chapter 1, Getting Started*.

How to do it...

Every core type field allows you to specify a custom analyzer for indexing and for searching as field parameters.

For example, if we want the name field to use a standard analyzer for indexing and a simple analyzer for searching, the mapping will be as follows:

```
{ "name": {
    "type": "string",
    "index_analyzer": "standard",
    "search_analyzer": "simple"
 } }
```

How it works...

The concept of the analyzer comes from Lucene (the core of Elasticsearch). An analyzer is a Lucene element that is composed of a tokenizer that splits text into tokens, as well as one or more token filters. These filters carry out token manipulation such as lowercasing, normalization, removing stop words, stemming, and so on.

During the indexing phase, when Elasticsearch processes a field that must be indexed, an analyzer is chosen. First, it checks whether it is defined in the index_analyzer field, then in the document, and finally, in the index.

Choosing the correct analyzer is essential to getting good results during the query phase.

Elasticsearch provides several analyzers in its standard installation. The following table shows the most common ones:

Name	Description
standard	This divides the text using a standard tokenizer: to normalize lowercase tokens, and remove unwanted tokens.
simple	This splits non-letter text and converts it into lowercase.
whitespace	This divides text on space separators.
stop	This processes the text with the standard analyzer, then applies custom stopwords.
keyword	This considers all text as a token.
pattern	This divides text using a regular expression.
snowball	This works as a standard analyzer, as well as a stemming at the end of processing.

Figure 2.4 – List of the most common general-purpose analyzers

For special language purposes, Elasticsearch supports a set of analyzers aimed at analyzing text in a specific language, such as Arabic, Armenian, Basque, Brazilian, Bulgarian, Catalan, Chinese, CJK, Czech, Danish, Dutch, English, Finnish, French, Galician, German, Greek, Hindi, Hungarian, Indonesian, Italian, Norwegian, Persian, Portuguese, Romanian, Russian, Spanish, Swedish, Turkish, and Thai.

See also

Several Elasticsearch plugins extend the list of available analyzers. The most famous ones are as follows:

- The ICU analysis plugin (`https://www.elastic.co/guide/en/elasticsearch/plugins/master/analysis-icu.html`)

- The Phonetic analysis plugin (`https://www.elastic.co/guide/en/elasticsearch/plugins/master/analysis-phonetic.html`)

- The Smart Chinese analysis plugin (`https://www.elastic.co/guide/en/elasticsearch/plugins/master/analysis-smartcn.html`)

- The Japanese (kuromoji) analysis plugin (`https://www.elastic.co/guide/en/elasticsearch/plugins/master/analysis-kuromoji.html`)

Using index components and templates

Real-world index mapping can be very complex and often, parts of it can be reused between different indices types. To be able to simplify this management, mappings can be divided into the following:

- **Components**: These will collect the reusable parts of the mapping.

- **Index templates**: These aggregate the components in a single template.

Using components is the most manageable way to scale on large index mappings because they can simplify large template management.

Getting ready

You will need an up-and-running Elasticsearch installation, as we described in the *Downloading and installing Elasticsearch* recipe of *Chapter 1*, *Getting Started*.

To execute the commands in this recipe, you can use any HTTP client, such as curl (`https://curl.haxx.se/`), Postman (`https://www.getpostman.com/`), or similar. I suggest using the Kibana console, which provides code completion and better character escaping for Elasticsearch.

How to do it...

We want to build an index mapping composed of two reusable components. To achieve this, follow these steps:

1. First, we will create three components for the timestamp, order, and items. These will store parts of our index mapping:

```
PUT _component_template/timestamp-management
{ "template": {
    "mappings": {
      "properties": {
        "@timestamp": { "type": "date"  } } } } }
```

```
PUT _component_template/order-data
{ "template": {
    "mappings": {
      "properties": {
        "id": { "type": "keyword" },
        "date": { "type": "date" },
        "customer_id": { "type": "keyword" },
        "sent": { "type": "boolean" } } } } }
```

```
PUT _component_template/items-data
{ "template": {
    "mappings": {
      "properties": {
        "item": {
          "type": "object",
          "properties": {
            "name": { "type": "keyword" },
            "quantity": { "type": "long" },
            "cost": { "type": "alias", "path": "item.
price" },
            "price": { "type": "double" },
            "vat": { "type": "double" } } } } } } }
```

2.   Now, we can create an index template that can sum them up:

```
PUT _index_template/order
{
  "index_patterns": ["order*"],
  "template": {
    "settings": { "number_of_shards": 1 },
    "mappings": {
      "properties": { "id": { "type": "keyword" } }
    },
    "aliases": { "order": { } }
  },
  "priority": 200,
  "composed_of": ["timestamp-management", "order-data",
"items-data"],
  "version": 1,
  "_meta": { "description": "My order index template" } }
```

How it works...

The process of using index components to build indices templates is very simple: you can register as many components as you wish (*Steps 1* and *2* in this recipe) and then aggregate them when you define the template (*Step 3*). By using this approach, your template is divided into blocks, and the index template is simpler to manage and easily reusable.

For simple use cases, using components to build indices template is too verbose. This approach shines when you need to manage different logs or documents in Elasticsearch that have common parts because you can refactorize them very quickly and reuse them.

Components are simple partial templates that are merged in an index template. Here, the parameters are as follows:

- `index_patterns`: This is a list of index glob patterns. When an index is created, if its name matches the glob patterns, the template is applied when the index is created.

- `aliases`: This is an optional alias definition to be applied to the created index.

- `template`: This is the template to be applied to the index.

- `priority`: This is an optional order of priority for applying this template. The standard priority of ELK components is 100, so if the value is set below 100, a custom template can override an ELK one.

- `version`: This is an optional incremental number that is managed by the user to keep track of the updates that are made to the template.

- `_meta`: This is an optional JSON object that contains metadata for the index.

- `composed_of`: This is an optional list of index components that are merged to build the final index mapping.

> **Note**
> This functionality is available from Elasticsearch version 7.8 and above.

See also

The *Adding metadata to a mapping* recipe in this chapter about using the _meta field.

3
Basic Operations

Before we start with indexing and searching in Elasticsearch, we need to cover how to manage indices and perform operations on documents. In this chapter, we'll start by discussing different operations that can be performed on indices, such as `create`, `delete`, `update`, `open`, and `close`. These operations are very important because they allow you to define the container (index) that will store your documents. The index `create`/`delete` actions are similar to the SQL `create`/`delete` database commands.

After that, we'll learn how to manage mappings to complete the discussion we started in the previous chapter and lay down the basis for the next chapter, which is mainly centered on searching.

A large portion of this chapter is dedicated to performing **create**, **read**, **update**, and **delete** (**CRUD**) operations on records, which are at the *core* of storing and managing records in Elasticsearch.

To improve indexing performance, it's also important to understand bulk operations and avoid their common pitfalls.

This chapter won't cover operations involving queries, as that is the main theme of *Chapter 4, Exploring Search Capabilities, Chapter 5, Text and Numeric Queries*, and *Chapter 6, Relationships and Geo Queries*, as well as cluster operations, which will be discussed in *Chapter 9, Managing Clusters*, because they are mainly related to controlling and monitoring the cluster.

In this chapter, we will cover the following recipes:

- Creating an index
- Deleting an index
- Opening or closing an index
- Putting a mapping in an index
- Getting a mapping
- Reindexing an index
- Refreshing an index
- Flushing an index
- Using ForceMerge on an index
- Shrinking an index
- Checking whether an index exists
- Managing index settings
- Using index aliases
- Managing dangling indices
- Resolving index names
- Rolling over an index
- Indexing a document
- Getting a document
- Deleting a document
- Updating a document
- Speeding up atomic operations (bulk operations)
- Speeding up GET operations (multi-GET)

Technical requirements

To follow along and test the commands in this chapter, you must have a working Elasticsearch cluster installed.

For recipes that are marked as (XPACK), the installed Elasticsearch version should have at least the free version of X-Pack installed.

To simplify command management and how they're executed, I suggest installing Kibana.

Creating an index

The first thing you must do before you can start indexing data in Elasticsearch is create an index – the main container of our data.

An index is similar to the concept of a database in SQL; it is a container for types (tables in SQL) and documents (records in SQL).

Getting ready

You will need an up-and-running Elasticsearch installation, as we described in the *Downloading and installing Elasticsearch* recipe of *Chapter 1, Getting Started*.

To execute the commands in this recipe, you can use any HTTP clients, such as curl (https://curl.haxx.se/), Postman (https://www.getpostman.com/), or others. I suggest using the Kibana console as it provides code completion and better character escaping for Elasticsearch.

How to do it...

The HTTP method for creating an index is PUT; the REST URL contains the index's name:

```
http://<server>/<index_name>
```

To create an index, follow these steps:

1. From the command line, execute a PUT call:

    ```
    PUT /myindex
    { "settings": {
        "index": {
          "number_of_shards": 2, "number_of_replicas": 1
        } } }
    ```

2. The following output will be returned by Elasticsearch:

    ```
    {   "acknowledged" : true,
      "shards_acknowledged" : true, "index" : "myindex" }
    ```

3. If the index already exists, a 400 error will be returned:

```
{ "error": {
    "root_cause": [
      { "type": "resource_already_exists_exception",
        "reason": "index [myindex/xaXAnnwcTUiTePcKGWJw3Q]
already exists",
        "index_uuid": "xaXAnnwcTUiTePcKGWJw3Q",
        "index": "myindex"} ],
    "type": "resource_already_exists_exception",
    "reason": "index [myindex/xaXAnnwcTUiTePcKGWJw3Q]
already exists",
    "index_uuid": "xaXAnnwcTUiTePcKGWJw3Q",
    "index": "myindex"
}, "status": 400 }
```

How it works...

During index creation, the replication process can be set with two parameters in the settings/index object:

- number_of_shards, which controls the number of shards that compose the index (every shard can store up to 2^{32} documents).
- number_of_replicas, which controls the number of replications (how many times your data is replicated in the cluster for high availability). A good practice is to set this value to at least 1.

The API call initializes a new index, which means the following:

- The index is created in a primary node first and then its status is propagated to all the nodes at the cluster level
- A default mapping (empty) is created
- All the shards that are required by the index are initialized and ready to accept data

The index creation API allows you to define the mapping when it's created. The parameter that's required to define a mapping is mapping, and it accepts multiple mappings. So, in a single call, it is possible to create an index and place the required mappings there.

There are also some limitations to the index's name; the only accepted characters are as follows:

- ASCII letters; that is, [a-z]

- Numbers; that is, [0-9]

- Period, ., minus, -, &, and _

There's more...

The create index command also allows you to pass the mappings section, which contains the mapping definitions. It is a shortcut for creating an index with mappings without executing an extra PUT mapping call.

A common example of this call, using the mapping from the *Putting a mapping in an index* recipe, is as follows:

```
PUT /myindex
{ "settings": {
    "number_of_shards": 2, "number_of_replicas": 1 },
  "mappings": {
    "properties": {
      "id": { "type": "keyword", "store": true },
      "date": { "type": "date", "store": false },
      "customer_id": { "type": "keyword", "store": true },
      "sent": { "type": "boolean" },
      "name": { "type": "text" },
      "quantity": { "type": "integer" },
      "vat": { "type": "double", "index": true }
    } } }
```

When you design your index, you need to choose the right number of shards with caution so that you don't waste resources. The best practice is to have shards that are no bigger than 20/25 GB.

Having more shards in the index increases the index rate but decreases the search time due to the cost of merging the results.

The right number of shards must be chosen if several gigabytes of data are involved: having small shards wastes different resources (that is, memory, threads, and file descriptors).

> **Note**
>
> Many shards that contain small data should be used mainly in the case of *script filter* usage or a similar query that isn't using the power of Lucene to execute queries but they are CPU-bound. For these scenarios, using a lot of nodes and shards can reduce the search time, but if possible, it's best to rewrite your models and queries and try to use searches only on pure Lucene functions.

See also

Please look at the following recipes for more information regarding this recipe:

- All the main concepts related to indexing were discussed in the *Understanding clusters, replication, and sharding* recipe in *Chapter 1, Getting Started*.

- After creating an index, you generally need to add a mapping, as described in the *Putting a mapping in an index* recipe in this chapter.

Deleting an index

The counterpart of creating an index is deleting one. Deleting an index means deleting its shards, settings, mappings, and data. There are many common scenarios where we need to delete an index, such as the following:

- Removing the index to clean unwanted or obsolete data (for example, old Logstash indices).

- Resetting an index for a scratch restart.

- Deleting an index that has some missing shards, mainly due to some failures, to bring the cluster back to a valid state. (If a node dies and it's storing a single replica shard of an index, this index will be missing a shard, so the cluster state becomes red. In this case, you'll need to bring the cluster back to a green state, but you will lose the data contained in the deleted index.)

Getting ready

You will need an up-and-running Elasticsearch installation, as we described in the *Downloading and installing Elasticsearch* recipe of *Chapter 1, Getting Started*.

To execute the commands in this recipe, you can use any HTTP client, such as curl (`https://curl.haxx.se/`), Postman (`https://www.getpostman.com/`), or others. I suggest using the Kibana console as it provides code completion and better character escaping for Elasticsearch.

Please ensure that you created the index in the previous recipe since we will need to delete it here.

How to do it...

The HTTP method for deleting an index is `DELETE`.

The following URL contains just the index's name:

```
http://<server>/<index_name>
```

To delete an index, follow these steps:

1. Execute a `DELETE` call by writing the following command:

```
DELETE /myindex
```

2. Then, check the result that's returned by Elasticsearch. If everything is correct, it should look as follows:

```
{ "acknowledged" : true }
```

3. If the index doesn't exist, a `404` error will be returned:

```
{ "error" : {
    "root_cause" : [ {
        "type" : "index_not_found_exception",
        "reason" : "no such index [myindex]",
        "resource.type" : "index_or_alias",
        "resource.id" : "myindex",
        "index_uuid" : "_na_","index" : "myindex"
      } ],
    "type" : "index_not_found_exception",
    "reason" : "no such index [myindex]",
    "resource.type" : "index_or_alias",
    "resource.id" : "myindex", "index_uuid" : "_na_",
    "index" : "myindex" },
  "status" : 404 }
```

How it works...

When an index is deleted, all the data related to the index is removed from the disk and is lost.

The deleting process is composed of two steps: first, the cluster is updated, and then the shards are deleted from storage. This operation is very quick: in a traditional filesystem, it is implemented as a recursive delete.

It's not possible to restore a deleted index if there is no backup.

Also, calling the delete API using the special `_all` value as an index name can be done to remove all the indices. In production, it is good practice to disable the ability to delete all the indices by adding the following line to `elasticsearch.yml`:

```
action.destructive_requires_name:true
```

See also

The previous recipe, *Creating an index*, is strongly related to this recipe.

Opening or closing an index

If you want to keep your data but save resources (memory or CPU), a good alternative to deleting indexes is to close them.

Elasticsearch allows you to open and close an index, putting it into online or offline mode, respectively.

Getting ready

You will need an up-and-running Elasticsearch installation, as we described in the *Downloading and installing Elasticsearch* recipe of *Chapter 1, Getting Started*.

To execute the commands in this recipe, you can use any HTTP client, such as curl (https://curl.haxx.se/), postman (https://www.getpostman.com/), or others. I suggest using the Kibana console as it provides code completion and better character escaping for Elasticsearch.

To execute the following commands correctly, you will need the index we created in the *Creating an index* recipe.

How to do it...

The HTTP method for opening/closing an index is POST.

The URL format for opening an index is as follows:

```
http://<server>/<index_name>/_open
```

The URL format for closing an index is as follows:

```
http://<server>/<index_name>/_close
```

To open and close an index, follow these steps:

1. From the command line, execute a POST call to close an index using the following command:

    ```
    POST /myindex/_close
    ```

2. If the call was successful, you should see the following output:

    ```
    { "acknowledged" : true,
      "shards_acknowledged" : true,
      "indices" : {"myindex" : { "closed" : true } } }
    ```

3. To open an index from the command line, use the following command:

    ```
    POST /myindex/_open
    ```

4. If the call was successful, you should see the following output:

    ```
    {"acknowledged" : true, "shards_acknowledged" : true }
    ```

How it works...

When an index is closed, there is no overhead on the cluster (except for the metadata's state): the index shards are switched off and they don't use file descriptors, memory, or threads.

The data is left unchanged in storage.

There are many use cases regarding closing an index:

*   It can disable date-based indices (indices that store their records by date) – for example, when you keep an index for a week, month, or day and you want to keep a fixed number of old indices (that is, 2 months old) online and some offline (that is, from 2 months to 6 months old).

- When you do searches on all the active indices of a cluster and don't want to search in some indices (in this case, using an alias is the best solution, but you can achieve the same with an alias with closed indices).

An alias cannot have the same name as an index.

There's more...

There is the possibility to freeze the indices with `_freeze` and unfreeze them with `_unfreeze`. When an index is frozen, it's in *read-only* mode: the frozen state is similar to the closed one, but the difference is that a frozen index can also be searched.

Since the frozen index queries run using a special thread pool with a single thread so that resources aren't wasted, their performance is very poor if the index is big.

The best use case for frozen indices is to search indices with few records where their performance isn't that important and there are no requirements for writing these indices.

See also

In the *Using index aliases* recipe in this chapter, we will discuss the advantages of using indices references in a time-based index to simplify how opened indices are managed.

Putting a mapping in an index

In the previous chapter, we learned how to build mappings by indexing documents. In this recipe, you will learn how to put a type mapping in an index. This kind of operation can be considered as the Elasticsearch version of an SQL-created table.

Getting ready

You will need an up-and-running Elasticsearch installation, as we described in the *Downloading and installing Elasticsearch* recipe of *Chapter 1, Getting Started.*

To execute the commands in this recipe, you can use any HTTP client, such as curl (`https://curl.haxx.se/`), Postman (`https://www.getpostman.com/`), or others. I suggest using the Kibana console as it provides code completion and better character escaping for Elasticsearch.

To execute the following commands correctly, you will need the index we created in the *Creating an index* recipe.

How to do it...

The HTTP method for putting a mapping in an index is PUT (POST also works).

The URL format for putting a mapping in an index is as follows:

```
http://<server>/<index_name>/_mapping
```

To put a mapping in an index, follow these steps:

1. After considering a possible order data model to be used as a mapping, the call will be as follows:

```
PUT /myindex/_mapping
{ "properties": {
    "id": { "type": "keyword", "store": true },
    "date": { "type": "date", "store": false },
    "customer_id": {"type": "keyword","store": true },
    "sent": { "type": "boolean" },
    "name": { "type": "text" },
    "quantity": { "type": "integer" },
    "vat": { "type": "double", "index": false } } }
```

2. In the case of success, you will see the following output:

```
{ "acknowledged" : true }
```

How it works...

This call checks if the index exists and it creates one or more types of mappings, as described in the definition. To learn how to define a mapping description, see *Chapter 2, Managing Mappings*.

When a mapping is inserted, if there is an existing mapping for this type, it is merged with the new one. If there is a field with a different type and the type could not be updated, an exception for expanding the fields property is raised. To prevent an exception during the merging mapping phase, it's possible to specify the ignore_conflicts parameter as true (the default is false).

The PUT mapping call allows you to set the mapping for several indices in one go; that is, you can list the indices separated by commas or, to apply all the indexes, use the _all alias.

For people who are used to SQL, the PUT mapping covers the functionality of altering a document by adding a new field.

There's more...

There is no delete operation for mapping. It's not possible to delete a single mapping from an index. To remove or change a mapping, you need to do the following:

1. Create a new index with the new or modified mapping.
2. Reindex all the records.
3. Delete the old index with an incorrect mapping.

In Elasticsearch 5.x and above, there is also a new operation to speed up this process known as the _reindex command, as we will see in the *Reindexing an index* recipe in this chapter.

See also

The *Getting a mapping* recipe is strongly related to this recipe, which allows you to control the result of the put mapping command.

Getting a mapping

After setting our mappings for processing types, we may need to control or analyze the mapping to prevent issues such as wrong type detection, new fields being created due to a data mismatch, and broken index template configurations.

How we get the mapping for a type helps us understand its structure or evolution due to some merging and implicit type guessing.

Getting ready

You will need an up-and-running Elasticsearch installation, as we described in the *Downloading and installing Elasticsearch* recipe of *Chapter 1, Getting Started*.

To execute the commands in this recipe, you can use any HTTP clients, such as curl (https://curl.haxx.se/), Postman (https://www.getpostman.com/), or others. I suggest using the Kibana console as it provides code completion and better character escaping for Elasticsearch.

To execute the following commands correctly, you will need the mapping we created in the *Putting a mapping in an index* recipe.

How to do it...

The HTTP method for getting a mapping is GET.

The URL formats for getting mappings are as follows:

```
http://<server>/_mapping
http://<server>/<index_name>/_mapping
```

To get a mapping from an index, follow these steps:

1. Considering the mapping for the previous recipe, the call will be as follows:

    ```
    GET /myindex/_mapping?pretty
    ```

    The pretty argument in the URL is optional but very handy for printing the output in a more comfortable way to be read by a user.

2. The output that's returned by Elasticsearch should look as follows:

    ```
    {
        "myindex" : {
          "mappings" : {
            "properties" : {
              "customer_id" : { "type" : "keyword" },
              "date" : { "type" : "date" },
              "vat" : { "type" : "double", "index" : false
            } } } }
    ```

How it works...

The mapping is stored at the cluster level in Elasticsearch. The call checks the existence of both the index and the type, and then it returns the stored mapping.

The returned mapping is in a reduced form, which means that the default values for a field are not returned. To reduce network and memory consumption, Elasticsearch only returns non-default values.

Retrieving a mapping is very useful for several purposes:

*   Debugging template-level mappings
*   Checking if the implicit mapping was derived correctly by guessing fields
*   Retrieving the mapping metadata so that it can be used to store type-related information

- Simply checking if the mapping is correct (often due to a query failing)

> **Tip**
> If you need to fetch several mappings, it is better to do it at the index level or the cluster level to reduce the number of API calls.

See also

Please take a look at the following recipes for more information regarding this recipe:

- To insert a mapping in an index, please refer to the *Putting a mapping in an index* recipe in this chapter.
- To manage dynamic mapping in an index, please refer to the *Using dynamic templates in document mapping* recipe in *Chapter 2, Managing Mapping*.

Reindexing an index

There are a lot of common scenarios that involve changing a mapping. Due to the limitations of Elasticsearch mapping (it's additive), it not possible to delete a defined one, so you often need to reindex index data in a new index with a new mapping. The most common scenarios are as follows:

- Changing the analyzer of the mapping
- Adding a new subfield to the mapping, where you need to reprocess all the records to search for the new subfield
- Removing unused mappings
- Changing a record structure that requires a new mapping

Getting ready

You will need an up-and-running Elasticsearch installation, as we described in the *Downloading and installing Elasticsearch* recipe of *Chapter 1, Getting Started*.

To execute the commands in this recipe, you can use any HTTP client, such as curl (https://curl.haxx.se/), Postman (https://www.getpostman.com/), or others. I suggest using the Kibana console as it provides code completion and better character escaping for Elasticsearch.

To execute the following commands correctly, you will need the index we created in the *Creating an index* recipe.

How to do it...

The HTTP method for reindexing an index is POST.

The URL format for getting a mapping is as follows:

```
http://<server>/_reindex
```

To reindex the data between two indices, follow these steps:

1. If we want to reindex data from myindex and place it in the myindex2 index, you can use the following call:

    ```
    POST /_reindex?pretty=true
    { "source": { "index": "myindex" },
      "dest": { "index": "myindex2" } }
    ```

2. The following output will be returned by Elasticsearch:

    ```
    {
        "took" : 20,   "timed_out" : false,
        "total" : 0, "updated" : 0, "created" : 0,
        "deleted": 0, "batches": 0, "version_conflicts" : 0,
        "noops": 0, "retries": { "bulk": 0, "search" : 0 },
        "throttled_millis": 0, "requests_per_second": -1.0,
        "throttled_until_millis" : 0, "failures" : [ ] }
    ```

How it works...

The reindex functionality, which was introduced in Elasticsearch 5.x and above, provides an efficient way to reindex a document.

In the previous versions of Elasticsearch, this functionality had to be implemented at the client level. The advantages of the new Elasticsearch implementation are as follows:

*   You can quickly copy data because it is completely managed on the server side.
*   You can manage the operation better due to the new task API.
*   Better error-handling support as it is done at the server level. This allows us to manage failovers better during reindex operations.
*   The source index can be a remote one, so this command lets you copy/back up data or part of a dataset from an Elasticsearch cluster to another one.

At the server level, this action is comprised of the following steps:

1. Initializing an Elasticsearch task to manage the operation
2. Creating the target index and copying the source mappings, if required
3. Executing a query to collect the documents to be reindexed
4. Reindexing all the documents using bulk operations until all the documents have been reindexed

The main parameters that can be provided for this action are as follows:

-   The `source` section manages how to select source documents. The most important subsections are as follows:
    -   `index`, which is the source index to be used. It can also be a list of indices.
    -   `query` (optional), which is the Elasticsearch query to be used to select parts of the document.
    -   `sort` (optional), which can be used to provide a way of sorting the documents.
-   The `dest` section manages how to control target written documents. The most important parameters in this section are as follows:
    -   `index`, which is the target index to be used. If it is not available, it must be created.
    -   `version_type` (optional) where, when it is set to `external`, the external version is preserved.
    -   `routing` (optional), which controls the routing in the destination index. It can be any of the following:
        -   `keep` (the default), which preserves the original routing
        -   `discard`, which discards the original routing
        -   `=<text>`, which uses the text value for the routing process
    -   `pipeline` (optional), which allows you to define a custom pipeline for ingestion. We will learn more about the ingestion pipeline in *Chapter 12, Using the Ingest Module.*
    -   `size` (optional), which specifies the number of documents to be reindexed.
    -   `script` (optional), which allows you to define a script for document manipulation. This will be discussed in the *Reindexing with a custom script* recipe in *Chapter 8, Scripting in Elasticsearch.*

See also

Please take a look at the following recipes for more information regarding this recipe:

- Check out the *Speeding up atomic operations (bulk operations)* recipe of this chapter, which will talk about using bulk operations to ingest data quickly. These bulk actions are used under the hood by the `reindex` functionality.

- To manage task execution, please refer to the *Using the task management API* recipe in *Chapter 9, Managing Clusters*.

- The *Reindexing with a custom script* recipe in *Chapter 8, Scripting in Elasticsearch*, will show several common scenarios for reindexing documents with a custom script.

- *Chapter 12, Using the Ingest Module*, will discuss how to use the ingestion pipeline.

Refreshing an index

When you send data to Elasticsearch, the data is not instantly searchable. This only happens after a time interval (generally a second) known as the **refresh rate**. This delayed approach to data reading/writing allows you to efficiently write large blocks of data by reducing small disk action and increasing the throughput.

Elasticsearch allows the user to control the state of the searcher by forcefully refreshing an index. If it's not forced, the newly indexed document will only be searchable after a fixed time interval (usually, 1 second).

Getting ready

You will need an up-and-running Elasticsearch installation, as we described in the *Downloading and installing Elasticsearch* recipe of *Chapter 1, Getting Started*.

To execute the commands in this recipe, you can use any HTTP client, such as curl (https://curl.haxx.se/), Postman (https://www.getpostman.com/), or others. I suggest using the Kibana console as it provides code completion and better character escaping for Elasticsearch.

To execute the following commands correctly, please use the index we created in the *Creating an index* recipe.

How to do it...

The HTTP method that's used for both operations is POST.

The URL format for refreshing an index is as follows:

```
http://<server>/<index_name(s)>/_refresh
```

The URL format for refreshing all the indices in a cluster is as follows:

```
http://<server>/_refresh
```

To refresh an index, follow these steps:

1. If we consider the order type from the previous chapter, the call will be as follows:

    ```
    POST /myindex/_refresh
    ```

2. If everything is fine, the following output will be returned by Elasticsearch:

    ```
    { "_shards" : { "total" : 2, "successful" : 1, "failed" :
    0 } }
    ```

How it works...

Near-real-time (**NRT**) capabilities are automatically managed by Elasticsearch, which automatically refreshes the indices every second if data is changed in them.

To force a refresh before the internal Elasticsearch interval, you can call the refresh API on one or more indices (more indices are comma-separated), or on all the indices.

Elasticsearch doesn't refresh the state of an index when every document is inserted to prevent poor performance due to the excessive I/O that's required in closing and reopening file descriptors.

You must force the refresh to have your last index data available for searching.

Generally, the best time to call the refresh is after indexing a lot of data, to ensure that your records can be searched instantly. It's also possible to force a refresh during document indexing by adding refresh=true as a query parameter.

For example, the following code block shows how to set `refresh` in the URL to guarantee that the new value is searchable:

```
POST /myindex/_doc/2qLrAfPVQvCRMe7Ku8r0Tw?refresh=true
{ "id": "1234",  "date": "2013-06-07T12:14:54",
  "customer_id": "customer1",  "sent": true,
  "in_stock_items": 0,
  "items": [
    { "name": "item1", "quantity": 3, "vat": 20 },
    { "name": "item2", "quantity": 2, "vat": 20 },
    { "name": "item3", "quantity": 1, "vat": 10 } ] }
```

See also

Please refer to the *Flushing an index* recipe in this chapter to learn how to force indexed data writing on disk and the *Using ForceMerge on an index* recipe in this chapter to learn how to optimize an index for searching.

Flushing an index

For performance reasons, Elasticsearch stores some data in memory and on a transaction log. If we want to free memory, we need to empty the transaction log, and to ensure that our data is safely written on disk, we need to flush an index.

Elasticsearch automatically provides periodic flushing on disk, but forcing flushing can be useful in the following situations:

- When we need to shut down a node to prevent stale data
- When we need to have all the data in a safe state (for example, after a big indexing operation so that all the data can be flushed and refreshed)

Getting ready

You will need an up-and-running Elasticsearch installation, as we described in the *Downloading and installing Elasticsearch* recipe of *Chapter 1, Getting Started*.

To execute the commands in this recipe, you can use any HTTP client, such as curl (`https://curl.haxx.se/`), Postman (`https://www.getpostman.com/`), or others. I suggest using the Kibana console as it provides code completion and better character escaping for Elasticsearch.

To execute the following commands correctly, please use the index we created in the *Creating an index* recipe.

How to do it...

The HTTP method that's used for both operations is POST.

The URL format for flushing an index is as follows:

```
http://<server>/<index_name(s)>/_flush[?refresh=True]
```

The URL format for flushing all the indices in a cluster is as follows:

```
http:///_flush[?refresh=True]
```

To flush an index, follow these steps:

1. if we consider the order type from the previous chapter, the call will be as follows:

   ```
   POST /myindex/_flush
   ```

2. If everything is fine, the following output will be returned by Elasticsearch:

   ```
   { "_shards" : { "total" : 2, "successful" : 1, "failed" :
   0 } }
   ```

   The result contains the shard operation status.

How it works...

To reduce I/O operations while writing data to disk, Elasticsearch caches some data in memory until a **refresh interval** occurs. This approach allows Elasticsearch to execute a multi-document write as a single atomic I/O operation, thereby reducing disk I/O and increasing throughput.

To clean up memory and force this data on disk, you can use the flush operation.

In the flush call, it is possible to give an extra request parameter, refresh, which is also used to force the index to refresh.

Flushing often affects index performance. Use it wisely!

See also

Please refer to the *Refreshing an index* recipe in this chapter to learn how to search for more recently indexed data and the *Using ForceMerge on an index* recipe in this chapter to learn how to optimize an index for searching.

Using ForceMerge on an index

The Elasticsearch core is based on Lucene, which stores data in segments on disk. During the life of an index, a lot of segments are created and changed. Since many other NoSQL systems (such as Cassandra, Accumulo, and HBase) prevent segments and part of the data from being rewritten, the records are not deleted in place, but they are put in a **tombstone** state. This means that the document is marked and deleted in metadata without the data being changed on disk. With the increasing number of segments, the speed of searching is decreased due to the time required to read all of them or skipping the records that aren't live (tombstones). The **ForceMerge** operation allows us to consolidate the index for quicker searching performance and reducing segments.

Getting ready

You will need an up-and-running Elasticsearch installation, as we described in the *Downloading and installing Elasticsearch* recipe of *Chapter 1, Getting Started*.

To execute the commands in this recipe, you can use any HTTP client, such as curl (`https://curl.haxx.se/`), Postman (`https://www.getpostman.com/`), or others. I suggest using the Kibana console as it provides code completion and better character escaping for Elasticsearch.

To execute the following commands correctly, please use the index we created in the *Creating an index* recipe.

How to do it...

The HTTP method that's used here is POST.

The URL format for optimizing one or more indices is as follows:

```
http://<server>/<index_name(s)>/_flush[?refresh=True]
```

The URL format for optimizing all the indices in a cluster is as follows:

```
http://<server>/_flush[?refresh=True]
```

To optimize or ForceMerge an index, follow these steps:

1. Considering the index we created in the *Creating an index* recipe, the call will be as follows:

    ```
    POST /myindex/_forcemerge
    ```

2. The following output will be returned by Elasticsearch:

    ```
    { "_shards" : { "total" : 2, "successful" : 1,
    "failed" : 0 } }
    ```

The result contains the shard's operation status.

How it works...

Lucene stores your data in several segments on disk. These segments are created when you index a new document or record, or when you delete a document.

In Elasticsearch, the deleted document is not removed from disk; instead, it is marked as deleted (and referred to as a tombstone). To free up space, you need to forcemerge to purge deleted documents.

Due to all these factors, the segment numbers can be large. (For this reason, in the setup, we have increased the file description number for Elasticsearch processes.)

Internally, Elasticsearch has a merger, which tries to reduce the number of segments, but it's designed to improve the index's performance rather than search performances. The forcemerge operation in Lucene tries to reduce the segments in an I/O-heavy way by removing unused ones, purging deleted documents, and rebuilding the index with a minimal number of segments.

The main advantages of this are as follows:

*   It reduces both file descriptors.
*   It frees up the memory that was used by the segment readers.
*   It improves performance during searches due to less segment management.

ForceMerge is a very I/O-heavy operation. The index can be unresponsive during this optimization. It is generally executed on indices that are rarely modified, such as the Logstash for previous days.

There's more...

You can pass several additional parameters to the ForceMerge call, such as the following:

- `max_num_segments`: The default value is `autodetect`. For full optimization, set this value to `1`.

- `only_expunge_deletes`: The default value is `false`. Lucene does not delete documents from segments; instead, it marks them as deleted. This flag only merges segments that have been deleted.

- `flush`: The default value is `true`. Elasticsearch performs a flush after a ForceMerge.

- `wait_for_merge`: The default value is `true`. If the request needs to wait, then the merge ends.

See also

Please refer to the *Refreshing an index* recipe in this chapter to learn how to search for more recent indexed data and the *Flushing an index* recipe in this chapter to learn how to force indexed data writing on disk.

Shrinking an index

The latest version of Elasticsearch provides us with a new way to optimize an index. By using the shrink API, it's possible to reduce the number of shards in an index.

This feature targets several common scenarios:

- A wrong number of shards will be provided during the initial design sizing. Often, sizing the shards without knowing the correct data or text distribution tends to oversize the number of shards.

- You should reduce the number of shards to reduce memory and resource usage.

- You should reduce the number of shards to speed up searching.

Getting ready

You will need an up-and-running Elasticsearch installation, as we described in the *Downloading and installing Elasticsearch* recipe of *Chapter 1, Getting Started*.

To execute the commands in this recipe, you can use any HTTP client, such as curl (https://curl.haxx.se/), Postman (https://www.getpostman.com/), or others. I suggest using the Kibana console as it provides code completion and better character escaping for Elasticsearch.

To execute the following commands correctly, please use the index we created in the *Creating an index* recipe.

How to do it...

The HTTP method that's used here is POST.

The URL format for optimizing one or more indices is as follows:

```
http://<server>/<source_index_name>/_shrink/<target_index_name>
```

To shrink an index, follow these steps:

1. We need all the primary shards of the index to be shrinking in the same node. We need the name of the node that will contain the shrink index. We can retrieve it using the _nodes API:

    ```
    GET /_nodes?pretty
    ```

    Within the result, you will see a section that's similar to the following:

    ```
    ... truncated ...
    "cluster_name" : "elastic-cookbook",
      "nodes" : {
        "9TiCStQuTDaTyMb4LgWDsg" : {
          "name" : "1e9840cf42df",
          "transport_address" : "172.18.0.2:9300",
          "host" : "172.18.0.2", "ip" : "172.18.0.2",
          "version" : "7.0.0", "build_flavor" : "default",
          "build_type" : "docker",
          "build_hash" : "f076a79",
          "total_indexing_buffer" : 103795916,
    ... truncated ...
    ```

    The name of my node is 1e9840cf42df.

2. Now, we can change the index settings, thus forcing allocation to a single node for our index and disabling the writing for the index. This can be done with the following code:

    ```
    PUT /myindex/_settings
    { "settings": {
        "index.routing.allocation.require._name":
    ```

```
    "1e9840cf42df",
        "index.blocks.write": true } }
```

3. We need to check if all the shards have been relocated. Let's check for their green status:

```
GET /_cluster/health?pretty
```

The result will be as follows:

```
{ "cluster_name" : "elastic-cookbook",
  "status" : "yellow", "timed_out" : false,
  "number_of_nodes" : 1, "number_of_data_nodes" : 1,
  "active_primary_shards" : 2, "active_shards" : 2,
  "relocating_shards" : 0, "initializing_shards" : 0,
  "unassigned_shards" : 1,
  "delayed_unassigned_shards" : 0,
  "number_of_pending_tasks" : 0,
  "number_of_in_flight_fetch" : 0,
  "task_max_waiting_in_queue_millis" : 0,
  "active_shards_percent_as_number" : 66.66666666666666 }
```

4. The index should be in a read-only state so that it can be shrunk. We need to disable the writing for the index by using the following code snippet:

```
PUT /myindex/_settings?
{"index.blocks.write":true}
```

5. If we consider the index we created in the *Creating an index* recipe, the shrink call for creating reduced_index will be as follows:

```
POST /myindex/_shrink/reduced_index
{ "settings": {
    "index.number_of_replicas": 1,
    "index.number_of_shards": 1,
    "index.codec": "best_compression" },
  "aliases": { "my_search_indices": {} } }
```

6.   The following output will be returned by Elasticsearch:

```
{"acknowledged":true}
```

7.   We can also wait for a `yellow` status if the index is ready to work:

```
GET /_cluster/health?wait_for_status=yellow
```

8.   Now, we can remove the read-only setting by changing the index settings:

```
PUT /myindex/_settings
{"index.blocks.write":false}
```

How it works...

The shrink API reduces the number of shards by executing the following steps:

1.   Elasticsearch creates a new target index with the same definition as the source index, but with a smaller number of primary shards.

2.   Elasticsearch hard-links (or copies) segments from the source index into the target index.

If the filesystem doesn't support hard-linking, then all the segments are copied into the new index, which is a much more time-consuming process. Elasticsearch recovers the target index as though it were a closed index that has just been reopened. On a Linux system, this process is very fast due to hard links.

The prerequisites for executing a shrink API call are as follows:

*   All the primary shards must be on the same node.

*   The target index must not exist.

*   The target number of shards must be a factor of the number of shards in the source index.

There's more...

This Elasticsearch functionality provides support for new scenarios in Elasticsearch usage.

The first scenario is when you overestimate the number of shards. If you don't know your data, it's difficult to choose the correct number of shards to be used. So, often, an Elasticsearch user tends to overestimate the number of shards.

Another interesting scenario is to use shrinking to provide a boost at indexing time. The main way to speed up Elasticsearch's writing capabilities to a high number of documents is to create indices with a lot of shards (in general, the ingestion speed is about equal to the number of shards multiplied for documents per second, as ingested by a single shard). The standard allocation moves the shards on different nodes, so generally, the more shards you have, the faster the writing speed will be. So, to achieve fast writing speeds, you should create 15 or 30 shards for an index. After the indexing phase, the index doesn't receive new records (such as time-based indices); the index is only searched. So, to speed up the search, you can shrink your shards.

See also

Please refer to the *Using ForceMerge on an index* recipe in this chapter to learn how to optimize your indices for searching.

Checking whether an index exists

A common pitfall error is to query for indices that don't exist. To prevent this issue, Elasticsearch allows you to check for an index's existence.

This check is often used during application startup to create indices that are required for the application to work correctly.

Getting ready

You will need an up-and-running Elasticsearch installation, as we described in the *Downloading and installing Elasticsearch* recipe of *Chapter 1, Getting Started*.

To execute the commands in this recipe, you can use any HTTP clients, such as curl (`https://curl.haxx.se/`), Postman (`https://www.getpostman.com/`), or others. I suggest using the Kibana console as it provides code completion and better character escaping for Elasticsearch.

To execute the following commands correctly, please use the index we created in the *Creating an index* recipe.

How to do it...

The HTTP method for checking an index's existence is HEAD.

The URL format for checking an index is as follows:

```
http://<server>/<index_name>/
```

To check if an index exists, follow these steps:

1. If we consider the index we created in the *Creating an index* recipe, the call will be as follows:

    ```
    HEAD /myindex/
    ```

2. If the index exists, an HTTP status code of 200 will be returned; if it is missing, a 404 code will be returned.

How it works...

This is a typical HEAD REST call to check for something's existence. It doesn't return a body response, only the status code, which is the result status of the operation.

The most common status codes are as follows:

*   The 20X family, if everything is okay
*   404, if the resource is not available
*   The 50X family, if there are server errors

Before you perform an action while indexing, generally on an application's startup, it's good practice to check if an index exists to prevent future failures.

Managing index settings

Index settings are more important because they allow you to control several important Elasticsearch functionalities, such as sharding or replication, caching, term management, routing, and analysis. The goal of this is to understand how to manage index settings.

Getting ready

You will need an up-and-running Elasticsearch installation, as we described in the *Downloading and installing Elasticsearch* recipe of *Chapter 1, Getting Started*.

To execute the commands in this recipe, you can use any HTTP client, such as curl (`https://curl.haxx.se/`), Postman (`https://www.getpostman.com/`), or others. I suggest using the Kibana console as it provides code completion and better character escaping for Elasticsearch.

To execute the following commands correctly, please use the index we created in the *Creating an index* recipe.

How to do it...

To retrieve the settings of your current index, use the following URL format:

```
http://<server>/<index_name>/_settings
```

To manage the index settings, follow these steps:

1. We are reading information using the REST API, so the method will be GET. The following is an example of a call that's using the index we created in the *Creating an index* recipe:

```
GET /myindex/_settings?pretty=true
```

2. The response will look something similar to the following:

```
{ "myindex" : {
    "settings" : {
      "index" : {
        "routing" : {
          "allocation" : {
            "require" : { "_name" : "1e9840cf42df" }
          } },
        "number_of_shards" : "1",
        "blocks" : { "write" : "true" },
        "provided_name" : "myindex",
        "creation_date" : "1554578317870",
        "number_of_replicas" : "1",
        "uuid" : "sDzB7n80SFi80f99IgLYtA",
        "version" : { "created" : "7000099" } } } } }
```

3. The response attributes depend on the index settings. In this case, the response will be the number of replicas (1), shards (2), and the index creation version (7000099). The UUID represents the unique ID of the index.

4. To modify the index settings, we need to use the PUT method. A typical way to change the settings is to increase the replica number:

```
PUT /myindex/_settings
{"index":{ "number_of_replicas": 2}}
```

How it works...

Elasticsearch provides a lot of options for tuning index behaviors, such as the following:

- **Replica management**, which can control how indices/shards are replicated in the cluster:

  - index.number_of_replicas: This is the number of replicas each shard has.

  - index.auto_expand_replicas: This allows you to define a dynamic number of replicas related to the number of shards.

  Setting index.auto_expand_replicas to 0-all allows an index to be created that is replicated in every node. (This is very useful for settings or cluster-propagated data, such as language options or stopwords.)

- **Refresh interval (default 1s)**: In the *Refreshing an index* recipe, we saw how to refresh an index manually. The index.refresh_interval index setting controls the rate of automatic refreshing.

- **Write management**: Elasticsearch provides several settings for blocking read or write operations in the index and for changing metadata. They live in the index. blocks settings.

- **Shard allocation management**: These settings control how the shards must be allocated. They live in the index.routing.allocation.* namespace.

Other index settings can be configured for very specific needs. In every new version of Elasticsearch, the community extends these settings to cover new scenarios and requirements.

There's more...

The refresh_interval parameter allows you to perform several tricks to optimize the indexing speed. It controls the rate of refresh and refreshing itself, and reduces the indices' performance due to files being opened or closed.

A good practice is to disable the refresh interval (by setting it to -1) during a big bulk indexing process and to restore the default behavior after it. This can be done by following these steps:

1. Disable the refresh:

```
PUT /myindex/_settings
{"index":{"refresh_interval": "-1"}}
```

2. Bulk-index millions of documents.

3. Restore the refresh:

```
PUT /myindex/_settings
{"index":{"refresh_interval": "1s"}}
```

4. Optionally, you can optimize an index for search performance:

```
POST /myindex/_forcemerge
```

See also

Please refer to the *Refreshing an index* recipe in this chapter to learn how to search for more recent indexed data and the *Using ForceMerge on an index* recipe in this chapter to learn how to optimize an index for searching.

Using index aliases

Real-world applications have a lot of indices and queries that span more indices. This scenario requires defining all the indices' names that the queries are based on; aliases allow you to group them under a common name/label.

Some common scenarios for this usage are as follows:

- Log indices divided by date (that is, `logstash-YYYY-MM-DD`) for which we want to create an alias for the last week, the last month, today, yesterday, and so on. This pattern is commonly used in log applications such as Logstash (`https://www.elastic.co/products/logstash`).

- Collecting a website's content in several indices (*New York Times*, *The Guardian*, and so on) for those we want to be referred to by the index alias sites.

Getting ready

You will need an up-and-running Elasticsearch installation, as we described in the *Downloading and installing Elasticsearch* recipe of *Chapter 1, Getting Started*.

To execute the commands in this recipe, you can use any HTTP client, such as curl (https://curl.haxx.se/), Postman (https://www.getpostman.com/), or others. I suggest using the Kibana console as it provides code completion and better character escaping for Elasticsearch.

How to do it...

The URL format for control aliases is as follows:

```
http://<server>/_aliases
http://<server>/<index>/_alias/<alias_name>
```

To manage the index aliases, follow these steps:

1. We are reading the aliases and statuses of all the indices using the REST API, so we will use the GET method here. The following is an example of a call:

   ```
   GET /_aliases
   ```

2. This will give us the following output:

   ```
   { ".monitoring-es-7-2019.04.06" : { "aliases" : { } },
     "myindex" : { "aliases" : { } } }
   ```

3. Aliases can be changed with the add and delete commands.

4. To read an alias for a single index, we can use the _alias endpoint:

   ```
   GET /myindex/_alias
   ```

   This will give us the following output:

   ```
   { "myindex" : { "aliases" : { } } }
   ```

5. To add an alias, type the following command:

   ```
   PUT /myindex/_alias/myalias1
   ```

   This will give us the following output:

   ```
   { "acknowledged" : true }
   ```

   This action adds the myindex index to the myalias1 alias.

6. To delete an alias, type the following command:

```
DELETE /myindex/_alias/myalias1
```

This will give us the following output:

```
{ "acknowledged" : true }
```

This action removed `myindex` from the `myalias1` alias.

How it works...

Elasticsearch, during search operations, automatically expands the alias, so the required indices are selected.

The alias metadata is kept in the cluster state. When an alias is added or deleted, all the changes are propagated to all the cluster nodes.

Aliases are mainly functional structures that simply manage indices when data is stored in multiple indices.

There's more...

Aliases can also be used to define a filter and routing parameter.

Filters are automatically added to the query to filter out data. Routing by using an alias allows us to control which shards must be hit during searching and indexing.

An example of this call is as follows:

```
POST /myindex/_aliases/user1alias
{ "filter": { "term": { "user": "user_1" } },
  "search_routing": "1,2", "index_routing": "2" }
```

In this case, we are adding a new alias, `user1alias`, to the `myindex` index, and are also adding the following:

- A filter to select only documents that match a field user with a `user_1` term.
- A list and a routing key to select the shards to be used during a search.
- A routing key to be used during indexing. The routing value is used to modify the destination shard of the document.

The `search_routing` parameter allows multi-value routing keys. The `index_routing` parameter is single-value only.

Managing dangling indices

In the case of a node failure, if there are not enough replicas, you can lose some shards (and the data within those shards).

Indices with missing shards are marked in red and they are put in read-only mode with issues in case you try to query the data.

In this situation, the only available option is to drop the broken index and recover them from the data or a backup. When the node that failed returns as active in the cluster, there will be some dangling indices (the orphan shards).

The APIs that we will look at in this recipe can be used to manage these indices.

Getting ready

You will need an up-and-running Elasticsearch installation, as we described in the *Downloading and installing Elasticsearch* recipe of *Chapter 1, Getting Started.*

To execute the commands in this recipe, you can use any HTTP client, such as curl (https://curl.haxx.se/), Postman (https://www.getpostman.com/), or others. I suggest using the Kibana console as it provides code completion and better character escaping for Elasticsearch.

How to do it...

The URL format for managing dangling indices is as follows:

```
http://<server>/_dangling
```

To manage a dangling index, follow these steps:

1. We need the list of dangling indices that are present in our cluster (we used GET to read here):

   ```
   GET /_dangling
   ```

   The output will be as follows:

   ```
   { "_nodes" : { "total" : 1, "successful" : 1,
       "failed" : 0 },
     "cluster_name" : "packtpub",
     "dangling_indices" : [
     { "index_name": "my-index-000001",
       "index_uuid": "zmM4eOJtBkeUjiHD-MihPQ",
   ```

```
      "creation_date_millis": 1589414451372,
      "node_ids": [ "pL47UN3dAb2d5RCWP61Q3e" ]
    } ] }
```

2. We can restore the data that's available in `index_uuid` (`zmM4e0JtBkeUjiHD`) of the previous response like so:

```
POST /_dangling/zmM4e0JtBkeUjiHD?accept_data_loss=true
```

The output will be as follows:

```
{ "acknowledged" : true }
```

3. If you wish to save space and remove the data of the dangling index with `index_uuid` (`zmM4e0JtBkeUjiHD`), you can execute the following command:

```
DELETE /_dangling/<index-uuid>?accept_data_loss=true
```

The output will be as follows:

```
{ "acknowledged" : true }
```

How it works...

Dangling indices are created when you delete indices whose shards are in offline nodes within your cluster. When the offline nodes go online, they have shards in a dangling state and they are marked as dangling indices in the cluster state.

If you want to delete or restore the data in existing orphan shards, you will only need partial data because the stale indices will be missing data shards. It is for this reason that the `accept_data_loss=true` flag must be provided.

It's not common to have dangling indices in your cluster but you can schedule a check weekly or monthly to check for them, as well as to check if you can do the required maintenance: generally, delete them to recover space.

See also

Please refer to the *Deleting an index* recipe in this chapter regarding deleting indices in Elasticsearch.

Resolving index names

In the previous recipe, we saw how to use a wildcard to select indices and their aliases.

If you have a large number of indices and aliases, when you try to select them using wildcards, some results won't be provided, so you'll need to understand why. It's also common to need to debug the slowness of a query (due to how much data has been queried) or an error because you are trying to query closed indices.

To help you solve such issues, you can use the resolve index API, which allows you to return all the information about the indices that can be queried.

Getting ready

You will need an up-and-running Elasticsearch installation, as we described in the *Downloading and installing Elasticsearch* recipe of *Chapter 1, Getting Started*.

To execute the commands in this recipe, you can use any HTTP client, such as curl (https://curl.haxx.se/), Postman (https://www.getpostman.com/), or others. I suggest using the Kibana console as it provides code completion and better character escaping for Elasticsearch.

You will also need the index we created in the *Creating an index* recipe with the aliases so that you receive a more detailed response (with alias information) from the API.

How to do it...

To resolve some indices, we need to create some indices and an alias that we will query for. Follow these steps:

1. Create some indices with an alias by running the following commands:

```
PUT /myindex
PUT /myindex/_alias/myalias1
POST /myindex/_aliases/user1alias
PUT /myindex/_alias/myindex1
```

2. We can add the rolling index to the `logs_write` alias like so:

```
GET /_resolve/index/myinde*
```

The output will be as follows:

```
{ "indices" : [
    { "name" : "myindex",
```

```
        "aliases" : [ "myalias1", "myindex1", "userlalias"
  ],
        "attributes" : [ "open" ] } ],
    "aliases" : [
      { "name" : "myindex1",
        "indices" : [ "myindex" ] } ],
    "data_streams" : [ ] }
```

How it works...

The resolve index API is a very useful API that can scan the cluster state and return all the indices, aliases, and datastreams that have been matched in the query wildcard.

If you need to resolve multiple patterns, they can be provided by adding a comma to separate the URL fragment (that is, `pattern1*`, `pattern*`, and so on).

This API is by far more efficient at reading the cluster's state and the indices' information; it automatically collects the following:

- `indices`: A list of indices that match the patterns, along with the respective alias and attributes. These attributes are very useful: you cannot query closed indices, but you can quickly check which are closed and fix possible bugs.

- `aliases`: A list of aliases that match the patterns. For every alias, the name is also returned in the list of referred indices.

- `data_streams`: A list of `data_streams` that match the patterns. Data Streams is an X-Pack special index designed for append-only data. The concept is very similar to rolling over indices, as we'll see in the next recipe.

See also

Please refer to the *Using index aliases* recipe in this chapter to learn how to manage aliases for indices.

Rolling over an index

When you're using a system that manages logs, it is very common to use rolling files for your log entries. By doing so, you can have indices that are similar to rolling files.

You can define some conditions that must be checked and leave it to Elasticsearch to roll new indices automatically and refer the use of an alias to a **virtual** index.

Getting ready

You will need an up-and-running Elasticsearch installation, as we described in *Downloading and installing Elasticsearch* recipe of *Chapter 1, Getting Started*.

To execute the commands in this recipe, you can use any HTTP client, such as curl (`https://curl.haxx.se/`), Postman (`https://www.getpostman.com/`), or others. I suggest using the Kibana console as it provides code completion and better character escaping for Elasticsearch.

How to do it...

To enable a rolling index, we need an index with an alias that points to it alone. For example, to set a log rolling index, we would follow these steps:

1. We need an index with a `logs_write` alias that points to it alone:

```
PUT /mylogs-000001
{ "aliases": { "logs_write": {} } }
```

The output will be an acknowledgment, as follows:

```
{ "acknowledged" : true,
  "shards_acknowledged" : true,
  "index" : "mylogs-000001" }
```

2. We can add the rolling index to the `logs_write` alias like so:

```
POST /logs_write/_rollover
{ "conditions": {
    "max_age": "7d", "max_docs": 100000,
    "max_size": "5g" },
  "settings": { "index.number_of_shards": 3 } }
```

The output will be as follows:

```
{ "acknowledged" : false,
  "shards_acknowledged" : false,
  "old_index" : "mylogs-000001",
  "new_index" : "mylogs-000002",
  "rolled_over" : false, "dry_run" : false,
  "conditions" : {
    "[max_docs: 100000]" : false,
    "[max_age: 7d]" : false } }
```

3. If your alias doesn't point to a single index, the following error will be returned:

```
{ "error" : {
    "root_cause" : [ {
        "type" : "illegal_argument_exception",
        "reason" : "source alias maps to multiple indices"
} ],
        "type" : "illegal_argument_exception",
        "reason" : "source alias maps to multiple indices"
    }, "status" : 400 }
```

How it works...

The rolling index is a special alias that manages the auto-creation of new indices when one of the conditions is matched.

This is a very convenient functionality because it is completely managed by Elasticsearch, reducing the need for a lot of custom backend user code.

The information for creating the new index is taken from the source, but you can also apply custom settings when the index is being created.

The naming convention is managed by Elasticsearch, which automatically increments the numeric part of the index's name (by default, it uses six ending digits).

You can define it by using different criteria to *roll over* your index:

*   max_age (Optional): The validity period for writing in this index.

*   max_docs (Optional): The maximum number of documents in an index.

*   max_size (Optional): The maximum size of the index. (Pay attention and divide it by the number of shards to get the real shard size.)

There's more...

Using rolling indices has several advantages, including the following:

*   You can have indices with a fixed number of documents/sizes, which can prevent you from having small indices that contain data.

*   You can automatically manage the time validity of the index, which can span different days.

If large data is stored in rolling indices, then the following disadvantages may occur:

- It's more difficult to filter indices for data and your queries should often hit all the indices of a rolling group (more time and resources will be needed for queries).
- There will be issues in guaranteeing a GDPR approach to the end of life of the indices because some days of data could be present in two indices.
- There will be more complexity in operational activities such as ForceMerge and cold/warm indices management.

See also

Please refer to the *Using index aliases* recipe in this chapter to learn how to manage aliases for indices.

Indexing a document

In Elasticsearch, there are two vital operations: **index** and **search**.

Indexing means storing one or more documents in an index; this is a similar concept to inserting records in a relational database.

In Lucene, the core engine of Elasticsearch, inserting or updating a document has the same cost: in Lucene and Elasticsearch, to update means to replace.

Getting ready

You will need an up-and-running Elasticsearch installation, as we described in the *Downloading and installing Elasticsearch* recipe of *Chapter 1, Getting Started*.

To execute the commands in this recipe, you can use any HTTP client, such as curl (https://curl.haxx.se/), Postman (https://www.getpostman.com/), or others. I suggest using the Kibana console as it provides code completion and better character escaping for Elasticsearch.

To execute the following commands correctly, please use the index and mapping we created in the *Putting a mapping in an index* recipe.

How to do it...

Several REST entry points can be used to index a document:

Method	URL
POST	http://<server>/<index_name>/_doc
PUT/POST	http://<server>/<index_name>/_doc /<id>
PUT/POST	http://<server>/<index_name>/_create /<id>

Table 3.1 – REST entry points

To index a document, follow these steps:

1. If we consider the `order` type from the previous chapter, the call to index a document will be as follows:

```
POST /myindex/_doc/2qLrAfPVQvCRMe7Ku8r0Tw
{ "id": "1234", "date": "2013-06-07T12:14:54",
  "customer_id": "customer1", "sent": true,
  "in_stock_items": 0,
  "items": [
    { "name": "item1", "quantity": 3, "vat": 20 },
    { "name": "item2", "quantity": 2, "vat": 20 },
    { "name": "item3", "quantity": 1, "vat": 10 } ] }
```

2. If the index operation was successful, the following output will be returned by Elasticsearch:

```
{ "_index" : "myindex",
  "_id" : "2qLrAfPVQvCRMe7Ku8r0Tw",
  "_version" : 1,  "result" : "created",
  "_shards" : { "total" : 2, "successful" : 1,
    "failed" : 0 },
  "_seq_no" : 0, "_primary_term" : 1 }
```

Some additional information is returned from the index operation, such as the following:

- An auto-generated ID, if it's not specified (in this example, this is `2qLrAfPVQvCRMe7Ku8r0Tw`)

- The version of the indexed document, as per the optimistic concurrency control (the version is `1` because it was the document's first time saving or updating)

- Whether the record has been created (in this example, this is `"result": "create"`)

How it works...

One of the most used APIs in Elasticsearch is the index API. Indexing a JSON document consists of the following steps:

1. Routing the call to the correct shard based on the ID, routing, or parent metadata. If the ID is not supplied by the client, a new one is created (see the *Managing your data* recipe of *Chapter 1, Getting Started*, for details).

2. Validating the sent JSON.

3. Processing the JSON according to the mapping. If new fields are present in the document (and the mapping can be updated), new fields will be added to the mapping.

4. Indexing the document in the shard. If the ID already exists, it is updated.

5. If it contains nested documents, it extracts and processes them separately.

6. Returning information about the saved document (ID and versioning).

It's important to choose the correct ID when you're indexing your data. If you don't provide an ID during the indexing phase, Elasticsearch will automatically associate a new one with your document. To improve performance, the ID should generally be of the same character length to improve the balancing of the data tree that stores them.

Due to the REST call's nature, it's better to pay attention when you're not using ASCII characters due to URL encoding and decoding (or to ensure that the client framework you use escapes them correctly).

Depending on the mappings, other actions take place during the indexing phase: propagation on the replica, nested processing, and percolator.

The document will be available for standard search calls after a refresh (forced with an API call or after the time slice of 1 second, near-real-time): not every GET API on the document requires a refresh, and these can be available instantly.

The refresh can also be forced by specifying the `refresh` parameter during indexing.

There's more...

Elasticsearch allows you to pass the index API's URL to several query parameters to control how the document is indexed. The most used ones are as follows:

- `routing`: This controls the shard to be used for indexing, as follows:

```
POST /myindex/_doc?routing=1
```

- `consistency(one/quorum/all)`: By default, an index operation succeeds if a quorum (`>replica/2+1`) of active shards is available. The right consistency value can be changed for index action:

```
POST /myindex/_doc?consistency=one
```

- `replication (sync/async)`: Elasticsearch returns from an index operation when all the shards of the current replication group have executed the index operation. Setting up `async` replication allows us to execute the index action synchronously on the primary shard and asynchronously on secondary shards. In this way, the API call returns the response action faster:

```
POST /myindex/_doc?replication=async
```

- `version`: The version allows us to use the **optimistic concurrency control** (`http://en.wikipedia.org/wiki/Optimistic_concurrency_control`). The first time a document is indexed, its version (1) is set on the document. Every time it's updated, this value is incremented. Optimistic concurrency control is a way to manage concurrency in every insert or update operation. The passed version value is the last seen version (usually, it's returned by a `GET` or a search). Indexing only happens if the current index version's value is equal to the passed one:

```
POST /myindex/_doc?version=2
```

- `op_type`: This can be used to force a `create` on a document. If a document with the same ID exists, the index will fail:

```
POST /myindex/_doc?op_type=create
```

- `refresh`: This forces a refresh once you've indexed the document. It allows documents to be ready for searching once they've been indexed:

```
POST /myindex/_doc?refresh=true
```

- `timeout`: This defines a time to wait for the primary shard to be available. Sometimes, the primary shard is not in a writable status (if it's relocating or recovering from a gateway) and a timeout for the write operation is raised after 1 minute:

```
POST /myindex/_doc?timeout=5m
```

See also

Please take a look at the following recipes for more information regarding this recipe:

- The *Getting a document* recipe in this chapter to learn how to retrieve a stored document
- The *Deleting a document* recipe in this chapter to learn how to delete a document
- The *Updating a document* recipe in this chapter to learn how to update the fields in a document

To learn more about *optimistic concurrency control* – that is, the Elasticsearch way to manage concurrency on a document – go to `http://en.wikipedia.org/wiki/Optimistic_concurrency_control`.

Getting a document

Once you've indexed a document, during your application's life, it will probably need to be retrieved.

The GET REST call allows us to get a document in real time without the need to refresh it.

Getting ready

You will need an up-and-running Elasticsearch installation, as we described in the *Downloading and installing Elasticsearch* recipe of *Chapter 1, Getting Started*.

To execute the commands in this recipe, you can use any HTTP clients, such as curl (`https://curl.haxx.se/`), Postman (`https://www.getpostman.com/`), or others. I suggest using the Kibana console as it provides code completion and better character escaping for Elasticsearch.

To execute the following commands correctly, please use the indexed document from the *Indexing a document* recipe.

How to do it...

The GET method allows us to return a document, given its index, type, and ID.

The REST API's URL is as follows:

```
http://<server>/<index_name>/_doc/<id>
```

To get a document, follow these steps:

1. If we consider the document that we indexed in the previous recipe, the call will be as follows:

    ```
    GET /myindex/_doc/2qLrAfPVQvCRMe7Ku8r0Tw
    ```

2. The indexed document should be returned by Elasticsearch:

    ```
    { "_index" : "myindex",
      "_id" : "2qLrAfPVQvCRMe7Ku8r0Tw",
      "_version" : 1, "_seq_no" : 0,
      "_primary_term" : 1, "found" : true,
      "_source" : {
        "id" : "1234", "date" : "2013-06-07T12:14:54",
        "customer_id" : "customer1" } }
    ```

3. Our indexed data is contained in the _source parameter, but other information is returned:

    *   _index: The index that stores the document
    *   _id: The ID of the document
    *   _version: The version of the document
    *   found: Whether the document has been found

    If the record is missing, a 404 error is returned as the status code.

How it works...

Using the Elasticsearch GET API on the document doesn't require you to refresh: all the GET calls are made in real time.

This call is very fast because Elasticsearch only redirects the search on the shard that contains the document without any other overhead, and the document IDs are often cached in memory for fast lookup.

The source of the document is only available if the _source field is stored (as per the default settings in Elasticsearch).

Several additional parameters can be used to control the GET call:

- _source allows us to retrieve only a subset of fields. This is very useful for reducing bandwidth or for retrieving calculated fields such as the attachment-mapping ones:

```
GET /myindex/_doc/2qLrAfPVQvCRMe7Ku8r0Tw?_
source=date,sent
```

- stored_fields, similar to source, allows us to retrieve only a subset of fields that are marked as stored in the mapping. Stored fields are kept in a separated memory portion of the index, and they can be retrieved without you having to parse the JSON source:

```
GET /myindex/_doc/2qLrAfPVQvCRMe7Ku8r0Tw?stored_
fields=date,sent
```

- routing allows us to specify the shard to be used for the GET operation. To retrieve a document, the routing that's used at indexing time must be the same as the one that was used at search time:

```
GET /myindex/_doc/2qLrAfPVQvCRMe7Ku8r0Tw?routing=customer_
id
```

- refresh allows us to refresh the current shard before performing the GET operation (it must be used with care because it slows down indexing and introduces some overhead):

```
GET /myindex/_doc/2qLrAfPVQvCRMe7Ku8r0Tw?refresh=true
```

- preference allows us to control which shard replica is chosen to execute the GET method. Generally, Elasticsearch chooses a random shard for the GET call. The possible values are as follows:

  - _primary for the primary shard.

  - _local, first trying the local shard and then falling back to a random choice. Using the local shard reduces the bandwidth usage and should generally be used with auto-replicating shards (replica set to 0-all).

  - custom value for selecting a shard-related value, such as customer_id or username.

There's more...

The GET API is very fast, so a good practice when developing applications is to try and use it as much as possible. Choosing the correct ID form during application development can bring a big boost in performance.

If the shard that contains the document is not bound to an ID, a query with an ID filter (we will learn more about them in *Chapter 5*, *Text and Numeric Queries*, in the *Using an IDS query* recipe) is required to fetch the document.

If you don't need to fetch the record and only want to check for its existence, you can replace GET with HEAD. The response will be a status code of 200 if it exists or 404 if it is missing.

The GET call also has a special endpoint, _source, that allows you to only fetch the source of the document.

The GET source's REST API URL is as follows:

```
http://<server>/<index_name>/_doc/<id>/_source
```

To fetch the source of the previous order, we can call the following code:

```
GET /myindex/_doc/2qLrAfPVQvCRMe7Ku8r0Tw/_source
```

See also

Please refer to the *Speeding up the GET operation* recipe in this chapter to learn how to execute multiple GET operations in one go to reduce fetching time.

Deleting a document

Deleting documents in Elasticsearch can be done in two ways: using the DELETE call or the delete_by_query call, which we'll look at in the next chapter.

Getting ready

You will need an up-and-running Elasticsearch installation, as we described in the *Downloading and installing Elasticsearch* recipe of *Chapter 1*, *Getting Started*.

To execute the commands in this recipe, you can use any HTTP client, such as curl (https://curl.haxx.se/), Postman (https://www.getpostman.com/), or others. I suggest using the Kibana console as it provides code completion and better character escaping for Elasticsearch.

To execute the following commands correctly, please use the indexed document from the *Indexing a document* recipe.

How to do it...

The REST API URL is the same as it is for GET calls, but the HTTP method is DELETE:

```
http://<server>/<index_name>/_doc/<id>
```

To delete a document, follow these steps:

1. If we consider the order indexed from the *Indexing a document* recipe, then the call to delete a document will be as follows:

   ```
   DELETE /myindex/_doc/2qLrAfPVQvCRMe7Ku8r0Tw
   ```

2. The following output will be returned by Elasticsearch:

   ```
   { "_index" : "myindex",
     "_id" : "2qLrAfPVQvCRMe7Ku8r0Tw",
     "_version" : 2,  "result" : "deleted",
     "_shards" : { "total" : 2, "successful" : 1,
       "failed" : 0 },
     "_seq_no" : 3, "_primary_term" : 1 }
   ```

3. If the record is missing, a 404 error is returned as the status code, and the return JSON will be as follows:

   ```
   { "_index" : "myindex",
     "_id" : "2qLrAfPVQvCRMe7Ku8r0Tw",
     "_version" : 3, "result" : "not_found",
     "_shards" : { "total" : 2, "successful" : 1,
       "failed" : 0 },
     "_seq_no" : 4, "_primary_term" : 1 }
   ```

How it works...

Deleting records only hits shards that contain documents, so there is no overhead. If the document is a child, the parent must be set to look for the correct shard.

Several additional parameters can be used to control the delete call. The most important ones are as follows:

- `routing`, which allows you to specify the shard to be used for the delete operation
- `version`, which allows you to define a version of the document to be deleted to prevent that document from being modified

The `DELETE` operation has to restore functionality. Every document that is deleted is lost forever.

Deleting a record is a fast operation and very easy to use if the IDs of the documents to delete are available. Otherwise, we must use the `delete_by_query` call, which we will look at in the next chapter.

See also

Please refer to the *Deleting by query* recipe in *Chapter 4, Exploring Search Capabilities*, to learn how to delete a bunch of documents that match a query.

Updating a document

Documents stored in Elasticsearch can be updated during their lives. There are two available solutions for performing this operation in Elasticsearch: adding a new document or using the update call.

The update call can work in two ways:

- By providing a script that uses the update strategy
- By providing a document that must be merged with the original one

The main advantage of updating versus using an index is the networking reduction and the increased possibility of reducing conflicts due to concurrent changes.

Getting ready

You will need an up-and-running Elasticsearch installation, as we described in the *Downloading and installing Elasticsearch* recipe of *Chapter 1, Getting Started*.

To execute the commands in this recipe, you can use any HTTP client, such as curl (https://curl.haxx.se/), Postman (https://www.getpostman.com/), or others. I suggest using the Kibana console as it provides code completion and better character escaping for Elasticsearch.

To execute the following commands correctly, please use the indexed document from the *Indexing a document* recipe.

To use dynamic scripting languages, they must be enabled. See *Chapter 9, Managing Clusters*, for more information.

How to do it...

Since we are changing the state of the data, the HTTP method we must use here is POST and the REST URL will be as follows:

```
http://<server>/<index_name>/_update/<id>
```

The REST format is changed by the previous version of Elasticsearch.

To update a document, follow these steps:

1. If we consider the order type from the previous recipe, the call to update a document will be as follows:

    ```
    POST /myindex/_update/2qLrAfPVQvCRMe7Ku8r0Tw
    { "script": {
        "source": "ctx._source.in_stock_items += params.
    count",
        "params": { "count": 4 } } }
    ```

2. If the request is successful, the following output will be returned by Elasticsearch:

    ```
    { "_index" : "myindex",
     "_id" : "2qLrAfPVQvCRMe7Ku8r0Tw",
     "_version" : 4, "result" : "updated",
     "_shards" : { "total" : 2, "successful" : 1,
       "failed" : 0 },
     "_seq_no" : 8, "_primary_term" : 1 }
    ```

3.   The record will look as follows:

```
{ "_index" : "myindex",
  "_id" : "2qLrAfPVQvCRMe7Ku8r0Tw",
  "_version" : 8, "_seq_no" : 12,
  "_primary_term" : 1, "found" : true,
  "_source" : {
    "id" : "1234", "date" : "2022-01-07T12:14:54",
    "customer_id" : "customer1", "sent" : true,
    "in_stock_items" : 4 } }
```

The visible changes are as follows:

- The scripted field has been changed.

- The version has been incremented.

How it works...

The update operation takes a document, applies the changes that are required to the script or the updated document, and then reindexes the changed document. In *Chapter 8, Scripting in Elasticsearch*, we will explore the scripting capabilities of Elasticsearch.

The standard language for scripting in Elasticsearch is **Painless**, and it's been used in these examples.

The script can operate on `ctx._source`, which is the source of the document (it must be stored to work), and it can change the document in situ. It's possible to pass parameters to a script by passing a JSON object. These parameters are available in the execution context.

A script can control Elasticsearch's behavior once the script has been executed by setting the `ctx.op` value of the context. The available values are as follows:

- `ctx.op="delete"`: Here, the document will be deleted after the script's execution.

- `ctx.op="none"`: Here, the document will skip the indexing process. A good practice to improve performance is to set `ctx.op="none"` so that the script doesn't update the document, thus preventing any reindexing overhead.

`ctx` also manages the timestamp of the record in `ctx._timestamp`. It's possible to pass an additional object in the `upsert` property, which will be used if the document is not available in the index:

```
POST /myindex/_update/2qLrAfPVQvCRMe7Ku8r0Tw
{ "script": {
    "source": "ctx._source.in_stock_items += params.count",
    "params": { "count": 4 } },
  "upsert": { "in_stock_items": 4 } }
```

If you need to replace some field values, a good solution is not to write a complex update script, but to use the special `doc` property, which allows us to overwrite the values of an object. The document that's provided in the `doc` parameter will be merged with the original one. This approach is easier to use, but it cannot set `ctx.op`, so if the update doesn't change the value of the original document, the next successive phase will always be executed:

```
POST /myindex/_update/2qLrAfPVQvCRMe7Ku8r0Tw
{ "doc": { "in_stock_items": 10 } }
```

If the original document is missing, it is possible to provide a `doc` value (the document to be created) for an `upsert` as a `doc_as_upsert` parameter:

```
POST /myindex/_update/2qLrAfPVQvCRMe7Ku8r0Tw
{ "doc": { "in_stock_items": 10 }, "doc_as_upsert" : true }
```

By using Painless scripting, it is possible to apply advanced operations on fields, such as the following:

- Removing a field, like so:

```
"script" : {"inline": "ctx._source.remove("myfield"}}
```

- Adding a new field, like so:

```
"script" : {"inline": "ctx._source.myfield=myvalue"}}
```

The update REST call is very useful because it has some advantages:

- It reduces bandwidth usage because the update operation doesn't need to make a round trip to the client of the data.

- It's safer because it automatically manages optimistic concurrent control: if a change is made during script execution, the script that it's re-executed with updates the data.

- It can be bulk executed.

See also

Please refer to the following recipe, *Speeding up atomic operations (bulk operations)*, to learn how to use bulk operations to reduce the networking load and speed up ingestion.

Speeding up atomic operations (bulk operations)

When we are inserting, deleting, or updating a large number of documents, the HTTP overhead is significant. To speed up this process, Elasticsearch allows us to execute the bulk of CRUD calls.

Getting ready

You will need an up-and-running Elasticsearch installation, as we described in the *Downloading and installing Elasticsearch* recipe of *Chapter 1, Getting Started*.

To execute the commands in this recipe, you can use any HTTP client, such as curl (https://curl.haxx.se/), Postman (https://www.getpostman.com/), or others. I suggest using the Kibana console as it provides code completion and better character escaping for Elasticsearch.

How to do it...

Since we are changing the state of the data, we must use the POST HTTP method. The REST URL will be as follows:

```
http://<server>/<index_name/_bulk
```

To execute a bulk action, we will perform the following steps via curl (because it's very common to prepare your data on files and send them to Elasticsearch via the command line):

1. We need to collect the create/index/delete/update commands in a structure made up of bulk JSON lines, composed of a line of action with metadata, and another optional line of data related to the action. Every line must end with a new line, \n. A bulk data file should be presented like this:

```
{ "index":{ "_index":"myindex", "_id":"1" } }
{ "field1" : "value1", "field2" : "value2" }
{ "delete":{ "_index":"myindex", "_id":"2" } }
{ "create":{ "_index":"myindex", "_id":"3" } }
{ "field1" : "value1", "field2" : "value2" }
{ "update":{ "_index":"myindex", "_id":"3" } }
{ "doc":{"field1" : "value1", "field2" : "value2" }}
```

2. This file can be sent with the following POST:

```
curl -s -XPOST localhost:9200/_bulk --data-binary @
bulkdata;
```

3. The output that's returned by Elasticsearch should collect all the responses from the actions.

    You can execute the previous commands in Kibana with the following call:

```
POST /_bulk
{ "index":{ "_index":"myindex", "_id":"1" } }
{ "field1" : "value1", "field2" : "value2" }
{ "delete":{ "_index":"myindex", "_id":"2" } }
{ "create":{ "_index":"myindex", "_id":"3" } }
{ "field1" : "value1", "field2" : "value2" }
{ "update":{ "_index":"myindex", "_id":"3" } }
{ "doc":{"field1" : "value1", "field2" : "value2" }}
```

How it works...

Using bulk operation allows you to aggregate different calls as a single one: a header part with the action to be performed, and a body for other operations such as index, create, and update.

The header is composed of the action's name and the object of its parameters. Looking at the previous index example, we have the following:

```
{ "index":{ "_index":"myindex", "_id":"1" } }
```

For indexing and creating, an extra body is required for the data:

```
{ "field1" : "value1", "field2" : "value2" }
```

The `delete` action doesn't require optional data, so only the header composes it:

```
{ "delete":{ "_index":"myindex", "_id":"1" } }
```

However, it is possible to use an update action in a bulk operation with a format similar to the `index` one:

```
{ "update":{ "_index":"myindex", "_id":"3" } }
```

The header accepts all the common parameters of the update action, such as `doc`, `upsert`, `doc_as_upsert`, `lang`, `script`, and `params`. To control the number of retries in the case of concurrency, the bulk update defines the `_retry_on_conflict` parameter, set to the number of retries to be performed, before raising an exception.

So, a possible body for the update would be as follows:

```
{ "doc":{"field1" : "value1", "field2" : "value2" }}
```

The bulk item can accept several parameters, such as the following:

- `routing`, to control the routing shard.
- `parent`, to select a parent item shard. This is required if you are indexing some child documents. Some global bulk parameters that can be passed using query arguments are as follows:
  - `consistency` (`one`, `quorum`, `all`) (default `quorum`), which controls the number of active shards before executing write operations.
  - `refresh` (default `false`), which forces a refresh in the shards involved in bulk operations. The newly indexed document will be available immediately, without having to wait for the standard refresh interval (1 second).
  - `pipeline`, which forces an index using the ingest pipeline provided.

Previous versions of Elasticsearch required users to pass the `_type` value, but this was removed in version 7.x due to type removal.

Usually, Elasticsearch client libraries that use the Elasticsearch REST API automatically implement serialization for bulk commands.

The correct number of commands to serialize in a bulk execution is the user's choice, but there are some things to consider:

- In the standard configuration, Elasticsearch limits the HTTP call to 100 MB in size. If the size is over that limit, the call is rejected.

- Multiple complex commands take a lot of time to be processed, so pay attention to client timeout.

- The small size of commands in a bulk operation doesn't improve performance.

If the documents aren't big, 500 commands in a bulk operation can be a good number to start with, and it can be tuned depending on your data structures (number of fields, number of nested objects, the complexity of fields, and so on).

Speeding up GET operations (multi-GET)

The standard GET operation is very fast, but if you need to fetch a lot of documents by ID, Elasticsearch provides the _mget operation.

Getting ready

You will need an up-and-running Elasticsearch installation, as we described in the *Downloading and installing Elasticsearch* recipe of *Chapter 1, Getting Started*.

To execute the commands in this recipe, you can use any HTTP client, such as curl (https://curl.haxx.se/), Postman (https://www.getpostman.com/), or others. I suggest using the Kibana console as it provides code completion and better character escaping for Elasticsearch.

To execute the following commands correctly, please use the indexed document we created in the *Indexing a document* recipe.

How to do it...

The multi-GET REST URLs are as follows:

```
http://<server</_mget
http://<server>/<index_name>/_mget
```

To execute a multi-GET action, follow these steps:

1. First, we must use the POST method with a body that contains a list of document IDs and the index or type if they are missing. As an example, using the first URL, we need to provide the index, type, and ID:

```
POST /_mget
{ "docs": [
      { "_index": "myindex",
        "_id":"2qLrAfPVQvCRMe7Ku8r0Tw" },
      { "_index": "myindex", "_id": "2" } ] }
```

This kind of call allows us to fetch documents in several different indices and types.

2. If the index and the type are fixed, the call should be in the following form:

```
GET /myindex/_mget
{ "ids" : ["1", "2"] }
```

The multi-GET result is an array of documents.

How it works...

Multi-GET calling is a shortcut for executing many GET commands in one shot.

Internally, Elasticsearch spreads the get object in parallel on several shards and collects the results to return to the user.

The get object can contain the following parameters:

*   _index: The index that contains the document. It can be omitted if it's passed in the URL.
*   _id: The document's ID.
*   stored_fields (optional): A list of fields to retrieve.
*   _source (optional): The source filter object.
*   routing (optional): The shard routing parameter.

The advantages of a multi-GET are as follows:

*   Reduced networking traffic, both internally and externally for Elasticsearch.
*   Increased speed if it's used in an application: the time for processing a multi-GET is quite similar to a standard get.

See also...

Please refer to the *Getting a document* recipe in this chapter to learn how to execute a simple `get` and the general parameters of a GET call.

4
Exploring Search Capabilities

Now that we have set the mappings and put the data inside the indices, we can start exploring the search capabilities of **Elasticsearch**. In this chapter, we will cover how to search using different factors: sorting, highlighting, scrolling, suggesting, counting, and deleting. These actions are the core part of Elasticsearch; ultimately, everything in Elasticsearch is about serving the query and returning good-quality results.

This chapter is divided into two parts: the first part shows how to perform an API call-related search, and the second part will look at two special query operators that are the basis for building complex queries in the upcoming chapters.

In this chapter, we will cover the following recipes:

- Executing a search
- Sorting results
- Highlighting results
- Executing a scrolling query
- Using the search_after functionality
- Returning inner hits in results
- Suggesting a correct query
- Counting matched results
- Explaining a query
- Query profiling
- Deleting by query
- Updating by query
- Matching all of the documents
- Using a Boolean query
- Using the search template

Technical requirements

All the recipes in this chapter require us to prepare and populate the required indices—the online code is available in the GitHub repository(`https://github.com/PacktPublishing/Elasticsearch-8.x-Cookbook`). Here, you can find scripts to initialize all of the required data.

Executing a search

Elasticsearch was born as a search engine; its main purpose is to process queries and give results as quickly as possible. In this recipe, we'll learn that a search in Elasticsearch is not only limited to matching documents—it can also calculate additional information that's required to improve the quality of the search.

Getting ready

You will need an up-and-running Elasticsearch installation, as described in the *Downloading and installing Elasticsearch* recipe of *Chapter 1, Getting Started*.

To execute these commands, any HTTP client can be used, such as Curl (`https://curl.haxx.se/`), Postman (`https://www.getpostman.com/`), or something similar. I suggest using the Kibana console as it provides code completion and better character escaping.

To correctly execute the following commands, you will need an index populated with the `ch04/populate_kibana.txt` commands or the `ch04/populate_kibana.sh` script, which is available in the online code. (Remember to set the user and password credentials in the `ES_USER/ES_PASSWORD` environment variables before executing the script.)

The mapping that's used in all the queries and searches in this chapter is similar to the following book representation:

Figure 4.1 – ML model of a Book entity

The command to create the schema is as follows:

```
PUT /mybooks
{ "mappings": {
    "properties": {
        "join_field": {
            "type": "join", "relations": {   "order": "item"
```

```
... truncated...
      "title": {
          "term_vector": "with_positions_offsets",
          "type": "text",
          "fields": {
              "keyword": { "type": "keyword", "ignore_above": 256
}
... truncated...
```

How to do it...

To execute the search and view the results, we will perform the following steps:

1. From the command line, we can execute a search, as follows:

    ```
    GET /mybooks/_search
    { "query": { "match_all": {} } }
    ```

    In this case, we have used a match_all query that returns all of the documents. We'll discuss this kind of query in the *Matching all of the documents* recipe.

2. If everything works as expected, the command will return the following:

    ```
    { "took" : 0,
        "timed_out" : false,
        "_shards" : {
          "total" : 1, "successful" : 1, "skipped" : 0,
    "failed" : 0
        },
        "hits" : {
          "total" : { "value" : 3, "relation" : "eq" },
          "max_score" : 1.0,
          "hits" : [
    {
              "_index" : "mybooks",
              "_id" : "3",
              "_score" : 1.0,
              "_source" : {...truncated...}
          }
        ... truncated...
    ```

The preceding results contain the following information:

- `took` is the milliseconds of time required to execute the query.

- `timed_out` indicates whether a timeout occurred during the search. This is related to the timeout parameter of the search. If a timeout occurs, you will either get partial or no results.

- `_shards` refers to the status of the shards, and this is divided into the following sections:

  - `total`: This refers to the number of shards.

  - `successful`: This refers to the number of shards in which the query was successful.

  - `skipped`: This refers to the number of shards that were skipped during the search (for example, if you were searching more than 720 shards simultaneously).

  - `failed`: This refers to the number of shards in which the query failed because some error or exception occurred during the query.

- `hits` refers to the results, and they are composed of the following:

  - `total` is the object that contains the number of documents that match the query. `value` refers to the number of results (in the case of results that are greater than 10,000, it's set to 10,000), and `relation` indicates whether the size value is `eq` (equal) to the count value or `gte` (greater than or equal to) the returned count. For a large number of results, by default, the correct result score is not computed to speed up the searches. (Please refer to the `track_total_hits` query parameter, in the next section, to learn how to return an exact match.)

  - `max_score` is the match score of the first document. Usually, it is 1 if no match scoring has been computed, for example, in sorting or filtering.

  - `hits` refers to the list of result documents.

The resulting document has lots of fields that are always available and others that depend on search parameters that are not available in the actual response. The most important fields are listed as follows:

- `_index`: This refers to the index that contains the document.

- `_id`: This refers to the ID of the document.

- `_source`: This refers to the document source—the original `json` file sent to Elasticsearch.

- `_score`: This is the query score of the document (if the query doesn't require a score, it's 1.0).

- `sort`: If the document has been sorted, this refers to the values that are used for sorting.

- `highlight`: This refers to the highlighted segments if highlighting was requested.

- `stored_fields`: Some fields can be retrieved without needing to fetch the source object.

- `script_fields`: Some fields can be computed using scripting.

How it works...

The search in Elasticsearch is a distributed computation composed of many steps. The main ones are as follows:

1. In the master or coordinator nodes, validation of the query body is required.

2. A selection of indices to be used in the query is needed; the shards are randomly chosen.

3. The execution of the query part in the data nodes collects the top hits of the query.

4. Next is the aggregation of results in the master and coordinator nodes, along with the scoring.

5. Finally, return the results to the user.

The following diagram shows how the query is distributed inside the cluster:

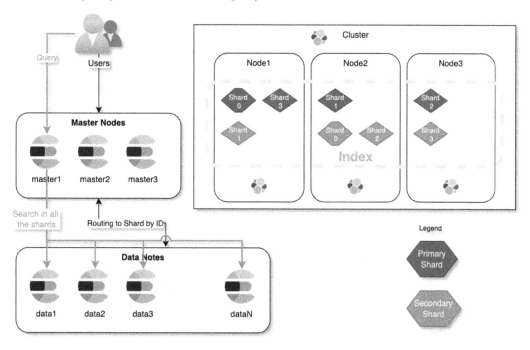

Figure 4.2 – Query distribution inside an Elasticsearch cluster

The HTTP method to execute a search is GET (although POST also works). The REST endpoints are as follows:

```
http://<server>/_search
http://<server>/<index_name(s)>/_search
```

Not all HTTP clients allow you to send data through a GET call. So, if you need to send body data, the best practice is to use the POST call.

Multi-indices and types are comma-separated. If an index or a type has been defined, the search is limited to only them. One or more aliases can be used as index names.

Usually the core query is contained in the body of the GET/POST call, but a lot of options can also be expressed as URI query parameters, such as the following:

- q: This is the query string to perform simple string queries, which can be done as follows:

```
GET /mybooks/_search?q=uuid:11111
```

- df: This sets the default field to be used within the query. This can be done as follows:

```
GET /mybooks/_search?df=uuid&q=11111
```

- from (the default value is 0): This refers to the start index of the hits.

- size (the default value is 10): This refers to the number of hits to be returned.

- analyzer: This refers to the default analyzer to be used. See the *Specifying different analyzers* recipe in *Chapter 2, Managing Mappings*.

- default_operator (the default value is OR): This can be set to AND or OR.

- explain: This allows the user to return information regarding how the score has been calculated. It is calculated as follows:

```
GET /mybooks/_search?q=title:joe&explain=true
```

- stored_fields: This parameter allows the user to define fields that must be returned. This can be done as follows:

```
GET /mybooks/_search?q=title:joe&stored_fields=title
```

- sort (the default value is score): This allows the user to change the order of the documents. By default, sorting is ascendant; if you need to change the order, add desc to the field, as follows:

```
GET /mybooks/_search?sort=title.keyword:desc
```

- timeout (by default, this is not active): This defines the timeout for the search. Elasticsearch tries to collect results until a timeout. If a timeout is fired, all the hits that have been accumulated are returned.

- search_type: This defines the search strategy. A reference is available in the online Elasticsearch documentation at https://www.elastic.co/guide/en/elasticsearch/reference/current/search-request-search-type.html.

- `track_scores` (the default value is `false`): If `true`, this tracks the score and allows it to be returned with the hits. It's used in conjunction with `sort` because sorting, by default, prevents the return of a match score.

- `pretty` (the default value is `false`): If `true`, the results will be pretty-printed by putting in the JSON output newlines and tabs to simplify readiness.

Generally, the query contained in the body of the search is a JSON object. The body of the search is the core of Elasticsearch's search functionalities; the list of search capabilities extends in every release. For the current version of Elasticsearch, the following parameters are available:

- `query`: This contains the query to be executed. Later in this chapter, we will see how to create different kinds of queries that cover several scenarios.

- `from`: This allows the user to control the pagination. The `from` parameter defines the start position of the hits to be returned (the default is `0`) and `size` (the default is `10`).

> **Note**
>
> Query argument parameters always have precedence over body arguments: if the same argument is defined as both a query argument and a body argument, only the query argument is evaluated.

The pagination is applied to the currently returned search results. Firing the same query can bring different results if a lot of records have the same score. Therefore, their order could change, or you could have live data and new documents ingested. If you need to process all the result documents without repetition, you need to execute the `scan` or `scroll` queries:

- `sort`: This allows the user to change the order of the matched documents. This option is covered, in full, in the *Sorting results* recipe.

- `post_filter`: This allows the user to filter out the query results without affecting the aggregation count. Usually, it's used for filtering by facet values.

- `_source`: This allows the user to control the returned source. It can be disabled (`false`), partially returned (`obj.*`), or use multiple exclude/include rules. This functionality can be used instead of fields to return values (for complete coverage of this, take a look at the online Elasticsearch reference page at `http://www.elasticsearch.org/guide/en/elasticsearch/reference/current/search-request-source-filtering.html`).

- `fielddata_fields`: This allows the user to return a field data representation of the field.

- `stored_fields`: This controls the fields to be returned.

Returning only the required fields reduces the network and memory usage, thus improving performance. The recommended way to retrieve custom fields is to use the `_source` filtering function because it doesn't need to use Elasticsearch's extra resources:

- `aggregations/aggs`: These control the analytics of the aggregation layer. These will be discussed in *Chapter 5, Text and Numeric Queries*.

- `index_boost`: This allows the user to define the per-index boost value. It is used to increase/decrease the score of results in boosted indices.

- `highlighting`: This allows the user to define the fields and settings to be used for calculating a query abstract (see the *Highlighting results* recipe).

- `Version` (the default value is `false`): This adds the version of a document in the results.

- `Rescore`: This allows the user to define an extra query to be used in the score to improve the quality of the results. The `rescore` query is executed on the hits that match the first query and the filter.

- `min_score`: If this is given, all of the result documents that have a score lower than this value are rejected.

- `explain`: This returns information regarding how the TD/IF score is calculated for a particular document.

- `script_fields`: This defines a script that computes extra fields via scripting to be returned with a hit. We'll look at Elasticsearch scripting in *Chapter 8, Scripting in Elasticsearch*.

- `suggest`: If given a query and a field, this returns the most significant terms related to this query. This parameter allows the user to implement the Google-like *do you mean* functionality (see the *Suggesting a correct query* recipe).

- `search_type`: This defines how Elasticsearch should process a query. We'll see the scrolling query in the *Executing a scrolling query* recipe.

- `scroll`: This controls the scrolling in the scroll/scan queries. `scroll` allows the user to have an Elasticsearch equivalent of the DBMS cursor.

- `collapse`: This allows the user to define a field that can be used to collapse results: generally, it's used in conjunction with the `sort` parameter. (You can find more references at `https://www.elastic.co/guide/en/elasticsearch/reference/8.1/collapse-search-results.html`.)

- `_name`: This allows returns for every hit that matches the named queries. It's very useful if you have a Boolean and you want the name of the matched query.

- `search_after`: This allows the user to skip results using the most efficient way of scrolling. We'll examine this functionality in the *Using the search_after functionality* recipe.

- `preference`: This allows the user to select which shard/s to use for executing the query.

- `track_total_hits` (the default is `false`): This allows you to return the exact number of matching hits at the cost of some performance.

> **Note**
>
> Multiple searches can be executed with a single REST call using the _msearch endpoint, which accepts multiple JSON files (one for every search) such as the _bulk command.

There's more...

To improve the quality of the results score, Elasticsearch provides the `rescore` functionality. This capability allows the user to reorder a top number of documents with one or more queries that are, generally, much more expensive (either CPU or time-consuming), for example, if the query contains a lot of match queries or scripting. This approach allows the user to execute the `rescore` query on just a small subset of results, reducing the overall computation time and number of resources.

As with every query, the `rescore` query is executed at the shard level, so it's automatically distributed.

The best candidates to be executed in the `rescore` query are complex queries with a lot of nested options, and everything that is used in scripting (due to the massive overhead of scripting languages).

The following example will show you how to execute a fast query (a Boolean one) in the first phase and then perform a `rescore` query on it with a `match` query in the `rescore` section:

```
POST /mybooks/_search
{"query": {
    "match": {
      "description": {
        "operator": "or",
        "query": "nice guy joe"
      } }
  },
  "rescore": {
    "window_size": 100,
    "query": {
      "rescore_query": {
        "match_phrase": {
          "description": {
            "query": "joe nice guy",
            "slop": 2
          } } },
      "query_weight": 0.8,
      "rescore_query_weight": 1.5
} } }
```

The `rescore` parameters are as follows:

- `window_size`: The example is `100`. This controls how many results per shard must be considered in the `rescore` functionality.

- `query_weight`: The default value is `1.0`, and the `rescore_query_weight` default value is `1.0`. These are used to compute the final score using the following formula:

$$final\_score = query\_score * query\_weight + rescore\_score * rescore\_query\_weight$$

If a user wants to only keep the `rescore` score, they can set the `query_weight` parameter to `0`.

See also

For further reference, take a look at the following recipes that are related to this recipe:

- The *Executing an aggregation* recipe, in *Chapter 7*, *Aggregations*, explains how to use the aggregation framework during queries.

- The *Highlighting results* recipe, in this chapter, explains how to use the highlighting functionality for improving the user experience in results.

- The *Executing a scrolling query* recipe, in this chapter, covers how to efficiently paginate results.

- The *Suggesting terms for a query* recipe, in this chapter, will help you to correct text queries.

- More information about multiple searches in a single REST call can be found at `https://www.elastic.co/guide/en/elasticsearch/reference/current/search-multi-search.html`.

Sorting results

When searching for results, the standard criterion for sorting in Elasticsearch is the relevance to a text query. Often, real-world applications need to control the sorting criteria in scenarios, such as the following:

- Sorting a user by their last name and their first name

- Sorting items by stock symbols and price (ascending and descending)

- Sorting documents by size, file type, source, and more

- Sorting items related to the maximum, the minimum, or the average of some of the children fields

Getting ready

You will need an up-and-running Elasticsearch installation, as described in the *Downloading and installing Elasticsearch* recipe of *Chapter 1*, *Getting Started*.

To execute these commands, any HTTP client can be used, such as Curl (`https://curl.haxx.se/`), Postman (`https://www.getpostman.com/`), or similar. I suggest using the Kibana console as it provides code completion and better character escaping for Elasticsearch.

To correctly execute the following commands, you will need an index populated with the ch04/populate_kibana.txt command, which is available in the online code.

How to do it...

In order to sort the results, we will perform the following steps:

1. Add a sort section to your query, as follows:

```
GET /mybooks/_search
{ "query": { "match_all": {} },
  "sort": [
    { "price": {
        "order": "asc",
        "mode": "avg",
        "unmapped_type": "double",
        "missing": "_last"
    } },
    "_score"
  ] }
```

The returned result should be similar to the following:

```
...truncated...
    "hits" : {
    "total" : { "value" : 3, "relation" : "eq" },
    "max_score" : null,
    "hits" : [
      {
        "_index" : "mybooks", "_id" : "1", "_score" : 1.0,
        "_source" : {
          ...truncated...
          "price" : 4.3, "quantity" : 50
        },
        "sort" : [ 4.3, 1.0 ]
...truncated...
```

The sort result is very special—an extra sort field is created to collect the value that's used for sorting.

How it works...

The `sort` parameter can be defined as a list that can contain both simple strings and JSON objects. The sort strings are the names of the fields (such as `field1`, `field2`, `field3`, `field4`, and more) that are used for sorting and are similar to the `order by` SQL function.

The JSON object allows users to have extra parameters, as follows:

- `unmapped_type` (`long`, `int`, `double`, or `string`): This defines the type of the `sort` parameter if the value is missing. It's a best practice to define it to prevent sorting errors due to missing values.

- `missing` (`_last` or `_first`): This defines how to manage missing values—whether to put them at the end (`_last`) of the results or the start (`_first`).

- `order` (`asc` or `desc`): This defines whether the order must be considered ascendant (which is the default) or descendent.

- `mode`: This defines how to manage multi-valued fields. The possible values are listed as follows:

  - `min`: The minimum value is chosen (that is to say that, in the case of multiple prices on an item, it chooses the lowest for comparison).

  - `max`: The maximum value is chosen.

  - `sum`: The `sort` value will be computed as the sum of all the values. This mode is only available in numeric array fields.

  - `avg`: The `sort` value will be the average of all the values. This mode is only available in numeric array fields.

  - `median`: The `sort` value will be the median of all the values. This mode is only available in numeric array fields.

If we want to add the relevance score value to the sort list, we must use the special `_score` sort field.

If you are sorting for a nested object, two extra parameters can be used, as follows:

- `nested_path`: This defines the nested object to be used for sorting. The field defined for sorting will be relative to the `nested_path` parameter. If it is not defined, then the sorting field is related to the document root.

- `nested_filter`: This defines a filter that is used to remove nested documents that don't match the sorting value extraction. This filter allows for a better selection of values to be used in sorting.

For example, if we have an `address` object nested in a `person` document and we can sort according to `city.name`, we can use the following:

- `address.city.name` without defining the `nested_path` parameter
- `city.name` if we define a `nested_path` parameter

The sorting process requires the sorting fields of all the matched query documents to be fetched and compared. To prevent high memory usage, it's better to sort numeric fields, and in the case of string sorting, choose short text fields processed with an analyzer that doesn't tokenize the text.

There's more...

If you are using `sort`, pay attention to the tokenized fields. This is because the sort order depends on the lower-order token if ascendant and the higher-order token if descendent. In the case of tokenized fields, this behavior is not the same as a common sort query because we execute it at the term level:

- For example, if we sort by the descending `title` field, we use the following:

```
GET /mybooks/_search?sort=title:desc
```

In the preceding example, the results are as follows:

```
{ ...truncated...
    "hits" : {
        "total" : { "value" : 3, "relation" : "eq" },
        "max_score" : null,
        "hits" : [
            {
                "_index" : "mybooks", "_id" : "1", "_score" :
null,
                "_source" : {
                    ...truncated...
                    "title" : "Joe Tester",
                    ...truncated...
                "sort" : [ "bill" ]
                ...truncated...
```

- The expected SQL results can be obtained using a not-tokenized keyword field, which, in this case, is `title.keyword`:

```
GET /mybooks/_search?sort=title.keyword:desc
```

The results are as follows:

```
{ ...truncated...
  "hits" : {
    "total" : 3, "max_score" : null,
    "hits" : [ {
        "_index" : "mybooks",
        "_id" : "2",
        ...truncated...
        "sort" : [ "Bill Baloney" ]
} ] } }
```

There are two special sorting types: `geo distance` and `scripting`.

Geo distance sorting uses the distance from a GeoPoint (location) as the metric to compute the ordering. A sorting example could be as follows:

```
...truncated...
  "sort" : [ {
    "_geo_distance" : {
      "pin.location" : [-70, 40],
      "order" : "asc", "unit" : "km"
    } } ],
...truncated...
```

The preceding example accepts special parameters, such as the following:

- `unit`: This defines the metric to be used to compute the distance.

- `distance_type` (`sloppy_arc`, `arc`, or `plane`): This defines the type of distance to be computed. The `_geo_distance` name for the field is mandatory.

The point of reference for the sorting can be defined in several ways, as we already discussed in the *Mapping a GeoPoint field* recipe of *Chapter 2, Managing Mapping*.

Using the scripting for sorting will be discussed in the *Sorting data using scripts* recipe of *Chapter 8, Scripting in Elasticsearch*, after we introduce the scripting capabilities of Elasticsearch.

See also

For further reference, take a look at the following recipes that are related to this recipe:

- The *Mapping a GeoPoint field* recipe, in *Chapter 2, Managing Mapping*, explains how to correctly create a mapping for a GeoPoint field.

- The *Sorting data using scripts* recipe, in *Chapter 8, Scripting in Elasticsearch*, will explain the use of a custom script for computing values to sort on.

Highlighting results

Elasticsearch performs a good job of finding matching results in big text documents. It's useful for searching text in very large blocks. However, to improve user experience, you need to show users the abstract—a small portion of the text part of the document that has matched the query. The abstract is a common way to help users understand how the matched document is relevant to them.

The highlight functionality in Elasticsearch is designed to do this job.

Getting ready

You will need an up-and-running Elasticsearch installation, as described in the *Downloading and installing Elasticsearch* recipe of *Chapter 1, Getting Started*.

To execute these commands, any HTTP client can be used, such as Curl (`https://curl.haxx.se/`), Postman (`https://www.getpostman.com/`), or similar. I suggest using the Kibana console as it provides code completion and better character escaping for Elasticsearch.

To correctly execute the following commands, you will need an index populated with the `ch04/populate_kibana.txt` command, which is available in the online code.

How to do it...

To search and highlight the results, we will need to perform the following steps:

1. From the command line, we can execute a search with a `highlight` parameter, as follows:

```
GET /mybooks/_search?from=0&size=10
  { "query": { "query_string": { "query": "joe" } },
    "highlight": {
      "pre_tags": [""],
      "fields": {
```

```
            "description": { "order": "score" },
            "title": { "order": "score" }
        },
        "post_tags": [""]
} }
```

2. If everything works as expected, the command will return the following result:

```
{ ...truncated...
    "hits" : {
        "total" : { "value" : 1, "relation" : "eq" ,
        "max_score" : 1.0126972,
        "hits" : [
            { "_index" : "mybooks",  "_id" : "1",
                "_score" : 1.0126973,
                ...truncated...
                "highlight" : {
                    "description" : ["<b>Joe</b> Testere nice guy"
],
                    "title" : ["<b>Joe</b> Tester" ]
...truncated...
```

As you can see, in the standard results, there is a new `highlight` field, which contains the highlighted fields within an array of fragments.

How it works...

When the `highlight` parameter is passed to the search object, Elasticsearch tries to execute the highlight parameter within the document results.

The highlighting phase, which occurs after the document fetchinf phase, tries to extract the highlight using the following steps:

1. It collects the terms that are available in the query.

2. It initializes the highlighter with the parameters that have been given during the query.

3. It extracts the interested fields and tries to load them if they are stored; otherwise, they are taken from the source.

4. It executes the query on single fields to detect the more relevant parts.

5. It adds the highlighted fragments found to the hit.

Using the highlighting functionality is very easy, but there are some important factors to pay attention to:

- The field that is used for highlighting must be available in one of these forms: stored field, inside the source, or inside the stored term vector.

  First, the Elasticsearch highlighter checks for the presence of the data field as the term vector (this is a faster way to execute the highlighting). If the field does not use the term vector (a special indexing parameter that allows you to store an index additional positional text data), it tries to load the field value from the stored fields. If the field is not stored, it finally loads the JSON source, interprets it, and extracts the data value, if available. Of course, the last approach is the slowest and most resource-intensive.

- If a special analyzer is used in the search, it should also be passed to the highlighter (this is often automatically managed).

When executing highlighting on a large number of fields, you can use the wildcard to multi-select them (that is to say, use `title*`).

The common properties for controlling highlighting field usage are listed as follows:

- `order`: This defines the selection order of the matched fragments.
- `force_source`: This skips the term vector or stored field and takes the field from the source (`false` is the default).
- `type` (the default is `unified`, and the valid values are `unified`, `plain`, and `fvh`): This is used to force a specific highlight type. For simple highlighting, the default value is the best one.
- `number_of_fragment`: The default value is 5. This parameter controls how many fragments will return. It can be configured globally or for a specific field.
- `fragment_size`: The default value is `100`. This is the number of characters that the fragments must contain. It can be configured globally or for a specific field.

There are several optional parameters that can be passed in the highlight object to control the highlighting markup, and these are as follows:

- `pre_tags/post_tags`: These parameters refer to a list of strings or tags that can be used for marking the highlighted text.
- `tags_schema="styled"`: This allows the user to define a tag schema that marks highlighting with different tags in order of importance. This is a helper to reduce the definition of a lot of the `pre_tags/post_tags` tags.

- encoder: The default value is html. If this is set to html, it will escape the HTML tags in the fragments.

- require_field_match: The default value is true. If this is set to false, it also allows you to highlight fields that don't match the query.

- boundary_chars: This is a list of characters that is used for phrase boundaries (these are characters such as , , ;, :, and /).

- boundary_max_scan: The default value is 20. This controls how many characters the highlighting must scan for boundaries in a match. It's used to provide better fragment extraction.

- matched_fields: This allows the user to combine multi-fields to execute the highlighting. This is very useful if the field that you use for highlighting is a multi-field that's been analyzed with different analyzers (such as standard, linguistic, and more). It can only be used when the highlighter is a **Fast Vector Highlighter** (**FVH**). An example of this usage could be as follows:

```
{ "query": {
    "query_string": {
        "query": "content.plain:some text",
        "fields": [ "content" ]
    } },
    "highlight": {
        "order": "score",
        "fields": {
            "content": {
                "matched_fields": [ "content", "content.plain"
],
                "type": "fvh"
} } } }
```

See also

For further reference, take a look at the following points that are related to this recipe:

- To understand how to structure a search, refer to the *Executing a search* recipe in this chapter.

- For more examples of border cases with highlighting, refer to the official documentation of Elasticsearch. This can be found at https://www.elastic.co/guide/en/elasticsearch/reference/master/search-request-highlighting.html.

Executing a scrolling query

Every time a query is executed, the results are calculated and returned to the user in real time. In Elasticsearch, there is no deterministic order for records—pagination on a big block of values can lead to inconsistency between results. This is due to added and deleted documents and also documents that have the same score.

The scrolling query tries to resolve this kind of problem by giving a special cursor that allows the user to uniquely iterate all of the documents.

Getting ready

You will need an up-and-running Elasticsearch installation, as described in the *Downloading and installing Elasticsearch* recipe of *Chapter 1, Getting Started*.

To execute these commands, any HTTP client can be used, such as Curl (https://curl.haxx.se/), Postman (https://www.getpostman.com/), or similar. I suggest using the Kibana console as it provides code completion and better character escaping for Elasticsearch.

To correctly execute the following commands, you will need an index populated with the ch04/populate_kibana.txt command, which is available in the online code.

How to do it...

In order to execute a scrolling query, we will perform the following steps:

1. From the command line, we can execute a search of the scan type, as follows:

    ```
    GET /mybooks/_search?scroll=10m&size=1
    { "query": { "match_all": {} } }
    ```

    If everything works as expected, the command will return a result that is similar to the following:

    ```
    { "_scroll_id" : "FGluY2x1ZGVfY29udGV4dF91dWlkDX-
    F1ZXJ5QW5kRmV0Y2gBFnhHblM3SnAxUWJxaFpKZ11hRYUtX-
    b0EAAAAAAAACBYzb3dHY1A2LVROaVduV0ZmYUJncF93",
        ...truncated...
      "hits" : {
        "total" : { "value" : 3, "relation" : "eq" },
        "max_score" : 1.0,
        "hits" : [
    ```

```
        {
            "_index" : "mybooks",  "_id" : "1",        "_score"
    : 1.0,
            ...truncated...
```

The preceding result is composed of the following:

- `scroll_id`: This refers to the value to be used for scrolling records.

- `took`: This is the time required to execute the query.

- `timed_out`: This tells us whether the query has been timed out.

- `_shards`: This query status contains information about the status of the shards during the query.

- `hits`: This refers to an object that contains the total count and the resulting hits.

2. With `scroll_id`, you can use `scroll` to get the results, as follows:

```
POST /_search/scroll
{ "scroll" : "10m",
     "scroll_id" : "FGluY2x1ZGVfY29udGV4dF91dWlk-
DXF1ZXJ5QW5kRmV0Y2gBFnhHblM3SnAxUWJxaFpKZ1hRYUtX-
b0EAAAAAAAACBYzb3dHY1A2LVROaVduV0ZmYUJncF93" }
```

The result should be something similar to the following:

```
{ "_scroll_id" : "FGluY2x1ZGVfY29udGV4dF91dWlk-
DXF1ZXJ5QW5kRmV0Y2gBFnhHblM3SnAxUWJxaFpKZ1hRYUtX-
b0EAAAAAAAACBYzb3dHY1A2LVROaVduV0ZmYUJncF93",
    ...truncated...
  "hits" : {
    "total" : { "value" : 3, "relation" : "eq" },
    "max_score" : 1.0,
    "hits" : [
      {
          "_index" : "mybooks",  "_id" : "2",
          "_score" : 1.0,
          ...truncated... } ] } }
```

For the most curious readers, `scroll_id` is a base64 format that contains information about the query type and an internal ID (you can check the decode value at `https://www.base64decode.org/`). In our case,

```
FGluY2x1ZGVfY29udGV4dF91dWlkDXF1ZXJ5QW5k  RmV0Y2gBFnhH-
blM3SnAxUWJxaFpKZlhRYUtXb0EAAA  AAAAAACBYzb3dHYlA2LVR0aVdu-
V0ZmYUJncF93corresponds to include_context_uuid queryAndFetch
xGnS7Jp1QbqhZJfXQaKWoA3owGbP6-TNiWnWFfaBgp_w.
```

How it works...

The scrolling query is interpreted as a standard search. This kind of search is designed to iterate on a large set of results, so the score and the order are not computed.

During the query phase, every shard stores the state of the IDs in memory until timeout. Processing a scrolling query is done in the following ways:

- The first part executes a query and returns `scroll_id`, which is used to fetch the results.

- The second part executes the document scrolling. You iterate the second step, getting the new `scroll_id` value, and fetch other documents.

If you need to iterate on a larger set of records, the scrolling query must be used; otherwise, you could have duplicated results.

The scrolling query is similar to every executed standard query, but there is a special parameter that must be passed in the query string.

The `scroll=(your timeout)` parameter allows the user to define how long the hits should live. The time can be expressed in seconds using the `s` postfix (for example, 5 seconds, 10 seconds, 15 seconds or more) or in minutes using the `m` postfix (for example, 5 minutes or 10 minutes). If you are using a long timeout, you must ensure that your nodes have a lot of RAM to keep the resulting ID live. This parameter is mandatory and must always be provided.

There's more...

Scrolling is very useful for executing re-indexing actions or iterating on very large result sets. The best approach for this kind of action is to use the sort query with the special _ doc field to obtain all of the matched documents and to be more efficient.

So, if you need to iterate on a large bucket of documents for re-indexing, you should execute a query that is similar to the following:

```
GET /mybooks/_search?scroll=10m&size=1
{ "query": { "match_all": {} },
  "sort": [ "_doc" ] }
```

The scroll result values are kept in memory until the scroll timeout. It's good practice to clean this memory if you no longer use the scroller; to delete a scroll from the Elasticsearch memory, the commands you need are as follows:

- If you know your scroll ID or IDs, you can provide them to the DELETE scroll API call, as follows:

```
DELETE /_search/scroll
  {"scroll_id": [ "DnF1ZXJ5VGhlbkZldGNoBQAA..." ] }
```

- If you want to clean all the scrolls, you can use the special _all keyword, as follows:

```
DELETE /_search/scroll/_all
```

See also

For further reference, take a look at the following points that are related to this recipe:

- The *Executing a search* recipe, in this chapter, explains how to structure a search.

- Refer to the official documentation about scrolling, which gives examples of how to use slices to map scrolling on multiple slices. This can be found at https://www.elastic.co/guide/en/elasticsearch/reference/master/search-request-scroll.html.

Using the search_after functionality

Elasticsearch's standard pagination, using from and size, performs very poorly on large datasets because, for every query, you need to compute and discard all of the results before the from value. Scrolling doesn't have this problem, but it consumes a lot due to memory search contexts; therefore, it cannot be used for frequent user queries.

To bypass these problems, Elasticsearch 5.x, and greater, provides the search_after functionality. This provides faster skipping for scrolling results.

Getting ready

You will need an up-and-running Elasticsearch installation, as described in the *Downloading and installing Elasticsearch* recipe of *Chapter 1, Getting Started*.

To execute these commands, any HTTP client can be used, such as Curl (https://curl.haxx.se/), Postman (https://www.getpostman.com/), or similar. I suggest using the Kibana console as it provides code completion and better character escaping for Elasticsearch.

To correctly execute the following commands, you will need an index populated with the ch04/populate_kibana.txt command, which is available in the online code.

How to do it...

In order to execute a scrolling query, we will perform the following steps:

1. From the command line, we can execute a search, which will provide a sort for your value. Then, we can use the _doc or _id fields of the document as the last sort parameter, as follows:

```
GET /mybooks/_search
{ "size": 1,
    "query": { "match_all": {} },
    "sort": [
        { "price": "asc" },
        { "_doc": "desc" } ] }
```

If everything works as expected, the command will return the following result:

```
{
    ...truncated...
    "hits" : {
        "total" : { "value" : 3, "relation" : "eq" },
        "max_score" : null,
        "hits" : [
            {
                "_index" : "mybooks", "_id" : "1",
                "_score" : null, "_source" : {
                    "uuid" : "11111",
                    "position" : 1,
```

```
...truncated...   },
        "sort" : [ 4.3, 0]
      } ] } }
```

2. To use the `search_after` functionality, you need to keep track of your last sort result. In this case, it is `[4.3, 0]`.

3. To fetch the next result, you must provide the `search_after` functionality with the last sort value of your last record, as follows:

```
GET /mybooks/_search
{ "size": 1,
    "query": { "match_all": {} },
    "search_after": [ 4.3, 0 ],
    "sort": [
        { "price": "asc" },
        { "_doc": "desc" } ] }
```

How it works...

Elasticsearch uses Lucene for indexing data. In Lucene indices, all the terms are sorted and stored in an ordered way, so it's natural for Lucene to be extremely fast at skipping to a term value. This operation is managed in the Lucene core with the `skipTo` method. This operation doesn't consume memory, and in the case of `search_after`, a query is built using the `search_after` values to quickly skip the Lucene search and speed up the result pagination.

The `search_after` functionality was introduced in Elasticsearch 5.x, but it must be kept as an important focal point to improve the user experience in search scrolling/pagination results.

See also

To learn how to structure a search for size pagination, refer to the *Executing a search* recipe in this chapter, and to learn how to use scrolling values in a query, please refer to the *Executing a scrolling query* recipe.

Returning inner hits in results

In Elasticsearch, when using nested and child documents, we can have complex data models. By default, Elasticsearch only returns documents that match the searched type and not the nested or children ones that match the query.

The `inner_hits` function was introduced in Elasticsearch 5.x to provide this functionality.

Getting ready

You will need an up-and-running Elasticsearch installation, as described in the *Downloading and installing Elasticsearch* recipe of *Chapter 1, Getting Started*.

To execute these commands, any HTTP client can be used, such as Curl (`https://curl.haxx.se/`), Postman (`https://www.getpostman.com/`), or similar. I suggest using the Kibana console as it provides code completion and better character escaping for Elasticsearch.

To correctly execute the following commands, you will need an index populated with the `ch04/populate_kibana.txt` command, which is available in the online code.

How to do it...

To return inner hits during a query, we will perform the following steps:

1. From the command line, we can execute a call by adding `inner_hits`, as follows:

    ```
    POST /mybooks-join/_search
    { "query": {
        "has_child": {
            "type": "author",
            "query": { "term": { "name": "peter" } },
            "inner_hits": {}, } } }
    ```

    If everything works as expected, the result returned by Elasticsearch should be as follows:

    ```
    { ...truncated...
      "hits" : {
          "total" : { "value" : 1, "relation" : "eq" },
          "max_score" : 1.0,
          "hits" : [
    ```

```
{ "_index" : "mybooks-join",
  "_id" : "1", "_score" : 1.0,
  "_source" : {
… truncated …
    "inner_hits" : {
      "author" : {
        "hits" : {
          "total" : 1, "max_score" : 1.2039728,
          "hits" : [
            {
              "_index" : "mybooks-join",
              "_id" : "a1", "_score" : 1.2039728,
              "_routing" : "1",
              "_source" : {
                "name" : "Peter",
                "surname" : "Doyle",
                "join" : {
                  "name" : "author",
                  "parent" : "1"
} } } ] } } } } ] } }
```

How it works...

When executing nested or children queries, Elasticsearch executes a two-step query, as follows:

1. It executes the nested query or the children query and returns the IDs of the referred values.

2. It executes the other part of the query filtering using the IDs returned from *step 1*.

Generally, the results of the nested query or the children query are not taken because they require memory. Using the inner_hits function, the nested query or the children query intermediate hits are kept and returned to the user.

To control the returned documents of the inner_hits function, standard parameters for the search are available, such as from, size, sort, highlight, _source, explain, scripted_fields, docvalues_fields, and version.

There is also a special query property name that is used to name `inner_hits`. This allows the user to easily determine it in the case of multiple `inner_hits` functions returning sections.

See also

For further reference, take a look at the following points that are related to this recipe:

- Please refer to the *Executing a search* recipe, in this chapter, for information about all the standard parameters in searches that are used for controlling returned hits.

- The *Using a has_child query, Using a top_children query, Using a has_parent query*, and *Using a nested query* recipes of *Chapter 6, Relationships and Geo Queries*, are useful when looking for queries that can be used for inner hits.

Suggesting a correct query

It's very common for users to commit typing errors or to require suggestions for the words that they are writing. These issues are resolved in Elasticsearch using the suggested functionality.

Getting ready

You will need an up-and-running Elasticsearch installation, as described in the *Downloading and installing Elasticsearch* recipe of *Chapter 1, Getting Started*.

To execute these commands, any HTTP client can be used, such as Curl (`https://curl.haxx.se/`), Postman (`https://www.getpostman.com/`), or similar. I suggest using the Kibana console as it provides code completion and better character escaping for Elasticsearch.

To correctly execute the following commands, you will need an index populated with the `ch04/populate_kibana.txt` command, which is available in the online code.

How to do it...

To suggest relevant terms by querying, we will perform the following steps:

1. From the command line, we can execute a `suggest` call, as follows:

```
GET /mybooks/_search
{ "suggest": {
     "suggest1": {
        "text": "we find tester",
        "term": { "field": "description" } } } }
```

2. If everything works as expected, the result returned by Elasticsearch should be as follows:

```
{ ...truncated...
    "suggest" : {
       "suggest1" : [
          { "text" : "we", "offset" : 0, "length" : 2,
             "options" : [ ]   },
   ... truncated ...
             {
                "text" : "testere",
                "score" : 0.8333333,
                "freq" : 2
             } ] } ] } }
```

The result is composed of the following:

- The status of the shards at the time of the query

- The list of tokens with their available candidates

How it works...

The `suggest` section works by collecting term stats on all the index shards. Using Lucene field statistics, it is possible to detect the correct term or the complete term. It's a statistical approach!

There are two types of suggesters, called *term* and *phrase*. They work as follows:

- The simpler suggester to use is the term suggester. It only requires the text and the field to work. Additionally, it also allows the user to set a lot of parameters, such as the minimum size for a word, learn how to sort results, and the suggester strategy. A complete reference is available on the Elasticsearch website at `https://www.elastic.co/guide/en/elasticsearch/reference/master/search-suggesters-term.html`.

- The phrase suggester is able to keep relations between the terms that it needs to suggest. The phrase suggester is less efficient than the term suggester, but it provides better results.

The `suggest` API features, parameters, and options often change between releases.

New suggesters can be added using plugins.

See also

For further reference, take a look at the following resources that are related to this recipe:

- Refer to the *Executing a search* recipe, in this chapter, for information about how to structure a search.

- The online documentation for phrase suggesters is available at `https://www.elastic.co/guide/en/elasticsearch/reference/current/search-suggesters-phrase.html`.

- The online documentation for the completion suggester is available at `https://www.elastic.co/guide/en/elasticsearch/reference/current/search-suggesters-completion.html`.

- The online documentation for the context suggester is available at `https://www.elastic.co/guide/en/elasticsearch/reference/current/suggester-context.html`.

Counting matched results

Often, it is necessary to only return the count of the matched results and not the results themselves.

There are a lot of scenarios involving counting, such as the following:

- To return the number of something (for example, the number of posts on a blog and the number of comments on a post).

- If you need to validate whether some items are available, For example, are there posts? Are there comments?

Getting ready

You will need an up-and-running Elasticsearch installation, as described in the *Downloading and installing Elasticsearch* recipe of *Chapter 1*, *Getting Started*.

To execute these commands, any HTTP client can be used, such as Curl (https://curl.haxx.se/), Postman (https://www.getpostman.com/), or similar. I suggest using the Kibana console as it provides code completion and better character escaping for Elasticsearch.

To correctly execute the following commands, you will need an index populated with the ch04/populate_kibana.txt command, which is available in the online code.

How to do it...

In order to execute a counting query, we will perform the following steps:

1. From the command line, we will execute a count query, as follows:

```
GET /mybooks/_count
{ "query": { "match_all": {} } }
```

2. If everything works as expected, the result returned by ElasticSearch should be as follows:

```
{ "count" : 3,
  "_shards" : {
    "total" : 1, "successful" : 1,
    "skipped" : 0, "failed" : 0 } }
```

The result is composed of the count result (a long type) and the shard status at the time of the query.

How it works...

The query is interpreted in the same way as it is interpreted for searching. The count action is processed and distributed in all the shards, which is executed as a low-level Lucene count call. Every shard hit returns a count that is aggregated and returned to the user.

In Elasticsearch, counting is faster than searching. If the resulting source hits are not required, it's good practice to use the count API because it's faster and requires fewer resources.

The HTTP method to execute a count is GET (but POST also works), and the REST endpoints are as follows:

```
http://<server>/_count
http://<server>/<index_name(s)>/_count
```

Multi-indices and types are comma-separated. If an index or a type is defined, the search is limited to only them. An alias can be used as an index name.

Typically, a body is used to express a query; however, for simple queries, q (the query argument) can be used. For example, look at the following code:

```
GET /mybooks/_count?q=uuid:11111
```

There's more...

In a previous version of Elasticsearch, the count API call (the _count REST entry point) was implemented as a custom action. However, since Elasticsearch version 5.x and greater, this has been removed. Internally, the actual count API is implemented as a standard search with the size set to 0.

This trick not only speeds up the searching process but reduces networking. You can use this approach to execute aggregations (we will explore this in *Chapter 7, Aggregations*) without returning hits.

The previous query can also be executed as follows:

```
GET /mybooks/_count?q=uuid:11111
```

If everything works as expected, the result returned by Elasticsearch should be as follows:

```
{ "count" : 1,
  "_shards" : {
    "total" : 1, "successful" : 1,
    "skipped" : 0,    "failed" : 0 } }
```

The count result (a long type) is also available in the standard _search result in `hits.total`.

See also

For further reference, take a look at the following resources that are related to this recipe:

- To learn how to use size to paginate, please refer to the *Executing a search* recipe in this chapter.

- To learn how to use aggregations, please refer to *Chapter 7, Aggregations*.

Explaining a query

When executing searches, it's very common to have documents that do match or don't match the query as expected. To easily debug these scenarios, Elasticsearch provides the explain query call. This allows you to check how the scores are computed against a document.

Getting ready

You will need an up-and-running Elasticsearch installation, as described in the *Downloading and installing Elasticsearch* recipe of *Chapter 1, Getting Started*.

To execute these commands, any HTTP client can be used, such as Curl (`https://curl.haxx.se/`), Postman (`https://www.getpostman.com/`), or similar. I suggest using the Kibana console as it provides code completion and better character escaping for Elasticsearch.

To correctly execute the following commands, you will need an index populated with the `ch04/populate_kibana.txt` command, which is available in the online code.

How to do it...

The steps that are required to execute the explain query call are as follows:

1. From the command line, we will execute an explain query against a document, as follows:

```
GET /mybooks/_explain/1?pretty
{ "query": { "term": { "uuid": "11111" } } }
```

2. If everything works as expected, the result returned by Elasticsearch should be as follows:

```
{   "_index" : "mybooks",   "_id" : "1",
   "matched" : true,
   "explanation" : {
     "value" : 0. 9808291,
     "description" : "weight(uuid:11111 in 0)
[PerFieldSimilarity], result of:",
     "details" : [
          {
               "value" : 3,
               "description" : "N, total number of
documents with field",
               "details" : [ ]
          } ]
... truncated ...
```

The important parts of the preceding result are listed as follows:

- matched: This refers to whether the documents match in the query.

- explanation: This section is composed of objects that are made up of the following:

- value: A double score of that query section.

- description: A string representation of the matching token (in the case of wildcards or multi-terms, it can provide information about the matched token).

- details: An optional list of explanation objects.

How it works...

The explain call is a view of how Lucene computes the results. In the description section of the explain object, there are Lucene representations of that part of the query.

A user doesn't need to be a Lucene expert to understand the explain descriptions, but they provide an insight into how the query is executed and how the terms are matched.

More complex queries with many subqueries are very hard to debug, particularly if you need to boost some special fields to obtain the desiderata sequence of documents. In these cases, using the explain API helps you to manage field boosting because it allows you to easily debug how they interact in your query or document.

There's more...

The explain query is one of the most commonly used APIs to understand why some results are not shown or why they do not have a wanted score. If a user only needs to check whether the query is valid, there is a special endpoint for this purpose:

```
GET http://<server>/_validate/query
```

This returns, as its result, a Boolean indicating whether the query is valid.

This API call is very useful for checking application generated code before executing the query on Elasticsearch. This is because, for simple checking, it does not have as high an overhead as the explain endpoint.

See also

For further reference, take a look at the following resources that are related to this recipe:

- The official documentation about the explain API can be found at `https://www.elastic.co/guide/en/elasticsearch/reference/current/search-explain.html`.

- The official documentation about the validate API can be found at `https://www.elastic.co/guide/en/elasticsearch/reference/current/search-validate.html`.

Query profiling

When you are executing queries, sometimes, they are not as fast as you think. The reasons why some queries need a lot of time to be executed vary, but using tools to profile them can solve your issues more quickly.

This feature is available from Elasticsearch 5.x, or greater, via the profile API. It allows the user to track the time spent by Elasticsearch in executing a search or an aggregation.

> **Note**
> This is only a debug tool because of the significant overhead involved in its execution: for each Lucene step, it computes the part of scores, which takes a significant amount of time during query execution.

Getting ready

You will need an up-and-running Elasticsearch installation, as described in the *Downloading and installing Elasticsearch* recipe of *Chapter 1, Getting Started*.

To execute these commands, any HTTP client can be used, such as Curl (https://curl.haxx.se/), Postman (https://www.getpostman.com/), or similar. I suggest using the Kibana console as it provides code completion and better character escaping for Elasticsearch.

To correctly execute the following commands, you will need an index populated with the ch04/populate_kibana.txt command, which is available in the online code.

How to do it...

The steps to profile a query are as follows:

1. From the command line, we will execute a search with the true profile set as follows:

    ```
    GET /mybooks/_search
    { "profile": true,
      "query": { "term": { "uuid": "11111" } } }
    ```

2. If everything works as expected, the result returned by Elasticsearch should be as follows:

    ```
    { "profile" : {
        "shards" : [
          {
            "id" : "[4pptZx4jSM-xcpiVT_d7Rw][mybooks][0]",
            "searches" : [ … truncated…
                "rewrite_time" : 5954,
                "collector" : [
                  {
                    "name" : "CancellableCollector",
                    "reason" : "search_cancelled",
                    "time_in_nanos" : 204857,
                    "children" : [
                        { "name" : "SimpleTopScoreDocCollector",
                          "reason" : "search_top_hits",
    ```

```
                "time_in_nanos" : 12288
                } ]    } ] ] } ],
        "aggregations" : [ ] } ] } }
```

The preceding output is very verbose. It's divided by shards and single hits.

The result exposes the type of query (for example, `TermQuery`) with details about the internal Lucene parameters. For every step, the time is tracked in a way that a user can easily detect the bottleneck in their query time.

How it works...

The profile APIs were introduced in Elasticsearch 5.x to help keep a track of timings when executing queries and aggregations. When a query is executed, if profiling is activated, all of the internal calls are tracked using the internal instrumental API. For this reason, the profile API adds an overhead to the computation.

The output is also very verbose and depends on the internal components of both Elasticsearch and Lucene, so the format of the result can change in the future. The typical usage for this feature is to reduce the level of time tracking execution, which refers to the slowest steps in the query, and try to optimize them.

Deleting by query

We saw how to delete a document in the *Deleting a document* recipe of *Chapter 3, Basic Operations*. Deleting a document is very quick, but it requires knowing the document ID for direct access and, in some cases, the routing value, too.

Elasticsearch provides a call to delete all of the documents that match a query using an additional module called `re-index`, which is installed by default.

Getting ready

You will need an up-and-running Elasticsearch installation, as described in the *Downloading and installing Elasticsearch* recipe of *Chapter 1, Getting Started*.

To execute these commands, any HTTP client can be used, such as Curl (https://curl.haxx.se/), Postman (https://www.getpostman.com/), or similar. I suggest using the Kibana console as it provides code completion and better character escaping for Elasticsearch.

To correctly execute the following commands, you will need an index populated with the ch04/populate_kibana.txt command, which is available in the online code.

How to do it...

In order to delete by query, we will perform the following steps:

1. From the command line, we will execute a query, as follows:

```
POST /mybooks/_delete_by_query
{ "query": { "match_all": {} } }
```

2. If everything works as expected, the result returned by Elasticsearch should be as follows:

```
{ "took" : 10, "timed_out" : false,
  "total" : 3, "deleted" : 3, "batches" : 1,
  "version_conflicts" : 0, "noops" : 0,
  "retries" : { "bulk" : 0, "search" : 0  },
  "throttled_millis" : 0,
  "requests_per_second" : -1.0,
  "throttled_until_millis" : 0,
  "failures" : [ ] }
```

The main components of the preceding result are listed as follows:

- Total: This refers to the number of documents that match the query.

- deleted: This refers to the number of documents that have been deleted.

- batches: This refers to the number of bulk actions executed to delete the documents.

- version_conflicts: This refers to the number of documents not deleted due to a version conflict during the bulk action.

- noops: This refers to the number of documents not executed to a noop event.

- retries.bulk: This refers to the number of bulk actions that have been retried.

- retries.search: This refers to the number of searches that have been retried.

- requests_per_second: This refers to the number of requests executed per second (which is -1.0 if this value is not set).

- throttled_millis: This refers to the length of sleep to conform to the request_per_second value.

- throttled_until_millis: This is generally 0, and it indicates the time for the next request if the request_per_second value is set.

- failures: This refers to an array of failures.

How it works...

The `delete_by_query` function is executed automatically using the following steps:

1. In a master node, the query is executed and the results are scrolled.
2. For every bulk size element (the default is 1,000), a bulk is executed.
3. The bulk results are checked for conflicts. If no conflicts exist, a new bulk is executed until all of the matched documents have been deleted.

The `delete_by_query` call automatically manages the back pressure (that is, it reduces the delete command rate if the server has a high load).

When you want to remove all the documents without re-indexing a new index, a `delete_by_query` command with a `match_all` query allows you to clean your mapping of all the documents. This call is analogous to the `truncate_table` command of the SQL language.

The `http` method to execute a `delete_by_query` command is `POST`. The REST endpoints are as follows:

```
http://<server>/_delete_by_query
http://<server>/<index_name(s)>/_delete_by_query
```

Multi-indices are defined as a unique comma-separated string. If an index or a type has been defined, the search is limited to only them. An alias can be used as the index name.

Typically, a body is used to express a query. However, for simple queries, q (the `query` argument) can be used. For example, look at the following code:

```
DELETE /mybooks/_delete_by_query?q=uuid:11111
```

There's more...

Further query arguments are listed as follows:

*   `conflicts`: If this is set to `proceed`, then, when there is a version conflict, the call doesn't exit; it skips the error and finishes the execution.
*   `routing`: This is only used to target some shards.
*   `scroll_size`: This controls the size of the scrolling and the bulk (the default is `1000`).
*   `request_per_seconds` (the default is `-1.0`): This controls how many requests can be executed in a second. The default value is unlimited.

See also

For further reference, take a look at the following resources that are related to this recipe:

- The *Deleting a document* recipe in *Chapter 3, Basic Operations*, is useful for learning how to execute a delete for a single document.

- The *Delete by query task* recipe in *Chapter 9, Managing Clusters*, is useful for learning how to monitor asynchronous deletion by using query actions.

Updating by query

In *Chapter 3, Basic Operations* we saw how to update a document in the *Update a document* recipe.

The `update_by_query` API call allows the user to execute an update on all the documents that match a query. It is very useful if you need to do the following:

- Reindex a subset of your records that match a query. This is very common if you change your document mapping and need the documents to be reprocessed.

- Update the values of your records that match a query.

This is the Elasticsearch version of the SQL update command.

This functionality is provided by an additional module, called reindex, which is installed by default.

Getting ready

You will need an up-and-running Elasticsearch installation, as described in the *Downloading and installing Elasticsearch* recipe of *Chapter 1, Getting Started*.

To execute these commands, any HTTP client can be used, such as Curl (`https://curl.haxx.se/`), Postman (`https://www.getpostman.com/`), or similar. I suggest using the Kibana console as it provides code completion and better character escaping for Elasticsearch.

To correctly execute the following commands, you will need an index populated with the `ch04/populate_kibana.txt` command, which is available in the online code.

How to do it...

In order to execute an update from a query that simply reindexes your documents, we will perform the following steps:

1. From the command line, we will execute a query, as follows:

```
POST /mybooks/_update_by_query
{ "query": { "match_all": {} },
  "script": { "source": "ctx._source.quantity=50" } }
```

2. If everything works as expected, the result returned by Elasticsearch should be as follows:

```
{ "took" : 7, "timed_out" : false,  "total" : 3,
  "updated" : 3, "deleted" : 0, "batches" : 1,
  "version_conflicts" : 0,  "noops" : 0,
  "retries" : { "bulk" : 0, "search" : 0 },
  "throttled_millis" : 0,
  "requests_per_second" : -1.0,
  "throttled_until_millis" : 0,  "failures" : [ ] }
```

The most important components of the preceding result are listed as follows:

- total: This refers to the number of documents that match the query.

- updated: This refers to the number of documents that have been updated.

- batches: This refers to the number of bulk actions that have been executed to update the documents.

- version_conflicts: This refers to the number of documents that have not been deleted due to a version conflict during the bulk actions.

- noops: This refers to the number of documents not changed due to a noop event.

- retries.bulk: This refers to the number of bulk actions that have been retried.

- retries.search: This refers to the number of searches that have been retried.

- requests_per_second: This refers to the number of requests executed per second (which is -1.0 if this value is not set).

- throttled_millis: This refers to the length of sleep to conform to the request_per_second value.

- throttled_until_millis: This is generally 0, and it indicates the time for the next request if the request_per_second value is set.

- failures: This refers to an array of failures.

How it works...

The update_by_query function works in a very similar way to the delete_by_query API. It is executed automatically using the following steps:

1. In a master node, the query is executed, and the results are scrolled.

2. For every bulk size element (the default is 1,000), a bulk with the update commands is executed.

3. The bulk results are checked for conflicts. If there are no conflicts, a new bulk is executed, and the search or bulk actions are executed until all of the matched documents have been deleted.

The HTTP method to execute an update_by_query call is POST, and the REST endpoints are as follows:

```
http://<server>/_update_by_query
http://<server>/<index_name(s)>/<type_name(s)>/_update_by_query
```

Multi-indices are defined via a comma-separated string. If an index or a type has been defined, the search is limited to only them. An alias can be used as an index name.

The additional query arguments are as follows:

- conflicts: If this is set to proceed, then, when there is a version conflict, the call doesn't exit; it skips the error and finishes the execution.

- routing: This is only used to target some shards.

- scroll_size: This controls the size of the scrolling and the bulk actions (the default size is 1000).

- request_per_seconds (the default is -1.0): This controls how many requests can be executed in a second. The default value is unlimited.

There's more...

The update_by_query API can accept a script section in its body. In this way, it can become a powerful tool for executing custom updates on a subset of documents. (We will examine scripting in more detail in *Chapter 8, Scripting in Elasticsearch*). It can be considered similar to the SQL update command.

With this facility, we can add a new field and initialize its value with a `script` section, as follows:

```
POST /mybooks/_update_by_query
{ "script": { "source": "ctx._source.hit=4" },
  "query": { "match_all": {} } }
```

In the preceding example, we add a `hit` field that is set to 4 for every document that matches the query. This is similar to the SQL command, which is as follows:

```
update mybooks set hit=4
```

The `update_by_query` API is one of the more powerful tools that Elasticsearch provides.

See also

The *Update a document* recipe in *Chapter 3, Basic Operations*, is useful for learning how to execute an update for a single document.

Matching all of the documents

One of the most common queries is the `match_all` query. This kind of query allows the user to return all of the documents that are available in an index. The `match_all` query and other query operators are part of the Elasticsearch query DSL.

Getting ready

You will need an up-and-running Elasticsearch installation, as described in the *Downloading and installing Elasticsearch* recipe in *Chapter 1, Getting Started*.

To execute these commands, any HTTP client can be used, such as Curl (https://curl.haxx.se/), Postman (https://www.getpostman.com/), or similar. I suggest using the Kibana console as it provides code completion and better character escaping for Elasticsearch.

To correctly execute the following commands, you will need an index populated with the `ch04/populate_kibana.txt` command, which is available in the online code.

How to do it...

In order to execute a `match_all` query, we will perform the following steps:

1. From the command line, we execute the query as follows:

```
POST /mybooks/_search
{ "query": { "match_all": {} } }
```

2. If everything works as expected, the result returned by Elasticsearch should be as follows:

```
{
    "took" : 0,   "timed_out" : false,
    "_shards" : { "total" : 1, "successful" : 1,
        "skipped" : 0, "failed" : 0 },
    "hits" : {
        "total" : { "value" : 3, "relation" : "eq" },
        "max_score" : 1.0,
        "hits" : [
            { "_index" : "mybooks", "_id" : "1",
                "_score" : 1.0,
                "_source" : {
                    "date" : "2021-10-22", "hit" : 4,
                    "quantity" : 50, "price" : 4.3,
                    "description" : "Joe Testere nice guy",
                    "position" : 1, "title" : "Joe Tester",
                    "uuid" : "11111" }
    ... truncated ...
```

The result is a standard query result, as we saw in the *Executing a search* recipe.

How it works...

The `match_all` query is one of the most common ones. It's faster because it doesn't require the score calculus (that is, it's wrapped in a Lucene, `ConstantScoreQuery`).

If no query is defined in the search object, `match_all` will be the default query.

See also

For further reference, refer to the *Executing a search* recipe.

Using a Boolean query

Most people who use a search engine have, at some point or another, used the minus (-) and plus (+) syntax to include or exclude query terms. A Boolean query allows the user to programmatically define queries to include, exclude, optionally include (`should`), or filter in the query.

This kind of query is one of the most important ones because it allows the user to aggregate a lot of simple queries or filters, as we will see in this chapter, to build a larger, more complex one.

There are two main concepts that are important in searches: `query` and `filter`. The query concept means that the matched results are scored using an internal Lucene scoring algorithm; in the filter concept, the results are matched without scoring. Because the filter doesn't need to compute the score, it is generally faster and can be cached.

Getting ready

You will need an up-and-running Elasticsearch installation, as described in the *Downloading and installing Elasticsearch* recipe of *Chapter 1, Getting Started*.

To execute these commands, any HTTP client can be used, such as Curl (`https://curl.haxx.se/`), Postman (`https://www.getpostman.com/`), or similar. I suggest using the Kibana console as it provides code completion and better character escaping for Elasticsearch.

To correctly execute the following commands, you will need an index populated with the `ch04/populate_kibana.txt` command, which is available in the online code.

How to do it...

To execute a Boolean query, we will perform the following steps:

1. We can execute a Boolean query from the command line as follows:

```
POST /mybooks/_search
{   "query": {
      "bool": {
        "must": [
            { "term": { "description": "joe" } }
        ],
… truncated …
        "filter": [
            { "term": { "description": "joe" } }
        ],
        "minimum_should_match": 1, "boost": 1 } } }
```

2. The result returned by Elasticsearch is similar to the previous recipes, but in this case, it should return one record (id:1).

How it works...

The bool query is often one of the most used queries because it allows the user to compose a large query using a lot of simpler ones. The inclusion of one of the following four parts is mandatory:

- must: This refers to a list of queries that must be satisfied. All the must queries must be verified to return the hits. It can be seen as an AND filter with all of its subqueries.

- must_not: This refers to a list of queries that must not be matched. It can be seen as not a filter of an AND query.

- should: This refers to a list of queries that can be verified. The minimum number of these queries must be verified, and this value is controlled by minimum_ should_match (the default is 1).

- filter: This refers to a list of queries that can be used as the filter. They allow the user to filter out results without changing the score and relevance. The filter queries are faster than the standard ones because they don't need to compute the score.

There's more...

If you define multiple subqueries in a Boolean query, understand that whatever query is hit by your result could be very important at the application level; generally, it is better to narrow your results. To obtain this result, you could use the special _name attribute, which can be defined in the query components:

- The previous query can be changed in the following way:

```
POST /mybooks/_search
{ "query": {
    "bool": {
        "should": [
            { "term": {
                "uuid": { "value": "11111", "_name":
"uuid:11111:matched" } } },
            {"term": {
                "uuid": { "value": "22222", "_name":
"uuid:22222:matched" } } }
        ],
        "filter": [
            { "term": {
                "description": {
                    "value": "joe", "_name": "fiter:term:joe" }
} } ],
        "minimum_should_match": 1,
        "boost": 1 } } }
```

- For every matched document, the result will contain the matched queries:

```
{
    ...truncated...   "hits" : {
    "total" : { "value" : 1, "relation" : "eq" },
    "max_score" : 0.9808292,
    "hits" : [
        {"_index" : "mybooks", "_id" : "1",
            "_score" : 0.9808292,
            "matched_queries" : [
                "uuid:11111:matched", "fiter:term:joe"
            ] } ] } }
```

Using the search template

Elasticsearch provides the capability of providing a template and some parameters to fill it. This functionality is very useful because it allows you to manage the query templates stored in the `.scripts` index and also allows you to change them without changing the application code.

Getting ready

You will need an up-and-running Elasticsearch installation, as described in the *Downloading and installing Elasticsearch* recipe of *Chapter 1, Getting Started*.

To execute these commands, any HTTP client can be used, such as Curl (`https://curl.haxx.se/`), Postman (`https://www.getpostman.com/`), or similar. I suggest using the Kibana console as it provides code completion and better character escaping for Elasticsearch.

To correctly execute the following commands, you will need an index populated with the `ch04/populate_kibana.txt` command, which is available in the online code.

How to do it...

The template query is composed of two components: the query and the parameters that must be filled in. We can execute a template query in several ways. In this recipe, we will look at some query types that we will explore in the upcoming chapters.

Using the new `_search/template` REST entrypoint is the best way to use the templates. To use it, perform the following steps:

1. We execute the query as follows:

```
POST /_search/template
{ "source": {
    "query": { "term": { "uuid": "{{value}}" } }
  },
  "params": { "value": "22222" } }
```

2. If everything works as expected, the result returned by Elasticsearch should be as follows:

```
{
    "took" : 3, "timed_out" : false,
    "_shards" : {
```

```
      "total" : 3, "successful" : 3,
      "skipped" : 0, "failed" : 0 },
   "hits" : {
      "total" : { "value" : 1, "relation" : "eq" },
      "max_score" : 0.9808292,
      "hits" : [
         { "_index" : "mybooks", "_id" : "2",
           "_score" : 0.9808292,
           "_source" : {
              "uuid" : "22222", "position" : 2,
              "title" : "Bill Baloney",
              "description" : "Bill Testere nice guy",
              "date" : "2016-06-12", "price" : 5,
              "quantity" : 34 } } ] } }
```

If we want to use an indexed stored template, we can perform the following steps:

1. We store the template in the .scripts index:

```
POST _scripts/myTemplate
{"script": {
    "lang": "mustache",
    "source": {
       "query": { "term": { "uuid": "{{value}}" } }
} } }
```

2. Now we can call the template with the following code:

```
POST /mybooks/_search/template
{ "id": "myTemplate", "params": { "value": "22222" } }
```

If you have a stored template and you want to validate it, you can use the REST render entrypoint.

The indexed templates and scripts are stored in the .script index. This is a normal index, and it can be managed as a standard data index.

If you want to render the query template, mainly for debugging purposes, follow these steps:

1. We render the template using the `_render/template` REST:

    ```
    POST /_render/template
    {"id": "myTemplate", "params": { "value": "22222" } }
    ```

2. The result will be shown as follows:

    ```
    { "template_output" : {
        "query" : { "term" : { "uuid" : "22222" } } } }
    ```

How it works...

A template query is composed of the following two components:

*   A template is a query object that is supported by Elasticsearch. The template uses the mustache (http://mustache.github.io/) syntax, which is a very common syntax to express templates.

*   An optional dictionary of parameters is used to fill the template.

When the search query is called, the template is loaded, populated with the parameter's data, and executed as a normal query. The template query is a shortcut so that you can use the same query with different values.

Typically, the template is generated by executing the query in the standard way and then by adding parameters, if required, during the process of creating the template. The mustache syntax is very rich and provides default values, JSON escaping, conditional parts, and more (the official documentation, which can be found at https://www. elastic.co/guide/en/elasticsearch/reference/master/search-template.html, covers all of these aspects).

It allows you to remove the query execution from the application code and put it on the filesystem or indices.

See also

For further reference, take a look at the following resources that are related to this recipe:

- Check the official mustache documentation at `http://mustache.github.io/` to learn about the template syntax.

- Check the official Elasticsearch documentation about search templates at `https://www.elastic.co/guide/en/elasticsearch/reference/master/search-template.html` for more examples of how to use the template syntax.

- Check the official Elasticsearch documentation about query templates at `https://www.elastic.co/guide/en/elasticsearch/reference/master/query-dsl-template-query.html`, which has some examples of query usage.

5
Text and Numeric Queries

In this chapter, we will examine the queries that are used for searching text and numeric values. They are simple and the most common ones used in Elasticsearch. The first part of this chapter covers the text queries from the simple term and terms query to the complex query string query. We'll gain an understanding of how the queries are strongly related to mapping when it comes to choosing the correct query based on mapping.

In the last part of this chapter, we will look at many special queries that cover fields, helpers for building complex queries from strings, and query templates.

In this chapter, we will cover the following recipes:

- Using a term query
- Using a terms query
- Using a terms set query
- Using a prefix query
- Using a wildcard query
- Using a regexp query
- Using span queries

- Using a match query

- Using a query string query

- Using a simple query string query

- Using the range query

- Using an IDs query

- Using the function score query

- Using the exists query

- Using a pinned query (XPACK)

Technical requirements

All the recipes in this chapter require us to prepare and populate the required indices—the online code is available from the PacktPub website or via GitHub (`https://github.com/PacktPublishing/Elasticsearch-8.x-Cookbook`). Here, you can find scripts to initialize all the required data.

Using a term query

Searching or filtering for a particular term is done frequently. Term queries work with exact value matches and, generally, are very fast.

Note that in the SQL world, term queries can be compared to the `equal` (`=`) query.

Getting ready

You will need an up-and-running Elasticsearch installation, as described in the *Downloading and installing Elasticsearch* recipe of *Chapter 1*, *Getting Started*.

To execute these commands, any HTTP client can be used, such as `curl` (`https://curl.haxx.se/`), Postman (`https://www.getpostman.com/`), or similar. I suggest you use the Kibana console, as it provides code completion and better character escaping for Elasticsearch.

To correctly execute the following commands, you will need an index populated with the `ch04/kibana_populate.txt` command, which is available in the online code.

How to do it...

To execute a term query, we will perform the following steps:

1. We will execute a `term` query from the command line, as follows:

```
POST /mybooks/_search
{ "query": { "term": { "uuid": "33333" } } }
```

If everything works as expected, the result returned by Elasticsearch should be as follows:

```
{ ... truncated ...
   "hits" : {
     "total" : { "value" : 1, "relation" : "eq" },
     "max_score" : 1.0296195,
     "hits" : [
       {
         "_index" : "mybooks",   "_id" : "3",
         "_score" : 1.0296195,
         "_source" : {
           "uuid" : "33333", "position" : 3,
   ... truncated ...
```

2. To execute a `term` query as a filter (so that the score is skipped to speed up the process without impacting the scoring for simple filtering), we need to use it wrapped in a Boolean query. The preceding `term` query will be executed in the following way:

```
POST /mybooks/_search
{ "query": { "bool": {
     "filter": { "term": { "uuid": "33333" } } } } }
```

If everything works as expected, the result returned by Elasticsearch should be as follows:

```
{... truncated ...
   "hits" : {
     "total" : { "value" : 1, "relation" : "eq" },
     "max_score" : 1.0296195,
     "hits" : [
       { "_index" : "mybooks",   "_id" : "3",
```

```
    "_score" : 1.0296195,
  "_source" : {
      "uuid" : "33333",  "position" : 3,
    ... truncated ...
```

The result is a standard query result, as we saw in the *Executing a search* recipe of *Chapter 4, Exploring Search Capabilities*.

How it works...

Due to its inverted index, Lucene is one of the fastest engines that you can use to search for a term or a value in a field. Every field that is indexed in Lucene is converted into a fast search structure for its particular type:

- The text is split into tokens if analyzed or saved as a single token.
- The numeric fields are converted into their fastest binary representation.
- The date and datetime fields are converted into binary forms.

In Elasticsearch, all of these conversion steps are automatically managed. Search for a term, independent from the value, and you will find it is archived by Elasticsearch using the correct format for the field.

Internally, during a `term` query execution, all of the documents that match the term are collected. Then, they are sorted by score (the scoring depends on the Lucene similarity algorithm; by default, BM25 is chosen).

For more details regarding Elasticsearch similarity algorithms, see `https://www.elastic.co/guide/en/elasticsearch/reference/master/index-modules-similarity.html`.

If we look for the results of the previous searches, the term query of the hit has `0.30685282` as its score, while the filter has `1.0`. The time required for scoring if the sample is very small is not relevant, but if you have thousands or millions of documents, it takes much longer.

If the score is not important, opt to use the term filter.

This filter is preferred to the query when the score is not important. Some typical scenarios are listed as follows:

- Filtering permissions
- Filtering numerical values
- Filtering ranges

In a filtered query, first, the filter is applied, narrowing down the number of documents to be matched against the query. Then, the query is applied.

There's more...

Matching a term is the basis of Lucene and Elasticsearch. To correctly use these queries, you need to pay attention to how the field is indexed.

As we saw in *Chapter 2*, *Managing Mapping*, the terms of an indexed field depend on the analyzer used to index it. To better understand this concept, a representation of a phrase depends on several analyzers, as shown in the following table. For standard string analyzers, if we have a similar phrase, for example, `Phrase: Peter's house is big`, the results will be similar to the following table:

Mapping index	Analyzer	Tokens
"index": false	(No index)	(No tokens)
"type": "keyword"	KeywordAnalyzer	["Peter's house is big"]
"type": "text"	StandardAnalyzer	["peter", "s", "house", "is", "big"]

Figure 5.1 – The results of Mapping index, Analyzer, and Tokens

The common pitfalls when it comes to searches include misunderstanding the analyzer and mapping configurations. `KeywordAnalyzer`, which is used as the default for the `not tokenized` field, saves the unchanged string as a single token.

`StandardAnalyzer`, which is used as the default for the `type="text"` field, tokenizes on whitespaces and punctuation; here, every token is converted into lowercase. You should use the same analyzer for indexing to analyze the query (at the default settings).

In the preceding example, if the phrase is analyzed with `StandardAnalyzer`, you cannot search for the term "Peter," but you can search for "peter" because the `StandardAnalyzer` analyzer executes terms in lowercase.

When the same field requires one or more search strategies, you need to use the `fields` property using the different analyzers that you need.

Using a terms query

The previous type of search works very well for searching a single term. If you want to search for multiple terms, you can process it in two ways: either using a Boolean query or using a multi-term query.

Getting ready

You will need an up-and-running Elasticsearch installation, as described in the *Downloading and installing Elasticsearch* recipe of *Chapter 1, Getting Started*.

To execute these commands, any HTTP client can be used, such as curl (https://curl.haxx.se/), Postman (https://www.getpostman.com/), or similar. I suggest you use the Kibana console, as it provides code completion and better character escaping for Elasticsearch.

To correctly execute the following commands, you will need an index populated with the ch04/populate_kibana.txt command, which is available in the online code.

How to do it...

To execute a terms query, we will perform the following steps:

1. We execute a terms query from the command line, as follows:

```
POST /mybooks/_search
{"query": { "terms": { "uuid": [ "33333", "32222" ]}}}
```

If everything works as expected, the result returned by Elasticsearch should be as follows:

```
{ ... truncated ...
  "hits" : {
    "total" : { "value" : 1, "relation" : "eq" },
    "max_score" : 1.0,
    "hits" : [
      { "_index" : "mybooks", "_id" : "3",
        "_score" : 1.0,
        "_source" : {
          "uuid" : "33333", "position" : 3,
          ... truncated ...
```

How it works...

The terms query is related to the previous kind of query; it extends the term query to support multi-values. This call is very useful because it is very common to the concept of filtering on multi-values. In traditional SQL, this operation is achieved with the in keyword inside the where clause, that is, Select * from *** where uuid in ("33333", "22222").

In the preceding examples, the query searches for uuid with the 33333 or 22222 values. The terms query is not merely a helper for the term matching function. The terms query allows you to define extra parameters to control the query behavior, such as the following:

- minimum_match/minimum_should_match: This controls how many matched terms are required to validate the query, as follows:

```
"terms": {
  "color": ["red", "blue", "white"],
    "minimum_should_match":2
}
```

- The preceding query matches all of the documents where the color field has at least two values from red, blue, and white.
- boost: This is the standard query boost value that is used to modify the query weight. This can be very useful if you want to give more relevance to the terms that have been matched to increase the final document score.

There's more...

Because terms filtering is very powerful, to enable you to speed up during searching, the terms can be fetched by other documents during the query.

This is a very common scenario. For example, imagine that a user contains the list of groups in which they have grants and you want to filter documents that can only be seen by some groups. The pseudocode should be as follows:

```
GET /my-index/document/_search
{ "query": {
    "terms": {
      "can_see_groups": {
        "index": "my-index", "type": "user",
        "id": "1bw71LaxSzSp_zV6NB_YGg", "path": "groups"
      } } } }
```

In the preceding example, the list of groups is fetched at runtime from a document (which is always identified by an index, type, and ID) and the path (field) that contains the values to put inside it. The routing parameter is also supported.

> **Note**
>
> Using a `terms` query with a lot of terms will be very slow. To prevent this, there is a limit of 65,536 terms. This value can be lifted, if required, by setting the `index.max_terms_count` index setting.

This is a similar pattern to SQL, as shown in the following example:

```
select * from xxx where can_see_group in (select groups from
user where user_id='1bw71LaxSzSp_zV6NB_YGg')
```

Generally, NoSQL datastores do not support joins, so the data must be optimized for searching using denormalization or other techniques.

Elasticsearch does not provide anything similar to the SQL joins, but it provides similar alternatives, such as the following:

- Child/parent queries via the join field
- Nested queries
- Terms filtered with external document term fetching

See also

For further reference, take a look at the following resources, all of which are related to this recipe:

- The *Executing a search* recipe in *Chapter 4, Exploring Search Capabilities*
- The *Using a term query* recipe in this chapter
- The *Using a boolean query* recipe in *Chapter 4, Exploring Search Capabilities*
- The *Using the nested query, Using the has_child query*, and *Using the has_parent query* recipes from *Chapter 6, Relationships and Geo Queries*

Using a terms set query

The previous query will work very well if you have a fixed number of values via the `minimum_should_match` parameter to look for.

The natural evolution of the previous type query is to be able to define the minimum number of terms that should be matched via a related field in the document or via scripting code: the terms set query is able to cover these scenarios.

Getting ready

You will need an up-and-running Elasticsearch installation, as described in the *Downloading and installing Elasticsearch* recipe of *Chapter 1, Getting Started*.

To execute these commands, any HTTP client can be used, such as curl (https://curl.haxx.se/), Postman (https://www.getpostman.com/), or similar. I suggest you use the Kibana console, as it provides code completion and better character escaping for Elasticsearch.

How to do it...

To execute a terms query, we will perform the following steps:

1. We will define an item mapping for an item entity:

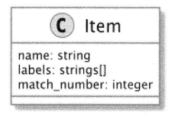

Figure 1.2 – The Item UML

The code is as follows:

```
PUT /ch05-item
{ "mappings": {
    "properties": {
      "name": { "type": "keyword" },
      "labels": { "type": "keyword" },
      "match_number": { "type": "integer" }
    } } }
```

2. We ingest some records via the following bulk command:

```
POST _bulk
{"index":{"_index":"ch05-item", "_id":"1"}}
{"name":"11111","labels":["one"],"match_number":2}
{"index":{"_index":"ch05-item", "_id":"2"}}
{"name":"22222","labels":["one", "two"],"match_number":2}
{"index":{"_index":"ch05-item", "_id":"3"}}
```

```
{"name":"33333","labels":["one", "two", "three"],"match_
number":3}
{"index":{"_index":"ch05-item", "_id":"4"}}
{"name":"44444","labels":["one", "two", "four"],"match_
number":3}
```

3. We want to select all the items that have the "one", "two", and "three" labels with the number of matches defined in the match_number field. The following terms_set query will achieve this:

```
GET /ch05-item/_search
{ "query": {
    "terms_set": {
      "labels": {
        "terms": [ "one", "two", "three"],
        "minimum_should_match_field": "match_number"
      } } } }
```

If everything works as expected, the result returned by Elasticsearch should be as follows:

```
{ ... truncated ...
  "hits" : {
    "total" : { "value" : 2, "relation" : "eq"},
    "max_score" : 2.1560106,
    "hits" : [
      {
        "_index" : "ch05-item",  "_id" : "3",
        "_score" : 2.1560106,
        "_source" : {
          "name" : "33333",
          "labels" : [ "one", "two", "three" ],
          "match_number" : 3
        }
      }... truncated ...
```

How it works...

This query type is very similar to the previous one, except that the number of values that need to be matched is dynamic: this parameter can be either another field that contains the number of matches or a script.

The field pointed in the `minimum_should_match_field` field must be a numeric one and should provide the number of matching terms required in the document.

If you need greater control of the matching term number, it's possible to provide a script in the `minimum_should_match_script` field that returns an integer: we will look at scripting in *Chapter 8, Scripting in Elasticsearch*.

See also

For further reference, take a look at the following points, all of which are related to this recipe:

- The *Executing a search* recipe in *Chapter 4, Exploring Search Capabilities*
- The *Using a terms query* recipe from this chapter
- The recipes in *Chapter 8, Scripting in Elasticsearch*, for additional details about how to use scripting in Elasticsearch

Using a prefix query

The `prefix` query is used when only the starting part of a term is known. It allows for the completion of truncated or partial terms.

Getting ready

You will need an up-and-running Elasticsearch installation, as described in the *Downloading and installing Elasticsearch* recipe of *Chapter 1, Getting Started*.

To execute these commands, any HTTP client can be used, such as curl (`https://curl.haxx.se/`), Postman (`https://www.getpostman.com/`), or similar. I suggest you use the Kibana console as it provides code completion and better character escaping for Elasticsearch.

To correctly execute the following commands, you will need an index populated with the `ch04/populate_kibana.txt` command, which is available in the online code.

How to do it...

To execute a `prefix` query, we will perform the following steps:

1. We execute a `prefix` query from the command line, as follows:

```
POST /mybooks/_search
{ "query": { "prefix": { "uuid": "222" } } }
```

If everything works as expected, the result returned by Elasticsearch should be as follows:

```
{ ... truncated ...
  "hits" : {
    "total" : { "value" : 1, "relation" : "eq" },
    "max_score" : 1.0,
    "hits" : [
      { "_index" : "mybooks", "_id" : "2",
        "_score" : 1.0,
        "_source" : {
          "uuid" : "22222", "position" : 2,
        ... truncated ...
```

How it works...

When a `prefix` query is executed, Lucene has a special method to skip to terms that start with a common: prefix. So, the execution of a `prefix` query is very fast.

In general, the `prefix` query is used in scenarios where term completion is required, as follows:

- Name completion
- Code completion
- Type completion

When you design a tree structure in Elasticsearch, if the ID of the item contains the hierarchical relationship (this approach is called **Materialized Path**), it can greatly speed up the application filtering. The following example shows how it's possible to model fruit and vegetable categories using a materialized path on the ID:

ID	Element
001	Fruit
00102	Apple
0010201	Green Apple
0010202	Red Apple
00103	Melon
0010301	White Melon
002	Vegetables

Figure 5.3 – Example of using a materialized path

In the preceding example, we have structured the ID that contains information about the tree structure. This allows us to create such queries:

- Filter by all the fruits, as follows:

```
"prefix": {"fruit_id": "001" }
```

- Filter by all the apple types, as follows:

```
"prefix": {"fruit_id": "001002" }
```

- Filter by all the vegetables, as follows:

```
"prefix": {"fruit_id": "002" }
```

If 'we compare this to a standard SQL `parent_id` table on a very large dataset, the reduction in joins and the fast search performance of Lucene can filter the results in milliseconds compared to some seconds or minutes.

Structuring the data in the correct way can give an impressive performance boost!

There's more...

The `prefix` query can be very handy when you are searching for ending text. For example, a user must match a document with a field, `filename`, with the `png` ending extension. Usually, users tend to execute a poor performance `regex` query that is similar to `.*png`. The regex needs to check every term of the fields, so the computation time is very long.

The best practice is to index the `filename` field with a reverse analyzer to convert a `suffix` query into a `prefix` one!

To achieve this, perform the following steps:

1. We define `reverse_analyzer` to the index level and put this into the settings, as follows:

    ```
    {
        "settings": {
          "analysis": {
            "analyzer": {
              "reverse_analyzer": {
                "type": "custom",
                "tokenizer": "keyword",
                "filter": [ "lowercase", "reverse" ]
    } } } } }
    ```

2. When we define the `filename` field, we use `reverse_analyzer` for its subfield, as follows:

    ```
    "filename": {
        "type": "keyword",
        "fields": {
            "rev": { "type": "text", "analyzer": "reverse_
    analyzer" } } }
    ```

3. Now we can search using a `prefix` query, using a similar query, as follows:

    ```
    "query": { "prefix": { "filename.rev": ".jpg" } }
    ```

If you were to use this approach, for example, when you index a file named `myTest.png`, the internal Elasticsearch data will be similar to the following ones:

```
filename:"myTest.jpg"
filename.rev:"gnp.tsetym"
```

Because the text analyzer is used both for indexing and searching the prefix text, `.png` will be automatically processed in `gnp` when the query is executed.

Moving from regex to prefix for the ending match can bring down your execution time from several seconds to several milliseconds!

See also

- Please refer to the *Using a term query* recipe, which is about full-term searches in Elasticsearch.

Using a wildcard query

The `wildcard` query is used when a part of a term is known. It allows the completion of truncated or partial terms. These queries are very famous because they are often used for commands on files within system shells (that is, `ls *.jpg`).

Getting ready

You will need an up-and-running Elasticsearch installation as described in the *Downloading and installing Elasticsearch* recipe of *Chapter 1, Getting Started*.

To execute the commands, any HTTP client can be used, such as Curl (`https://curl.haxx.se/`), Postman (`https://www.getpostman.com/`), or similar. I suggest you use the Kibana console as it provides code completion and better character escaping for Elasticsearch.

To correctly execute the following commands, you will need an index populated with the `ch04/populate_kibana.txt` command, which is available in the online code.

How to do it...

To execute a `wildcard` query, we will perform the following steps:

1. We will execute a `wildcard` query from the command line, as follows:

```
POST /mybooks/_search
{ "query": { "wildcard": { "uuid": "22?2*" } } }
```

If everything is alright, the result returned by Elasticsearch should be as follows:

```
{ ... truncated ...
  "hits" : {
    "total" : { "value" : 1, "relation" : "eq" },
    "max_score" : 1.0,
    "hits" : [
       { "_index" : "mybooks",   "_id" : "2",
          "_score" : 1.0,
```

```
    "_source" : {
        "uuid" : "22222", "position" : 2,
        ... truncated ...
```

How it works...

The wildcard is very similar to a regular expression, but it only has two special characters:

- *: This means you need to match zero or more characters.
- ?: This means you need to match one character.

During the query execution, all the terms of the searched field are matched against the wildcard query. So, the performance of the wildcard query depends on the cardinality of your terms.

To improve performance, it's suggested that you do not execute a wildcard query that starts with * or ?. To speed up a search, it's good practice to have some starting characters that use the skipTo Lucene method in order to reduce the number of processed terms.

See also

For further reference, take a look at the following resources, all of which are related to this recipe:

- Refer to the *Using a regexp query* recipe for more complex rules than wildcard ones.
- Refer to the *Using a prefix query* recipe for learning how to create a query with terms that start with a prefix.

Using a regexp query

In the previous recipes, we have explored different term queries (such as terms, prefix, and wildcard). Another powerful term query is the regexp (**regular expression**) query.

Getting ready

You will need an up-and-running Elasticsearch installation, as described in the *Downloading and installing Elasticsearch* recipe of *Chapter 1, Getting Started*.

To execute the commands, any HTTP client can be used, such as Curl (`https://curl.haxx.se/`), Postman (`https://www.getpostman.com/`), or similar. I suggest you use a Kibana console as it provides code completion and better character escaping for Elasticsearch.

To correctly execute the following commands, you will need an index populated with the `ch04/populate_kibana.txt` command, which is available in the online code.

How to do it...

To execute a `regexp` query, we will perform the following steps:

1. We can execute a `regexp` term query from the command line, as follows:

```
POST /mybooks/_search
{ "query": {
    "regexp": {
      "description": {
        "value": "j.*",
        "flags": "INTERSECTION|COMPLEMENT|EMPTY"
      } } } }
```

The query result will be as follows:

```
{ ... truncated ...
  "hits" : {
    "total" : { "value" : 1, "relation" : "eq" },
    "max_score" : 1.0,
    "hits" : [
      { "_index" : "mybooks",  "_id" : "1",
        "_score" : 1.0,
        "_source" : {
          "uuid" : "11111",  "position" : 1,
          ... truncated ...
```

The score for a matched regex result is always `1.0`. This is because the regex computation disables score computing.

How it works...

The `regexp` query executes the regular expression against all terms of the documents. Internally, Lucene compiles the regular expression automatically to improve performance. Therefore, generally, the performance of this query is not fast, as the performance depends on the regular expression used.

The parameters that are used to control this process are listed as follows:

- `boost` (the default is `1.0`): This includes the values used for boosting the score for this query.

- `flags`: This is a list of one or more flags (pipe | delimiter. The available flags are listed as follows:

  a) `ALL`: This enables all of the optional `regexp` syntaxes.

  b) `ANYSTRING`: This enables any (`@`) string.

  c) `AUTOMATON`: This enables named automation (`<identifier>`).

  d) `COMPLEMENT`: This enables complements (`~`).

  e) `EMPTY`: This enables empty language (`#`).

  f) `INTERSECTION`: This enables intersections (`&`).

  g) `INTERVAL`: This enables numerical intervals (`<n-m>`).

  h) `NONE`: This enables no optional `regexp` syntax.

> **Tip**
>
> To avoid poor performance in a search, don't execute `regex` starting with `.*`. Instead, use a `prefix` query on a string that is processed with a reverse analyzer.

See also

For further reference, take a look at the following points that are related to this recipe:

- Refer to the official documentation for the `regexp` query, which can be found at `https://www.elastic.co/guide/en/elasticsearch/reference/master/query-dsl-regexp-query.html`, for the regular expression syntax used by Lucene.

- Refer to the *Using a prefix query* recipe for a subset of a `regex` query that begins the term.

- Refer to the *Using a wildcard query* recipe if your regex can be rewritten in a `wildcard` query.

Using span queries

The big difference between standard databases (SQL and also many NoSQL databases) and Elasticsearch is the number of facilities used to express text queries. The `span` query family is a group of queries that controls a sequence of text tokens using their positions: standard queries don't take care of the positional presence of text tokens.

Span queries allow you to define several kinds of queries:

- The exact phrase query.

- The exact fragment query (that is, take off and give up).

- Partial exact phrase with a slop (other tokens between the searched terms, that is, the man with slop 2 can also match the strong man, the old wise man, and more).

Getting ready

You will need an up-and-running Elasticsearch installation, as described in the *Downloading and installing Elasticsearch* recipe of *Chapter 1, Getting Started*.

To execute the commands, any HTTP client can be used, such as Curl (`https://curl.haxx.se/`), Postman (`https://www.getpostman.com/`), or similar. I suggest you use the Kibana console as it provides code completion and better character escaping for Elasticsearch.

To correctly execute the following commands, you will need an index populated with the `ch04/populate_kibana.txt` command, which is available in the online code.

How to do it...

To execute span queries, we will perform the following steps:

1. The main element in span queries is span_term whose usage is similar to the term of the standard query. It is possible to aggregate more than one span_term value to formulate a span query.

2. The span_first query defines a query in which the span_term value must either match the first token or be close to it. The following code is an example of this:

```
POST /mybooks/_search
{ "query": {
    "span_first": {
      "match": { "span_term": {"description": "joe"}},
      "end": 5
} } }
```

3. The span_or query is used to define multi-values in a span query. This is very handy for simple synonym searches, as shown in the following example:

```
POST /mybooks/_search
{ "query": {
    "span_or": {
      "clauses": [
          { "span_term": { "description": "nice" } },
          { "span_term": { "description": "cool" } },
          { "span_term": { "description": "wonderful"}}
      ] } } }
```

The list of clauses is the core of the span_or query because it contains the span terms that should match.

4. Similar to span_or, there is a span_multi query that wraps multi-term queries such as prefix, wildcard, and more. For example, consider the following code:

```
POST /mybooks/_search
{ "query": {
    "span_multi": {
      "match": {
          "prefix": { "description": { "value": "jo" } }
} } } }
```

5. Queries can be used to create the `span_near` query. This allows you to control the token sequence of the query, such as the ordering and amount of distance between the terms (`slop`), as follows:

```
POST /mybooks/_search
{ "query": {
    "span_near": {
      "clauses": [
        { "span_term": { "description": "nice" } },
        { "span_term": { "description": "joe" } },
        { "span_term": { "description": "guy" } }
      ],
      "slop": 3, "in_order": false } } }
```

6. For complex queries, skipping matching positional tokens is very important. This can be achieved with the `span_not` query, as shown in the following example:

```
POST /mybooks/_search
{ "query": {
    "span_not": {
      "include":{"span_term":{"description": "nice"}},
      "exclude": {
        "span_near": {
          "clauses": [
            { "span_term": { "description": "not" } },
            { "span_term": { "description": "nice" }}
          ],
          "slop": 1, "in_order": true
        } } } } }
```

The `include` section contains the span that must be matched, while `exclude` contains the span that must not be matched. It matches documents with the term `nice` but not `not  nice`. This can be very useful for excluding negative phrases!

7. To search with a span query that is surrounded by other terms, we can use the `span_containing` variable, as follows:

```
POST /mybooks/_search
{ "query": {
    "span_containing": {
```

```
      "little": {
        "span_term": { "description": "nice"}
      },
        "big": {
          "span_near": {
            "clauses": [
                { "span_term": { "description": "not" } },
                { "span_term": { "description": "guy" } }
            ],
            "slop": 5, "in_order": true
          } } } } }
```

The `little` section contains the span that must be matched. The `big` section contains the span that contains the `little` matches. In the preceding case, the matched expression will be similar to not * nice * guy.

8. To search with a `span` query that is enclosed by other span terms, we can use the `span_within` variable, as follows:

```
POST /mybooks/_search
{"query": {
    "span_within": {
      "little": {
        "span_term": { "description": "nice" }
      },
        "big": {
          "span_near": {
            "clauses": [
                { "span_term": { "description": "not" } },
                { "span_term": { "description": "guy" } }
            ],
            "slop": 5, "in_order": true
          } } } } }
```

The `little` section contains the span that must be matched. The `big` section contains the span that contains the `little` matches.

How it works...

Lucene provides the span queries that are available in Elasticsearch. The base span query is the span_term query, which works exactly as the term query. The goal of this span query is to match an exact term (that is, the field plus text). Additionally, it can be composed to formulate other kinds of span queries.

The main usage of span query is a proximity search, that is, terms that are close to each other.

Using span_term in span_first refers to matching a term that must be in the first position. If the end parameter (integer) has been defined, it extends the first token to match the passed value.

One of the most powerful span queries is span_or, which allows you to define multiple terms in the same position. It covers several scenarios, such as the following:

- Multinames
- Synonyms
- Several verbal forms

The span_or query does not have a span_and counterpart, because span queries are positional.

If the number of terms that must be passed to a span_or query is too large (that is, 100 or more), it can be reduced with a span_multi query with a prefix or a wildcard. This approach allows the matching of, for example, all the play, playing, plays, player, and players terms by using a prefix query with play.

Otherwise, the most powerful span query is span_near, which allows you to define a list of span queries (clauses) to be matched in sequence or not. The parameters that can be passed to this span query are listed as follows:

- in_order: This tells us that the terms that are matched in the clauses must be executed in order. If you define two span near queries with two span terms to match joe and black, and in_order is true, you will not be able to match the black joe text (the default is true).
- slop: This defines the distance between the terms that must be matched to the clauses (the default is 0).

If you set the values of slop to 0 and in_order to true, you are creating an exact phrase match query, which we will explore in the next recipe.

The `span_near` query and slop can be used to create a phrase matching that is able to have some terms that are unknown. For example, consider matching an expression such as `the house`. If you need to execute an exact match, you need to write a similar query, as shown in the following example:

```
{ "query": {
    "span_near": {
      "clauses": [
          { "span_term": { "description": "the" } },
          { "span_term": { "description": "house" } }
      ],
      "slop": 0, "in_order": true   } } }
```

Now, for example, if you have an adjective between the `the` article and `house` (that is, the wonderful house or the big house), the previous query will never match them. To achieve this goal, it is necessary to set the slop to `1`.

Usually, slop is set to 1, 2, or 3 as values: high values (> 10) have no meaning.

See also

Please refer to the *Using a match query* recipe for a simplified way to create simple span queries.

Using a match query

Elasticsearch provides a helper to build complex span queries that depend on simple preconfigured settings. This helper is called a **match query**.

Getting ready

You will need an up-and-running Elasticsearch installation, as described in the *Downloading and installing Elasticsearch* recipe of *Chapter 1, Getting Started*.

To execute these commands, any HTTP client can be used, such as Curl (https://curl.haxx.se/), Postman (https://www.getpostman.com/), or similar. I suggest you use the Kibana console as it provides code completion and better character escaping for Elasticsearch.

To correctly execute the following commands, you will need an index populated with the `ch04/populate_kibana.txt` command, which is available in the online code.

How to do it...

To execute match queries, we will perform the following steps:

1. The standard usage of a `match` query simply requires the field name and the query text. Consider the following example:

```
POST /mybooks/_search
{ "query": {
    "match": {
      "description": {
        "query": "nice guy", "operator": "and"
      } } } }
```

2. If you need to execute the same query as a phrase query, the type changes from `match` to `match_phrase`, as shown in the following example:

```
POST /mybooks/_search
{ "query": {
    "match_phrase": { "description": "nice guy" } } }
```

3. An extension of the previous query that is used in text completion or in `search as you type` functionality is `match_phrase_prefix`, as follows:

```
POST /mybooks/_search
{ "query": { "match_phrase_prefix": {"description": "nice
gu" } } }
```

4. A common requirement is searching for several fields with the same `query` call. The `multi_match` parameter provides this capability, as shown in the following example:

```
POST /mybooks/_search
{ "query": {
    "multi_match": {
      "fields": [ "description", "name" ],
      "query": "Bill", "operator": "and" } } }
```

How it works...

The `match` query aggregates several frequently used query types that cover standard query scenarios.

The standard match query creates a Boolean query that can be controlled by the following parameters:

- `operator`: This defines how to store and process the terms. If it's set to `OR`, all the terms are converted into a Boolean query with all of the terms in `should` clauses. If it's set to `AND`, the terms build a list of must clauses (the default is `OR`).

- `analyzer`: This allows the overriding of the default analyzer of the field (the default is either based on mapping or set in searcher).

- `Fuzziness`: This allows you to define a fuzzy term. Similar to this parameter, both `prefix_length` and `max_expansion` are available.

- `zero_terms_query (none/all)`: This allows you to define a tokenizer filter that removes all terms from the query. The default behavior is to either return nothing or all of the documents. This is the case when you build an English query search for `the` or `a`, which means it could match all of the documents (the default is none).

- `cutoff_frequency`: This allows you to handle dynamic stopwords (very common terms in text) at runtime. During query execution, terms in the `cutoff_frequency` parameter are considered stopwords. This approach is very useful as it allows you to convert a general query into a domain-specific query because the terms to skip depend on text statistics. The correct value must be defined empirically.

- `auto_generate_synonyms_phrase_query` (the default is `true`): This happens if the match query uses the multi-terms synonym expansion with the synonym_graph token filter (for more information, look at `https://www.elastic.co/guide/en/elasticsearch/reference/current/analysis-synonym-graph-tokenfilter.html`).

The Boolean query created from the match query is very handy, but it suffers from some common problems that relate to Boolean queries, such as term position. If the term position matters, you need to use another family of match queries: the phrase query.

The `match_phrase` type in a match query builds long-spanning queries from the query text. The parameters that can be used to improve the quality of phrase queries are the analyzer, for text processing, and the `slop` parameter, which controls the distance between the terms (refer to the *Using span queries* recipe).

If the last term is partially complete, and you want to provide your user's query while writing functionality, the phrase type can be set to `match_phrase_prefix`. This type builds a span near query in which the last clause is a span prefix term. Often, his functionality is used for `typehead` widgets such as the one shown in the following screenshot:

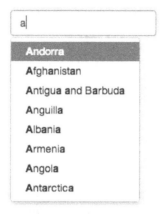

Figure 5.4 – An example of the autocompleter

The `match` query is a very useful query type, or, as I previously defined, it is a helper that allows you to build several common queries internally.

The `multi_match` parameter is similar to a `match` query in that it allows you to define multiple fields to search on. To define these fields, you can use several helpers, such as the following:

- **Wildcards field definition**: Using wildcards is a simple way in which to define multiple fields in one shot. For example, if you have fields for languages such as `name_en`, `name_es`, and `name_it`, you can define the search field as `name_*` to automatically search all of the name fields.

- **Boosting some fields**: Not all fields have the same importance. You can boost your fields using the `^` operator. For example, if you have title and content fields, and the title is more important than the content, you can define the fields in this way: `"fields":["title^3", "content"]`.

See also

For further reference, take a look at the following points, which are related to this recipe:

- The *Using span queries* recipe to build more complex text queries
- The *Using prefix query* recipe for simple initial typehead

Using a query string query

In the previous recipes, we looked at several types of queries that use text to match the results. The query string query is a special type of query that allows us to define complex queries by mixing the field rules.

It uses the Lucene query parser to parse text to complex queries.

Getting ready

You will need an up-and-running Elasticsearch installation, as described in the *Downloading and installing Elasticsearch* recipe of *Chapter 1*, *Getting Started*.

To execute the commands, any HTTP client can be used, such as Curl (https://curl.haxx.se/), Postman (https://www.getpostman.com/), or similar. I suggest you use the Kibana console as it provides code completion and better character escaping for Elasticsearch.

To correctly execute the following commands, you will need an index populated with the ch04/populate_kibana.txt command, which is available in the online code.

How to do it...

To execute a query_string query, we will perform the following steps:

1. We want to search for the text of nice guy, but with a condition of discarding the not term and displaying a price that is less than 5. The query will be as follows:

    ```
    POST /mybooks/_search
    { "query": {
        "query_string": {
            "query": """"nice guy" -description:not price:{ *
    TO 5 } """,
            "fields": [ "description^5" ],
            "default_operator": "and" } } }
    ```

If everything is alright, the result returned by Elasticsearch should be as follows:

```
{ … truncated …
   "hits" : {
     "total" : { "value" : 1, "relation" : "eq" },
     "max_score" : 2.3786995,
     "hits" : [
       {
         "_index" : "mybooks", "_id" : "1",
         "_score" : 2.3786995,
         "_source" : {
            … truncated …
            "description" : "Joe Testere nice guy",
            "price" : 4.3,
   … truncated …
```

How it works...

The `query_string` query is one of the most powerful types of query. The only required field is `query`, which contains the query that must be parsed with the Lucene query parser. For more information, please refer to `http://lucene.apache.org/core/7_0_0/queryparser/org/apache/lucene/queryparser/classic/package-summary.html`.

The Lucene query parser is able to analyze complex query syntax and convert it into many of the query types that we have seen in the previous recipes.

The optional parameters that can be passed to the query string query are as follows:

- `default_field`: This defines the default field to be used for the query. Additionally, it can be set at an index level that defines the `index` property `index.query.default_field` (the default is `_all`).

- `fields`: This defines the list of fields to be used. It replaces the `default_field` parameter. The `fields` parameter also allows us to use wildcards as values (that is, `city.*`).

- `default_operator`: This is the default operator to be used for text in the `query` parameter (the default is `OR`; the available values are `AND` and `OR`).

- `analyzer`: This is the analyzer that must be used for a query string.

- `allow_leading_wildcard`: Here, the * and ? wildcards are allowed as first characters. Using similar wildcards results in performance penalties (the default is `true`).

- `lowercase_expanded_terms`: This controls whether all of the expansion terms (generated by `fuzzy`, `range`, `wildcard`, and `prefix`) must be lowercase (the default is `true`).

- `Enable_position_increments`: This enables position increments in the queries. For every query token, the positional value is incremented by 1 (the default is `true`).

- `fuzzy_max_expansions`: This controls the number of terms that can be used in the fuzzy term expansion (the default is `50`).

- `fuzziness`: This sets the fuzziness value for fuzzy queries (the default is `AUTO`).

- `fuzzy_prefix_length`: This sets the prefix length for fuzzy queries (the default is `0`).

- `phrase_slop`: This sets the default slop (that is, the number of optional terms that can be present in the middle of the given terms) for phrases. If it is set to zero, the query is an exact phrase match (the default is `0`).

- `boost`: This defines the boost value of the query (the default is `1.0`).

- `analyze_wildcard`: This enables the processing of wildcard terms in the query (the default is `false`).

- `Auto_generate_phrase_queries`: This enables the autogeneration of phrase queries from the query string (the default is `false`).

- `minimum_should_match`: This controls how many `should` clauses should be verified to match the result. The value could be an integer value (that is, 3), a percentage (that is, 40%), or a combination of both (the default is `1`).

- `Lenient`: If it's set to `true`, the parser will ignore all format-based failures (such as text to number during date conversion) (the default is `false`).

- `locale`: This is the locale used for string conversion (the default is `ROOT`).

There's more...

The query parser is a very powerful tool that is used to support a wide range of complex queries. The most common cases are listed as follows:

- `field:text`: This is used to match a field that contains some text. It's mapped on a term query.

- `field:(term1 OR term2)`: This is used to match some terms in OR. It's mapped on a terms query.

- `field:"text"`: This is used to match the exact text. It's mapped on a match query.

- `_exists_:field`: This is used to match documents that have a field. It's mapped on an exists filter.

- `_missing_:field`: This is used to match documents that don't have a field. It's mapped on a missing filter.

- `field:[start TO end]`: This is used to match a range from the `start` value to the end value. The `start` and end values could be terms, numbers, or a valid datetime value. The `start` and end values are included in the range; if you want to exclude a range, you must replace the `[]` delimiters with `{}`.

- `field:/regex/`: This is used to match a regular expression.

The query parser also supports the text modifier and is used to manipulate the functionalities of the text. The most used ones are listed as follows:

- Fuzziness using the form of `text~`. The default fuzziness value is 2, which allows a Damerau-Levenshtein edit-distance algorithm (`http://en.wikipedia.org/wiki/Damerau%E2%80%93Levenshtein_distance`) of 2.

- Wildcards with ? that replace a single character or with * to replace zero or more characters (for example, `b?ll` or `bi*` to match bill).

- The proximity search of `"term1 term2"~3` allows you to match phrase terms with a defined slop. (For example, `"my umbrella"~3` matches `"my green umbrella"`, `"my new umbrella"`, and more.)

See also

For further reference, take a look at the following points that are related to this recipe:

- Refer to the Lucene official query parser syntax, which can be found at `http://lucene.apache.org/core/8_0_0/queryparser/org/apache/lucene/queryparser/classic/package-summary.html`. It provides a complete description of all the syntax.

- The official Elasticsearch documentation about the query string query can be found at `https://www.elastic.co/guide/en/elasticsearch/reference/master/query-dsl-query-string-query.html`.

Using a simple query string query

Typically, the programmer has the control to build complex queries using Boolean queries and the other query types. Thus, Elasticsearch provides two kinds of queries that give the user the ability to create string queries with several operators in them.

These kinds of queries are very common in advanced search engine usage, such as Google, which allows us to use the + and - operators for terms.

Getting ready

You will need an up-and-running Elasticsearch installation, as described in the *Downloading and installing Elasticsearch* recipe of *Chapter 1, Getting Started*.

To execute the commands, any HTTP client can be used, such as Curl (`https://curl.haxx.se/`), Postman (`https://www.getpostman.com/`), or similar. I suggest you use a Kibana console as it provides code completion and better character escaping for Elasticsearch.

To correctly execute the following commands, you will need an index populated with the `ch04/populate_kibana.txt` command, which is available in the online code.

How to do it...

To execute a `simple_query_string` query, we will perform the following steps:

1. We want to search for the `nice guy` text but not exclude the term `not`. The query will be as follows:

    ```
    POST /mybooks/_search
    { "query": {
        "simple_query_string": {
            "query": """"nice guy" -not""",
            "fields": [ "description^5", "_all" ],
            "default_operator": "and" } } }
    ```

2. If everything works as expected, the result returned by Elasticsearch should be as follows:

    ```
    { ... truncated ...
      "hits" : {
        "total" : { "value" : 2, "relation" : "eq"  },
        "max_score" : 2.3786995,
        "hits" : [
          { "_index" : "mybooks", "_id" : "1",
            "_score" : 2.3786995,
            "_source" : {
    ... truncated ...
              "description" : "Joe Testere nice guy",
    ... truncated ...
            } },
          { "_index" : "mybooks", "_id" : "2",
            "_score" : 2.3786995,
            "_source" : {
    ... truncated ...
              "description" : "Bill Testere nice guy",
    ... truncated ...
    ```

How it works...

The `simple_query_string` query takes the query text, tokenizes it, and builds a Boolean query by applying the rules provided in your text query.

If given to the final user, it's a good tool to express simple and advanced queries. Its parser is very complex, so it's able to extract fragments for an exact match to be interpreted as span queries.

The advantage of a simple query string query is that the parser always gives you a valid query.

Let's imagine you are using the previous query, the `query_string` query. If the user gives you, as input, a malformed query, it will throw an error. Additionally, if you use the simple query string, if malformed, it will be fixed and executed without errors.

See also

For a complete reference of these query types of syntaxes, the official documentation is available at `https://www.elastic.co/guide/en/elasticsearch/reference/master/query-dsl-simple-query-string-query.html` for the `simple_query_string` query.

Using the range query

All the previous queries work with defined or partially defined values, but it's very common in a real-world application to work for a range of values. The most common standard scenarios are as follows:

- Filtering by numeric value range (that is, price, size, and age).
- Filtering by date (that is, the events of `03/07/12` can be a range query from `03/07/12 00:00:00` to `03/07/12 24:59:59`).
- Filtering by term range (that is, from A to D).

Getting ready

You will need an up-and-running Elasticsearch installation, as described in the *Downloading and installing Elasticsearch* recipe of *Chapter 1, Getting Started*.

To execute the commands, any HTTP client can be used, such as Curl (`https://curl.haxx.se/`), Postman (`https://www.getpostman.com/`), or similar. I suggest you use the Kibana console as it provides code completion and better character escaping for Elasticsearch.

To correctly execute the following commands, you will need an index populated with the
ch04/populate_kibana.txt command, which is available in the online code.

How to do it...

To execute a range query, we will perform the following steps:

1. Consider the sample data of the previous examples, which contains a position
    integer field. By using it to execute a query for filtering positions between 3 and 5,
    we will get the following output:

```
POST /mybooks/_search
{ "query": {
    "range": {
      "position": {
        "from": 3, "to": 4,
        "include_lower": true, "include_upper": false
     } } } }
```

2. If everything works as expected, the result returned by Elasticsearch should be as
    follows:

```
{ ... truncated ...
  "hits" : {
    "total" : { "value" : 1, "relation" : "eq" },
    "max_score" : 1.0,
    "hits" : [
      { "_index" : "mybooks", "_id" : "3",
        "_score" : 1.0,
        "_source" : {
          "uuid" : "33333", "position" : 3,
... truncated ...
```

How it works...

The `range` query is used because scoring results can cover several interesting scenarios, such as the following:

- Items with high availability in stocks should be presented first.

- New items should be boosted.

- Most bought items should be boosted.

The `range` query is very handy with numeric values, as the preceding example shows. The parameters that a range query accepts are listed as follows:

- `from`: This is the starting value for the range (optional).

- `to`: This is the ending value for the range (optional).

- `include_in_lower`: This includes the starting value in the range (optional; the default is `true`).

- `include_in_upper`: This includes the ending value in the range (optional; the default is `true`).

In a `range` query, other helper parameters are available to simplify searches. They are listed as follows:

- `gt` (greater than): This has the same functionality to set both the `from` parameter and `include_in_lower` to `false`.

- `gte` (greater than or equal): This has the same functionality to set both the `from` parameter and `include_in_lower` to `true`.

- `lt` (less than): This has the same functionality to set both the `to` parameter and `include_in_upper` to `false`.

- `lte` (less than or equal to): This has the same functionality to set both the `to` parameter and `include_in_upper` to `true`.

There's more...

In Elasticsearch, what kind of query covers several types of SQL range queries, such as <, <=, >, and >=, on numeric values? In Elasticsearch, the date or time fields are managed internally as numeric fields, so it's possible to use the range queries or filters with date values. If the field is a `date` field, every value in the range query is automatically converted into a numeric value. For example, if you need to filter the documents for this year, the range fragment will be as follows:

```
"range": {
  "timestamp": {
    "from": "2014-01-01", "to": "2015-01-01",
    "include_lower": true, "include_upper": false } }
```

For `date` fields, it is also possible to specify a `time_zone` value to be used in order to correctly compute the matches.

If you are using a date value, you can use date math (`https://www.elastic.co/guide/en/elasticsearch/reference/master/common-options.html#date-math`) to round the values.

Using an IDs query

It's a common scenario to search by ID on a document that is part of a parent/child relationship or a join one. In this case, the children are not stored in shards based on their ID hash but based on their parent ID hash: the standard `GET` document doesn't work because a parent ID or routing value is required.

The `ids` query allows you to match documents by their IDs, spreading the query in all the searchable shards.

Getting ready

You will need an up-and-running Elasticsearch installation, as described in the *Downloading and installing Elasticsearch* recipe of *Chapter 1, Getting Started*.

To execute the commands, any HTTP client can be used, such as Curl (`https://curl.haxx.se/`), Postman (`https://www.getpostman.com/`), or similar. I suggest you use the Kibana console as it provides code completion and better character escaping for Elasticsearch.

To correctly execute the following commands, you will need an index populated with the `ch04/populate_kibana.txt` command, which is available in the online code.

How to do it...

To execute the `ids` queries or filters, we will perform the following steps:

1. The `ids` query for fetching the `"1"`, `"2"`, and `"3"` IDs of the `test-type` type is in the following form:

```
POST /mybooks/_search
{ "query": { "ids": { "values": [ "1", "2", "3" ] } }}
```

2. If everything works as expected, the result returned by Elasticsearch should be as follows:

```
{ ... truncated ...
  "hits" : {
    "total" : { "value" : 3, "relation" : "eq" },
    "max_score" : 1.0,
    "hits" : [
      { "_index" : "mybooks", "_id" : "1",
        "_score" : 1.0, "_source" : ... truncated... },
      { "_index" : "mybooks", "_id" : "2",
        "_score" : 1.0, "_source" : ... truncated... },
      { "_index" : "mybooks", "_id" : "3",
        "_score" : 1.0,  "_source" : ... truncated... }
    ] } }
```

In the preceding results, the request ID order is not respected.

How it works...

Querying by ID is a very fast operation because IDs are often cached in memory for faster lookup.

The parameters used in this query are listed as follows:

- `ids`: This includes a list of IDs that must be matched (required).
- `type`: This is a string, or a list of strings, that defines the types in which we need to search. If not defined, they are taken from the URL of the call (optional).

Internally, Elasticsearch stores the ID of a document in a special field called `_id`. An `_id` field is unique in an index.

Usually, the standard way of using the `ids` query is to select documents; this query allows you to fetch documents without knowing the shard that contains the documents. The documents are stored in shards, which are chosen based on a modulo operation computed on the document ID. If a parent ID, or a routing, is defined, they are used to select the shard. In this case, the only way to fetch the document while knowing its ID is to use the `ids` query.

If you need to fetch multiple IDs and there are no routing changes (due to the `routing` parameter at index time), it's better not to use this kind of query. Instead, use get or multi-get API calls to get documents, as they are much faster and also work in real time.

See also

For further reference, take a look at the following points that are related to this recipe:

- The *Getting a document* recipe from *Chapter 3, Basic Operations*
- The *Speeding up GET operations (Multi GET)* recipe from *Chapter 3, Basic Operations*

Using the function score query

This kind of query is one of the most powerful queries that are available. This is because it allows extensive customization of a scoring algorithm. The `function_score` query allows us to define a function that controls the score of the documents that are returned by a query.

Generally, these functions are CPU-intensive, and executing them on a large dataset (for instance, millions of records) requires a lot of memory and time, but computing them on a small subset can significantly improve the search quality.

The common scenarios used for this query are listed as follows:

- Creating a custom score function (for example, with the decay function)
- Creating a custom boost factor, for example, based on another field (that is, boosting a document by distance from a point)
- Creating a custom filter score function, for example, based on scripting Elasticsearch capabilities
- Ordering the documents randomly

Getting ready

You will need an up-and-running Elasticsearch installation, as described in the *Downloading and installing Elasticsearch* recipe of *Chapter 1, Getting Started.*

To execute the commands, any HTTP client can be used, such as Curl (https://curl.haxx.se/), Postman (https://www.getpostman.com/), or similar. I suggest you use the Kibana console as it provides code completion and better character escaping for Elasticsearch.

To correctly execute the following commands, you will need an index populated with the ch04/populate_kibana.txt command, which is available in the online code.

How to do it...

To execute a function_score query, we will perform the following steps:

1. We can execute a function_score query from the command line, as follows:

```
POST /mybooks/_search
{ "query": {
    "function_score": {
        "query": { "query_string":{ "query": "bill" } },
        "functions": [
            { "linear": {
                "position":{ "origin": "0", "scale": "20"}
            } } ],
        "score_mode": "multiply" } } }
```

2. Here, we execute a query that searches for bill, and we score the result with the linear function inside the position field.

The result should be as follows:

```
{ ... truncated ...
    "hits" : {
        "total" : { "value" : 2, "relation" : "eq" },
        "max_score" : 0.46101075,
        "hits" : [
            { "_index" : "mybooks", "_id" : "2",
                "_score" : 0.46101075, "_source" : ... truncated ...
        },
            { "_index" : "mybooks", "_id" : "3",
```

```
            "_score" : 0.43475333, "_source" : … truncated …
    } ] } }
```

How it works...

The function_score query is probably the most complex query type to master due to the natural complexity of mathematical algorithms involved in the scoring.

The generic full form of the function_score query is as follows:

```
"function_score": {
  "(query|filter)": {},
  "boost": "boost for the whole query",
  "functions": [
     { "filter": {}, "FUNCTION": {}  },
     { "FUNCTION": {} }
  ],
  "max_boost": number,
  "boost_mode": "(multiply|replace|...)",
  "score_mode": "(multiply|max|...)",
  "script_score": {},
  "random_score": { "seed ": number }
}
```

The parameters that are used are listed as follows:

- query or filter: This is the query used to match the required documents (optional; the default is to match all queries).

- boost: This is the boost to apply to the whole query (the default is 1.0).

- functions: This is a list of functions used to score the queries. In a simple case, use only one function. In the function object, a filter parameter can be provided to only apply the function to a subset of documents because the filter is applied first.

- max_boost: This sets the maximum allowed value for the boost score (the default is java FLT_MAX).

- boost_mode: This parameter defines how the function score is combined with the query score (the default is multiply). The possible values are listed as follows:

  - multiply (default): Here, the query score and function score are multiplied.

  - replace: Here, only the function score is used; the query score is ignored.

- Sum: Here, the query score and the function score are added.

- avg: Here, the average between the query score and the function score is taken.

- max: This is the maximum of the query score and the function score.

- min: This is the minimum of the query score and the function score.

- score_mode (the default is multiply): This parameter defines how the resulting function scores (when multiple functions are defined) are combined. The possible values are listed as follows:

  - multiply: The scores are multiplied.

  - Sum: The scores are summed.

  - avg: The scores are averaged.

  - first: The first function has a matching filter that is applied.

  - max: The maximum score is used.

  - Min: The minimum score is used.

- script_score: This allows you to define a script score function to be used to compute the score (optional). (Elasticsearch scripting will be discussed in *Chapter 8, Scripting in Elasticsearch.*) This parameter is very useful for implementing simple script algorithms. The original score value is in the _score function scope. This allows you to define similar algorithms, as follows:

```
"script_score": {
    "script": {
        "params": { "param1": 2, "param2": 3.1 },
        "source": "_score * doc['my_numeric_field'].value /
pow(param1, param2)"
    } }
```

- random_score: This allows you to randomly score the documents. It is very useful for retrieving records randomly (optional).

Elasticsearch provides native support for the most common scoring decay distribution algorithms, such as the following:

- **Linear**: This is used to linearly distribute the scores based on the distance from a value.
- **Exponential (exp)**: This is used for an exponential decay function.
- **Gaussian (gauss)**: This is used for the Gaussian decay function.

Choosing the correct function distribution depends on the context and data distribution.

See also

For further reference, take a look at the following points, all of which are related to this recipe:

- Refer to the official Elasticsearch documentation, which can be found at `https://www.elastic.co/guide/en/elasticsearch/reference/current/query-dsl-function-score-query.html`, for a complete reference on all the function score query parameters.
- Refer to the blog post at `https://www.elastic.co/blog/found-function-scoring` for several scenarios that use this kind of query.
- Refer to the experimental script score query provided by Elasticsearch 7.x, which can be found at `https://www.elastic.co/guide/en/elasticsearch/reference/7.10/query-dsl-script-score-query.html`.

Using the exists query

One of the main characteristics of Elasticsearch is its schema-less indexing capability. Records in Elasticsearch can have missing values. Due to its schema-less nature, two kinds of queries are required:

- **Exists field**: This is used to check whether a field exists in a document.
- **Missing field**: This is used to check whether a field is missing in a document.

Getting ready

You will need an up-and-running Elasticsearch installation, as described in the *Downloading and installing Elasticsearch* recipe of *Chapter 1, Getting Started*.

To execute the commands, any HTTP client can be used, such as curl (https://curl.haxx.se/), postman (https://www.getpostman.com/), or similar. I suggest you use the Kibana console as it provides code completion and better character escaping for Elasticsearch.

To correctly execute the following commands, you will need an index populated with the ch04/populate_kibana.txt command, which is available in the online code.

How to do it...

To execute existing and missing filters, we will perform the following steps:

1. To search all of the test-type documents that have a field called description, the query will be as follows:

    ```
    POST /mybooks/_search
    { "query": { "exists": { "field": "description" } } }
    ```

2. We can search all the test-type documents that do not have a field called description because there is no missing query, and we can obtain it using the Boolean must_ not query; the query will be as follows:

    ```
    POST /mybooks/_search
    { "query": {
        "bool": {
          "must_not": { "exists": { "field": "description" }
    } } } }
    ```

How it works...

The exists and missing filters only take a field parameter, which contains the name of the field to be checked. Using simple fields, there are no pitfalls. However, if you are using a single embedded object or a list of them, you need to use a subobject field because of the way Elasticsearch and Lucene work.

The following example helps you to understand how Elasticsearch maps JSON objects to Lucene documents internally if you are trying to index a JSON document:

```
{
  "name": "Paul",
  "address": {
    "city": "Sydney",
    "street": "Opera House Road",
    "number": "44"
  } }
```

Elasticsearch will index it internally, as follows:

```
name:paul
address.city:Sydney
address.street:Opera House Road
address.number:44
```

As we can see, there is no `address` field indexed, so the exists filter on `address` fails. To match documents with an address, you must search for a subfield (that is, `address.city`).

See also

For further reference, take a look at the following points, all of which are related to this recipe:

- Refer to the official Elasticsearch documentation, which can be found at `https://www.elastic.co/guide/en/elasticsearch/reference/current/query-dsl-function-score-query.html`, for a complete reference on all the function score query parameters.

- Refer to the blog post at `https://www.elastic.co/blog/found-function-scoring` for several scenarios that use this kind of query.

- Refer to the experimental script score query provided by Elasticsearch 7.x, which can be found at `https://www.elastic.co/guide/en/elasticsearch/reference/7.10/query-dsl-script-score-query.html`.

- Refer to the *Using a pinned query (XPACK)* recipe.

Using a pinned query (XPACK)

The `pinned` query is a special query provided by the X-Pack extension to provide a similar service to Google's promoted results. It's used to return results ranked based on a list of IDs.

Getting ready

You will need an up-and-running Elasticsearch installation, as described in the *Downloading and installing Elasticsearch* recipe of *Chapter 1, Getting Started*.

To execute the commands, any HTTP client can be used, such as curl (`https://curl.haxx.se/`), postman (`https://www.getpostman.com/`), or similar.
I suggest you use the Kibana console as it provides code completion and better character escaping for Elasticsearch.

To correctly execute the following commands, you will need an index populated with the `ch04/populate_kibana.txt` command, which is available in the online code.

How to do it...

To execute the `pinned` query, we will perform the following steps:

1. To rank the pinned documents, we will provide them as an `ids` list and a query will be provided in the `organic` field. The query will be as follows:

```
POST /mybooks/_search
{ "query": {
    "pinned": {
        "ids": ["1","2","3"],
        "organic": { "term": { "description": "bill" } }
    } } }
```

How it works...

The `pinned` query is a special query that will rank the matching results based on the order provided by the `ids` list.

Its parameters are listed as follows:

- `ids`: This is the list of IDs to be promoted.

- `organic`: This contains the query to be executed. It can be any kind of query.

During query execution, if an ID that is part of the `ids` parameter is matched, it's ranked above any other `organic` query result.

See also

For further reference, take a look at the following point, which relates to this recipe:

- Refer to the official Elasticsearch documentation at `https://www.elastic.co/guide/en/elasticsearch/reference/current/query-dsl-pinned-query.html`.

6
Relationships and Geo Queries

In this chapter, we will explore special queries that can be used to search for relationships between Elasticsearch and geolocation documents.

When we have a parent-child relationship (based on a `join` field mapping), we can use special queries to query for a similar relationship. Elasticsearch doesn't provide a SQL join, but it lets you search child/parent-related documents; it makes it possible to retrieve child documents via parent selection or by matching parent documents and filtering them by their children. Elasticsearch is very powerful tool and can help resolve many common data relationship issues. It is also used to easily solve issues in traditional relational databases. These features, in my experience, aren't popular among new Elasticsearch users but are valuable if you manage intelligence data sources.

In this chapter, we will also look at how to query nested objects using a nested query. The last part of this chapter is related to geolocation queries that provide queries based on distance, boxes, and polygons to match documents that meet these criteria.

In this chapter, we will cover the following recipes:

- Using the `has_child` query
- Using the `has_parent` query
- Using the `nested` query
- Using the `geo_bounding_box` query
- Using the `geo_shape` query
- Using the `geo_distance` query

Technical requirements

All the recipes in this chapter require us to prepare and populate the required indices – the online code is available on the PacktPub website or via GitHub (`https://github.com/PacktPublishing/Elasticsearch-8.x-Cookbook`). Here, you can find scripts to initialize all the required data.

Using the has_child query

Elasticsearch does not only support simple unrelated documents but also lets you define a hierarchy based on parents and children. The `has_child` query allows you to query parent documents of children by matching other queries.

Getting ready

As we described in the *Downloading and installing Elasticsearch* recipe in *Chapter 1, Getting Started*, you need an up-and-running Elasticsearch installation to execute the current recipe code.

To execute the code in the following section, any HTTP client can be used. This includes curl (`https://curl.haxx.se/`), Postman (`https://www.getpostman.com/`), or other similar versions. I suggest using the Kibana console, as this provides code completion and better character escaping for Elasticsearch.

To correctly execute the following commands, you will need an index populated with the `ch04/populate_kibana.txt` commands, available in the online code.

The index used in this recipe is `mybooks-join`, and the **Unified Modeling Language (UML)** of the data model is as follows:

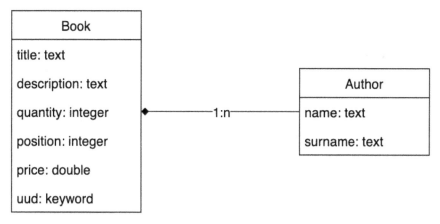

Figure 6.1 – The UML representation of the Book/Author relationship

The preceding UML representation describes a relationship between a **Book** entity and an **Author** one. **1:n** means that for every book, there is at least one author, but that can be more than one.

How to do it...

To execute the `has_child` queries, we will perform the following steps:

1. We want to search the parent (`book`) of the children (`author`), which has a term in the `name` field called `martin`. We can create this kind of query using the following code:

```
POST /mybooks-join/_search
{ "query": {
    "has_child": {
        "type": "author",
        "query": { "term": { "name": "martin" } },
        "inner_hits" : {}
    } } }
```

2. If everything is working well, the result returned by Elasticsearch should be as follows:

```
{ ... truncated ...
  "hits" : {
    "total" : { "value" : 1, "relation" : "eq"  },
    "max_score" : 1.0,
    "hits" : [
      { "_index" : "mybooks-join", "_id" : "3",
        "_score" : 1.0, "_source" : ...truncated...,
        "inner_hits" : {
          "author" : {
            "hits" : {
              "total" : { "value" : 1, "relation" : "eq" },
              "max_score" : 1.2039728,
              "hits" : [
                { "_index" : "mybooks-join", "_id" : "a3",
                  "_score" : 1.2039728, "_routing" : "3",
                  "_source" : {
                    "name" : "Martin",
                    "surname" : "Twisted",
                    "rating" : 3.2,
                    "join" : {
                      "name" : "author",
                      "parent" : "3"
} } } ] } } } } ] } }
```

For this example, we have used `inner_hits` to return matched children.

How it works...

This kind of query works by returning parent documents whose children match the query. The query can be of any type. The prerequisite of this kind of query is that the children must be correctly indexed in the shard of their parent. Internally, this kind of query is executed on the children, and all the IDs of the children are used to filter the parent. A system must have enough memory to store the children IDs.

The parameters that are used to control this process are as follows:

- The `type` parameter describes the type of children. This type is part of the same index as the parent; it's the name provided in the `join` field parameter at index time.

- The `query` parameter can be executed for the selection of the children. Any kind of query can be used.

- If defined, the `score_mode` parameter (the default is `none`; available values are `max`, `sum`, `avg`, and `none`) allows you to aggregate the children's scores with the parent's ones.

- `min_children` and `max_children` are optional parameters. This is the minimum/maximum number of children that are required to match the parent document.

- `ignore_unmapped` (`false` by default), when set to `true`, will ignore unmapped types. This is very useful when executing a query on multiple indices and some types are missing. The default behavior is to throw an exception if there is a mapping error.

In Elasticsearch, a document must have only one parent because the parent ID is used to choose the shard to put the children in. When working with child documents, it is important to remember that they must be stored in the same shard as their parents. Special precautions must be taken when fetching, modifying, and deleting them if the routing is unknown.

As the parent-child relationship can be considered similar to a foreign key in standard SQL, there are some limitations due to the distributed nature of Elasticsearch. This includes the following:

- There must be only one parent for each type.

- The `join` part of the child or parent is done in a shard and not distributed on all the nodes to reduce networking and increase performance; for this reason, the child and the parent should be in the same shard.

There's more...

Sometimes, you need to sort the parents according to their child field. In order to do this, you need to sort the parents by looking at the maximum score of the child field. To execute this kind of query, you can use the `function_score` query in the following way:

```
POST /mybooks-join/_search
{ "query": {
    "has_child": {
      "type": "author",
      "score_mode": "max",
      "query": {
        "function_score": {
          "script_score": {
            "script": """doc['rating'].value""" } } },
      "inner_hits": {} } } }
```

By executing this query for every child of a parent, the maximum score is taken (using the `function_score`), which is the value of the field that we want to sort.

In the preceding example, we have used scripting, which will be discussed in *Chapter 9, Managing Clusters*. This needs to be active so that it can be used.

See also

You can refer to the following recipes for further information related to this recipe:

- The *Indexing a document* recipe in *Chapter 3, Basic Operations*
- The *Mapping a child document with a join field* recipe in *Chapter 2, Managing Mapping*
- The *Using a function score query* recipe in *Chapter 6, Text and Numeric Queries*

Using the has_parent query

In the previous recipe, we saw the `has_child` query. Elasticsearch also provides a query to search child documents based on the parent query, `has_parent`, which complements the previous query.

Getting ready

You need an up-and-running Elasticsearch installation, as we described in the *Downloading and installing Elasticsearch* recipe in *Chapter 1, Getting Started*.

To execute these commands, I suggest using the Kibana console, as this provides code completion and better character escaping for Elasticsearch.

To correctly execute the following commands, you will need an index populated with the ch04/populate_kibana.txt commands, which are available in the online code. The index that's used in this recipe is mybooks-join.

How to do it...

To execute the has_parent query, we will perform the following steps:

1. We want to search for the children (author) of the parent (book) that has the bill term in the description field. We can create this kind of query using the following code:

```
POST /mybooks-join/_search
{ "query": {
    "has_parent": {
        "parent_type": "book",
        "query": { "term": {"description": "bill" }}}}}
```

2. If everything is fine here, the result returned by Elasticsearch should be as follows:

```
{
    ... truncated ...
    "hits" : {
        "total" : { "value" : 2, "relation" : "eq" },
        "max_score" : 1.0,
        "hits" : [
            { "_index" : "mybooks-join", "_id" : "a2",
                "_score" : 1.0, "_routing" : "2",
                "_source" : {
                    "name" : "Agatha", "surname" : "Princeton",
                    ... truncated ... },
            { "_index" : "mybooks-join", "_id" : "a3",
```

```
        "_score" : 1.0, "_routing" : "3",
        "_source" : {
            "name" : "Martin", "surname" : "Twisted",
        ... truncated ...
```

How it works...

This kind of query works by returning child documents whose parent ones match a subquery. Internally, this subquery is executed on the parents, and all the IDs of the matching parents are used to filter the children. The system must have enough memory to store all the parent IDs.

The parameters that are used to control this process are as follows:

- `parent_type`: This is the type of the parent.
- `query`: This is the query that can be executed to select the parents. Every kind of query can be used.
- `score`: The default value is `false`. Using the default configuration of `false`, Elasticsearch ignores the scores for parent documents, thus reducing memory and increasing performance. If it's set to `true`, the parent query score is aggregated into the children's.

Using the computed score, you can sort the resulting hits based on a parent with the same approach that was shown in the previous recipe using `function_score`.

See also

You can refer to the following recipes for further information related to this recipe:

- The *Indexing a document* recipe in *Chapter 4, Exploring Search Capabilities*
- The *Mapping a child document with a join field* recipe in *Chapter 3, Basic Operations*

Using the nested query

For queries based on nested objects, as we saw in *Chapter 2, Managing Mapping*, there is a special nested query. This kind of query is required because nested objects are indexed in a special way in Elasticsearch.

Getting ready

You need an up-and-running Elasticsearch installation, as we described in
the *Downloading and installing Elasticsearch* recipe in *Chapter 1, Getting Started*.

To execute these commands, any HTTP client can be used, such as curl (`https://curl.haxx.se/`), Postman (`https://www.getpostman.com/`), or similar.
I suggest using the Kibana console, as it provides code completion and better character escaping for querying text.

To correctly execute the following commands, you will need an index populated with
the `ch04/populate_kibana.txt` commands, which are available in the online code.

The index that's used in this recipe is `mybooks-join`.

How to do it...

To execute the `nested` query, we will perform the following steps:

1. We want to search the document for nested objects that are `blue` and whose size is greater than `10`. The `nested` query will be as follows:

```
POST /mybooks-join/_search
{ "query": {
    "nested": {
        "path": "versions",  "score_mode": "avg",
        "query": {
          "bool": {
            "must": [
                { "term": { "versions.color": "blue" } },
                { "range":{ "versions.size": { "gt": 10 }}}
            } ] } } } } }
```

The result returned by Elasticsearch, if everything is alright, should be as follows:

```
{ ... truncated ...
  "hits" : {
    "total" : { "value" : 1, "relation" : "eq" },
    "max_score" : 1.8754687,
    "hits" : [
        { "_index" : "mybooks-join", "_id" : "1",
          "_score" : 1.8754687,
          "_source" : {
```

```
        ... truncated ...
        "versions" : [
            { "color" : "yellow", "size" : 5 },
            { "color" : "blue", "size" : 15 }
        ] } } ] } }
```

How it works...

Elasticsearch manages nested objects in a special way. During indexing, they are extracted from the main document and indexed as a separate document, which is saved in the same Lucene chunk of the main document.

The `nested` query executes the first query on the nested documents, and after gathering the result IDs, they are used to filter the main document. The parameters that are used to control this process are as follows:

- `path`: This is the path of the parent document that contains the nested objects.
- `query`: This is the query that can be executed to select the nested objects. Every kind of query can be used.
- `score_mode`: The default value is `avg`. The valid values are `avg`, `sum`, `min`, `max`, and `none`, which control how to use the score of the nested document matches to improve the query.

Using `score_mode`, you can sort the result documents based on a nested object using the `function_score` query.

See also

You can refer to the following recipes for further information related to this recipe:

- The *Managing nested objects* recipe in *Chapter 2, Managing Mappings*
- The *Using the has_child query* recipe in this chapter

Using the geo_bounding_box query

One of the most common operations in geo-localization is searching for a box (square). The square is usually an approximation of the shape of a shop, a building, or a city.

This kind of query can be used in a percolator for real-time monitoring if users, documents, or events are entering a special place.

Getting ready

You need an up-and-running Elasticsearch installation, as we described in the *Downloading and installing Elasticsearch* recipe in *Chapter 1, Getting Started*.

To execute these commands, any HTTP client can be used, such as curl (https://curl.haxx.se/), Postman (https://www.getpostman.com/), or similar. I prefer using the Kibana console, as it provides code completion, formatting, and better character escaping for Elasticsearch.

To correctly execute the following commands, you will need an index populated with the ch04/populate_kibana.txt commands, which are available in the online code.

The index that's used in this section is mygeo-index.

How to do it...

To execute a geo_bounding_box query, we will perform the following steps:

1. A search to filter documents related to a bounding box with the 40.03, 72.0 40.717, and 70.99 coordinates can be achieved with the following query:

```
POST /mygeo-index/_search?pretty
{ "query": {
    "geo_bounding_box": {
      "pin.location": {
        "bottom_right": { "lat": 40.03, "lon": 72 },
        "top_left": { "lat": 40.717, "lon": 70.99 }
      } } } }
```

The result returned by Elasticsearch, if everything is alright, should be as follows:

```
{ … truncated …
  "hits" : {
… truncated …
    "hits" : [
      {
        "_index" : "mygeo-index",   "_id" : "2",
        "_score" : 1.0,
        "_source" : {
          "pin" : {
            "location" : { "lat" : 40.12, "lon" : 71.34 }
} } } ] } }
```

How it works...

Elasticsearch has a lot of optimizations to facilitate searching for a box shape. Latitude and longitude are indexed for fast-range checks, so this kind of filter is executed very quickly.

The parameters that are required to execute a geo-bounding box filter are the following:

- `top_left` (the top and left coordinates of the box).
- `bottom_right` (the bottom and right coordinates of the box) geo points.
- `validation_method` (default `STRICT`) is used for validating the geo point. The valid values are as follows:

  - `IGNORE_MALFORMED` is used to accept invalid values for latitude and longitude.
  - `COERCE` is used to try to correct wrong values.
  - `STRICT` is used to reject invalid values.

- `type` (`memory` by default) if the query should be executed in `memory` or `indexed`.

It's possible to use several representations of geo points, as described in the *Mapping a GeoPoint field* recipe in *Chapter 2, Managing Mapping*.

See also

You can refer to the following recipe for further information related to this recipe:

- The *Mapping a GeoPoint* field recipe in *Chapter 2, Managing Mapping*, for details on how to define a geo point

Using the geo_shape query

The *Using the geo_bounding_box query* recipe shows how to filter the square section, which is the most common case; Elasticsearch provides a way to filter user-defined shapes using the geo_shape query.

The common scenario when using this kind of query is searching for complex shapes such as countries, regions, or districts.

Getting ready

You need an up-and-running Elasticsearch installation, as we described in the *Downloading and installing Elasticsearch* recipe in *Chapter 1, Getting Started*.

To execute these commands, I suggest using the Kibana console, as this provides code completion, code formatting, and better character escaping for Elasticsearch.

To correctly execute the following commands, you will need an index populated with the ch04/populate_kibana.txt commands, which are available in the online code.

The index that's used in this section is mygeo-index.

How to do it...

To execute a geo_shape query, we will perform the following steps:

1. Searching documents in which pin.location is part of a triangle (its shape is made up of three geo points) is done with a query similar to the following:

```
POST /mygeo-index/_search
{ "query": {
    "bool": {
      "must": { "match_all": {} },
      "filter": {
        "geo_shape": {
          "pin.location": {
            "shape": {
              "type": "polygon",
              "coordinates": [
                [[-30,50],[-80,30],[-90,80],[-30,50]]
              ] },
            "relation": "within"} } } } } }
```

The result returned by Elasticsearch, if everything is alright, should be as follows:

```
{ ... truncated ...
  "hits" : {
    "total" : { "value" : 1, "relation" : "eq" },
    "max_score" : 1.0,
    "hits" : [
      { "_index" : "mygeo-index", "_id" : "1",
```

```
        "_score" : 1.0,
        "_source" : {
          "pin" : {
            "location" : { "lat" : 40.12, "lon" : -71.34
} } } } ] } }
```

How it works...

The `geo_shape` query allows you to define your own shape using the GeoJSON standards (`https://datatracker.ietf.org/doc/html/rfc7946`). In the preceding example, we chose to use a list of geo points to define our shape so that Elasticsearch can filter documents that are used in the polygon. This can be considered as an extension of a geo-bounding box for generic polygonal forms.

The `geo_shape` query allows the usage of the `ignore_unmapped` parameter, which helps to safely execute a search in the case of multi-indices or types where the field is not defined (the GeoPoint field is not defined for some indices or shards, and thus fails silently without giving errors).

See also

You can refer to the following recipes for further information related to this recipe:

- The *Mapping a GeoPoint field* recipe in *Chapter 2, Managing Mapping*
- The *Using the geo_bounding_box query* recipe in this chapter
- The **Rational Functional Tester** (**RFT**) documentation at `https://datatracker.ietf.org/doc/html/rfc7946`, which explains how to define different shape types
- This blog post, `https://www.factweavers.com/blog/advanced-geo-operations-in-elasticsearch/`, about the advanced usage of `geo_shape`

Using the geo_distance query

When you are working with geolocations, one common task is to filter results based on the distance from a location. This scenario covers very common site requirements, such as the following:

- Finding the nearest restaurant within a distance of 20 km
- Finding my nearest friends within a range of 10 km

The `geo_distance` query is used to achieve this goal.

Getting ready

You need an up-and-running Elasticsearch installation, as we described in the *Downloading and installing Elasticsearch* recipe in *Chapter 1, Getting Started*.

To execute these commands, any HTTP client can be used, such as curl (`https://curl.haxx.se/`), Postman (`https://www.getpostman.com/`), or similar. I suggest using the Kibana console, as it provides code completion and better character escaping for Elasticsearch.

To correctly execute the following commands, you will need an index populated with the `ch04/populate_kibana.txt` commands, which are available in the online code.

The index that's used in this recipe is `mygeo-index`.

How to do it...

To execute a `geo_distance` query, we will perform the following steps:

1. Searching documents in which `pin.location` is `200km` away from `lat` as `40` and `lon` as `70` is done with a query similar to the following:

```
GET /mygeo-index/_search
{ "query": {
    "geo_distance": {
      "pin.location": { "lat": 40, "lon": 70 },
      "distance": "200km" } } }
```

The result returned by Elasticsearch, if everything is alright, should be as follows:

```
{ … truncated …
  "hits" : {
    "total" : { "value" : 1, "relation" : "eq" },
    "max_score" : 1.0,
    "hits" : [
      { "_index" : "mygeo-index", "_id" : "2",
        "_score" : 1.0,
        "_source" : {
          "pin" : {
            "location" : { "lat" : 40.12, "lon" : 71.34 }
} } } ] } }
```

How it works...

As we discussed in the *Mapping a GeoPoint field* recipe, there are several ways to define a geo point to internally save searched items in an optimized way. The `distance` query executes a distance calculation between a given geo point and the points in the documents, returning hits that satisfy the distance requirement.

The parameters that control the distance query are as follows:

- The field and point of reference used to calculate the distance. In the preceding example, we have `pin.location` and `(40,70)`.

- `distance` defines the distance to be considered. It is usually expressed as a string by a number plus a unit.

- `unit` (optional) can be the unit of the distance value, if the distance is defined as a number. The valid values are as follows:

  - `in` or `inch`

  - `yd` or `yards`

  - `m` or `miles`

  - `km` or `kilometers`

  - `m` or `meters`

  - `mm` or `millimeters`

  - `cm` or `centimeters`

- `distance_type` (`arc` by default; valid choices are `arc`, which considers the roundness of the globe, or `plane`, which simplifies the distance in a linear way) defines the type of algorithm to calculate the distance.

- `validation_method` (`STRICT` by default) is used for validating the geo point. The valid values are as follows:

  - `IGNORE_MALFORMED` is used to accept invalid values for latitude and longitude.

  - `COERCE` is used to try to correct wrong values.

  - `STRICT` is used to reject invalid values.

- `ignore_unmapped` is used to safely execute the query in the case of multi-indices, which can have a missing definition of a geo point.

See also

You can refer to the following recipes for further information related to this recipe:

- The *Mapping a GeoPoint field* recipe in *Chapter 2*, *Managing Mapping*, about geo point definition

- The *Using the range query* recipe in *Chapter 5*, *Text and Numeric Queries*, with further details about using `range` queries

7
Aggregations

In developing search solutions, not only are the results important, but they also help us to improve the quality and the search focus. Elasticsearch provides a powerful tool to achieve these goals: **aggregations**. The main usage of aggregations is to provide additional data to the search results to improve their quality or to augment them with additional information.

For example, in a search for news articles, some facets that could be interesting to calculate could be the authors who wrote the articles and the date histogram of the publishing date; thus, aggregations are used, not only to improve the results' focus, but also to provide insight into stored data (analytics). This is the way that a lot of tools such as **Kibana** (`https://www.elastic.co/products/kibana`) are born.

Generally, aggregations are displayed to the end user with graphs or a group of filtering options (for example, a list of categories for the search results). Because the Elasticsearch aggregation framework provides scripting functionalities, it is able to cover a wide spectrum of scenarios. In this chapter, some simple scripting functionalities are shown relating to aggregations, but we will cover scripting in depth in the next chapter.

The aggregation framework is also the base for advanced analytics, as shown in software such as Kibana. It's very important to understand how the various types of aggregations work and when to choose them.

In this chapter, we will cover the following recipes:

- Executing an aggregation
- Executing a stats aggregation
- Executing a terms aggregation
- Executing a significant terms aggregation
- Executing a range aggregation
- Executing a histogram aggregation
- Executing a date histogram aggregation
- Executing a filter aggregation
- Executing a filters aggregation
- Executing a global aggregation
- Executing a geo distance aggregation
- Executing a children aggregation
- Executing a nested aggregation
- Executing a top hit aggregation
- Executing a matrix stats aggregation
- Executing a geo bounds aggregation
- Executing a geo centroid aggregation
- Executing a geotile grid aggregation
- Executing a sampler aggregation
- Executing a pipeline aggregation

Executing an aggregation

Elasticsearch provides several functionalities other than **Search**; this allows you to execute statistics and real-time analytics on searches using the aggregations.

Getting ready

You need an up-and-running Elasticsearch installation, as we described in the *Downloading and installing Elasticsearch* recipe in *Chapter 1*, *Getting Started*.

To execute these commands, any HTTP client can be used, such as cURL (https://curl.haxx.se/), Postman (https://www.getpostman.com/), or similar. Using the Kibana console is recommended, as it provides code completion and better character escaping for Elasticsearch.

To correctly execute the following commands, you will need an index populated with the ch07/populate_aggregation.sh commands available in the online code.

The index that's used in this recipe is index-agg.

How to do it...

To execute an aggregation, we will perform the following steps:

1. Compute the top 10 tags by name using the command line, executing a similar query with aggregations, as follows (we set the size to 0 because we are not interested in result hits):

    ```
    POST /index-agg/_search?size=0
    { "aggregations": {
        "tag": {
          "terms": { "field": "tag", "size": 10 } } } }
    ```

    In this case, we have used a term aggregation to count the terms.

2. The result returned by Elasticsearch, if everything is okay, should be as follows:

    ```
    { ... truncated...
      "aggregations" : {
        "tag" : {
          "doc_count_error_upper_bound" : 0,
          "sum_other_doc_count" : 2640,
          "buckets" : [
              { "key" : "laborum", "doc_count" : 31 },
              { "key" : "facilis", "doc_count" : 25 },
              { "key" : "maiores", "doc_count" : 25 },
              { "key" : "ipsam", "doc_count" : 24 },
              ... truncated... ] } } }
    ```

We have fixed the result size to 0 (default 10) in query arguments because we are not interested in result hits. The aggregation result is contained in the `aggregations` field. Each type of aggregation has its own result format (the explanation of this kind of result is given in the *Executing a term aggregation* recipe in this chapter).

> **Tip**
> It's possible to execute only an aggregation calculation without returning search results to reduce the bandwidth. This is done by passing the search `size` parameter set to 0 in the body of the JSON or as query arguments.

How it works...

Every search can return an aggregation calculation computed on the query results; the aggregation phase is an additional step in query post-processing, for example, highlighting. To activate the aggregation phase, an aggregation must be defined using the `aggs` or `aggregations` keyword.

There are several types of aggregation that can be used in Elasticsearch. In this chapter, we'll cover all the standard aggregations that are available; additional aggregation types can be provided with plugins and scripting.

Aggregations are the basis for real-time analytics. They allow us to execute the following:

- **Counting**
- **Histogram**
- **Range aggregation**
- **Statistics**
- **Geo distance aggregation**

The following are examples of the graphs that are generated by histogram aggregations:

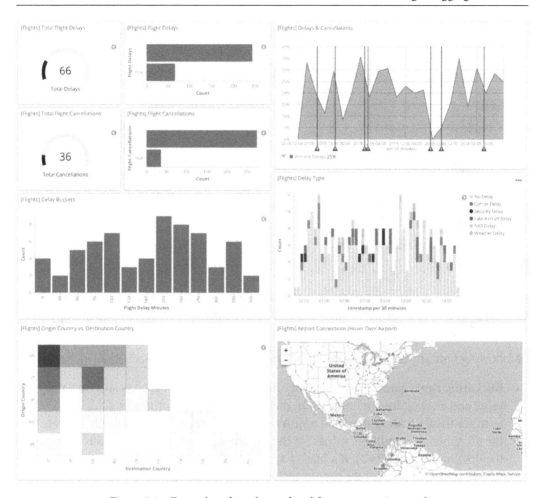

Figure 7.1 – Examples of graphs rendered from aggregation results

The aggregations are always executed on search hits; they are computed in a map or reduce way. The map step is distributed in shards, while the reduce step is done in the called node. During aggregation computation, a lot of data should be kept in memory and it can, therefore, be very memory-intensive.

For example, when executing a term aggregation, it requires that all the unique terms in the field that are used for aggregating are kept in memory. Executing this operation on millions of documents requires storing a large number of values in memory.

The aggregation framework was introduced in Elasticsearch 1.x with the possibility to execute analytics with several nesting levels of sub-aggregations. Aggregations keep information on which documents go into an aggregation bucket, and an aggregation output can be the input of the next aggregation.

Aggregations can be composed in a complex tree of sub-aggregations without depth limits.

The generic form for an aggregation is as follows:

```
"aggregations" : {
  "<aggregation_name>" : {
    "<aggregation_type>" : {
      <aggregation_body>
    }
    [,"aggregations" : { [<sub_aggregation>]+ } ]?
  }
  [,"<aggregation_name_2>" : { ... } ]*
}
```

Aggregation nesting allows for covering very advanced scenarios in executing analytics, such as aggregating data by country, by region, and by persons' ages, where age groups are ordered in descending order. There are no more limits to mastering analytics.

The following four kinds of aggregators can be used in Elasticsearch 8.x:

- **Bucketing aggregators**: These produce buckets, where a bucket has an associated value and a set of documents (that is, the terms aggregator produces a bucket per term for the field it's aggregating on). A document can end up in multiple buckets if the document has multiple values for the field being aggregated on (in our example, the document with id=3). If a bucket aggregator has one or more downstream (child) aggregators, they are run on each generated bucket.

- **Metric aggregators**: These receive a set of documents as input and produce statistical results that have been computed for the specified field. The output of metric aggregators does not include any information that's linked to individual documents, just the statistical data.

- **Matrix aggregators**: These operate on multiple fields and produce a matrix result based on the values extracted from the requested document fields.

- **Pipeline aggregators**: These aggregate the output of other aggregations and their associated metrics.

Generally, the order of buckets depends on the bucket aggregator that's used.

For example, using the terms aggregator, the buckets are, by default, ordered by count. The aggregation framework allows ordering by sub-aggregation metrics (that is, the preceding example can be ordered by the stats.avg value).

It's easy to create complex nested sub-aggregations that return huge numbers of results. Developers need to pay attention to the cardinality of returned aggregation results; it's very easy to return thousands of values!

See also

Refer to the *Executing a terms aggregation* recipe in this chapter for a more detailed explanation of aggregations that are used in this example. For pipeline aggregations, the official documentation can be found at `https://www.elastic.co/guide/en/elasticsearch/reference/master/search-aggregations-pipeline.html`. Since, as the official Elasticsearch documentation says, it's not safe to use, this feature could be removed in the future, and it is, therefore, not described in detail in the book.

Executing a stats aggregation

The most used metric aggregations are stats aggregations, which are able to compute several metrics of a bucket of documents in one go.

They are generally used as a terminal aggregation step to compute values that will be used directly or for further sorting.

Getting ready

You need an up and running Elasticsearch installation, as we described in the *Downloading and installing Elasticsearch* recipe in *Chapter 1, Getting Started*.

To execute these commands, any HTTP client can be used, such as curl (`https://curl.haxx.se/`), Postman (`https://www.getpostman.com/`), or similar. Using Kibana Console is recommended, as it provides code completion and better character escaping for Elasticsearch.

To correctly execute the following commands, you will need an index populated with the `ch07/populate_aggregation.sh` commands available in the online code.

The index that's used in this recipe is `index-agg`.

How to do it...

To execute a stats aggregation on our data sample, we will perform the following steps:

1. We want to calculate all statistics values of a matched query on the `age` field. The REST call should be as follows:

```
POST index-agg/_search?size=0
{ "aggs": {
    "age_stats": {
        "extended_stats": { "field": "age" } } } }
```

2. The result, if everything is okay, should be as follows:

```
{ ... truncated...
    "aggregations" : {
        "age_stats" : {
            "count" : 1000, "min" : 1.0, "max" : 100.0,
            "avg" : 53.243, "sum" : 53243.0,
            "sum_of_squares" : 3653701.0,
            "variance" : 818.8839509999999,
            "variance_population" : 818.8839509999999,
            "variance_sampling" : 819.7036546546545,
            "std_deviation" : 28.616148430562767,
            "std_deviation_population" : 28.616148430562767,
            "std_deviation_sampling" : 28.63046724478409,
            "std_deviation_bounds" : {
                "upper" : 110.47529686112554,
                "lower" : -3.9892968611255313,
                "upper_population" : 110.47529686112554,
                "lower_population" : -3.9892968611255313,
                "upper_sampling" : 110.50393448956818,
                "lower_sampling" : -4.017934489568177
            } } } }
```

In the answer, under the `aggregations` field, we have the statistical results of our aggregation under the defined field, `age_stats`.

How it works...

After the search phase, if any aggregations are defined, they are computed. In this case, we have requested an `extended_stats` aggregation labeled `age_stats`, which computes a lot of statistical indicators.

The available metric aggregators are as follows:

- `min`: Computes the minimum value for a group of buckets
- `max`: Computes the maximum value for a group of buckets
- `avg`: Computes the average value for a group of buckets
- `sum`: Computes the sum of all the buckets
- `value_count`: Computes the count of values in the bucket
- `stats`: Computes all the base metrics, such as `min`, `max`, `avg`, `count`, and `sum`
- `extended_stats`: Computes the `stats` metric plus `variance`, standard deviation (`std_deviation`), bounds of standard deviation (`std_deviation_bounds`), and sum of squares (`sum_of_squares`)
- `percentiles`: Computes the percentiles (the point at which a certain percentage of observed values occur) of some values (see Wikipedia at http://en.wikipedia.org/wiki/Percentile for more information about percentiles)
- `percentile_ranks`: Computes the rank of values that hit a percentile range
- `cardinality`: Computes an approximate count of distinct values in a field
- `geo_bounds`: Computes the maximum geo-bounds in the document where the geopoints are
- `geo_centroid`: Computes the centroid in the document where GeoPoints are

Every metric requires different computational needs, so it is good practice to limit the indicators only to the required one, so as not to waste CPU, memory, and performance.

In the preceding listing, I cited the most used metric aggregation available natively in Elasticsearch. Other metrics can be provided using custom plugins.

The syntax of all the metric aggregations has the same pattern, independent of the level of nesting, as in the **Domain Specific Language** (**DSL**) aggregation; they follow these patterns:

```
"aggs" : {
  "<name_of_aggregation>" : {
    "<metric_name>" : {
      "field" : "<field_name>"
    }
  }
}
```

See also

Refer to the official Elasticsearch documentation about stats aggregation at https://www.elastic.co/guide/en/elasticsearch/reference/current/search-aggregations-metrics-stats-aggregation.html, and extended stats aggregation at https://www.elastic.co/guide/en/elasticsearch/reference/current/search-aggregations-metrics-extendedstats-aggregation.html.

Executing a terms aggregation

The most used bucket aggregation is the terms one, which groups the documents into buckets based on a single term value. This aggregation is often used to narrow down the search using the computed values as filters for the queries.

Getting ready

You need an up and running Elasticsearch installation, as we described in the *Downloading and installing Elasticsearch* recipe in *Chapter 1, Getting Started.*

To execute the commands, any HTTP client can be used, such as cURL (https://curl.haxx.se/), Postman (https://www.getpostman.com/), or similar. Using Kibana Console is recommended, as it provides code completion and better character escaping for Elasticsearch.

To correctly execute the following commands, you will need an index populated with the ch07/populate_aggregation.sh commands available in the online code.

The index that's used in this recipe is index-agg.

How to do it...

To execute a terms aggregation, we will perform the following steps:

1. Calculate the top 10 tags of all the documents; the REST call should be as follows:

```
POST /index-agg/_search?size=0
{ "aggs": {
    "tag": {
      "terms": { "field": "tag", "size": 3 } } } }
```

2. The result returned by Elasticsearch, if everything is okay, should be as follows:

```
{ ... truncated...
   "aggregations" : {
      "tag" : {
        "doc_count_error_upper_bound" : 0,
        "sum_other_doc_count" : 2803,
        "buckets" : [
            { "key" : "laborum", "doc_count" : 31 },
            { "key" : "facilis", "doc_count" : 25 },
            { "key" : "maiores", "doc_count" : 25 } ] }}}
```

The aggregation result is composed of several buckets with terms, as follows:

- key: The term that's used to populate the bucket
- doc_count: The number of results with the key term

How it works...

During a search, there are a lot of phases that Elasticsearch executes; after the query execution, the aggregations are calculated and returned, along with the results.

In this recipe, we have seen the following parameters that the terms aggregation supports:

- field: This is the field to be used to extract the facets data. The field value can be a single string (as in the example tag) or a list of fields (that is, field1 and field2).
- size: This controls the number of term values to be returned (default 10).
- min_doc_count: This returns terms that have at least a minimum number of documents (optional).

- include: This defines the valid value to be aggregated using a regular expression (optional). This is evaluated before the exclude parameter. The regular expressions are controlled by the flags parameter, as follows:

```
"include" : {
   "pattern" : ".*labor.*",
   "flags" : "CANON_EQ|CASE_INSENSITIVE"
},
```

- exclude: This removes the terms that are contained in the exclude list (optional). The regular expressions are controlled by the flags parameter.

- order: This controls how to calculate the top *n* bucket values to be returned (optional, default doc_count). The order parameter can be one of the following types:

  - _count: Returns the aggregation values ordered by count (default)

  - _term: Returns the aggregation values ordered by term value (that is, "order" : { "_term" : "asc" })

- A sub-aggregation name, such as the following example:

```
{ "aggs": {
    "genders": {
      "terms": {
         "field": "tag",
         "order": { "avg_val": "desc" }  },
      "aggs": {
         "avg_age": { "avg": { "field": "age" }}}}}}
```

Term aggregation is very useful for representing an overview of values used for further filtering.

In the following figure, the term aggregation results are shown as a bar chart:

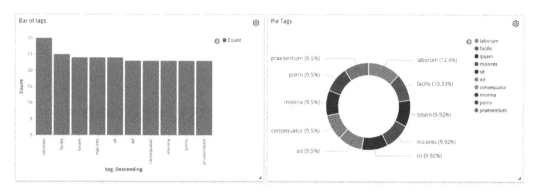

Figure 7.2 – Example of charts used to show term aggregation results

There's more...

Sometimes, we need to have much more control of terms aggregation; this can be achieved by adding an Elasticsearch script in the `script` field.

With scripting, it is possible to modify the term being used for the aggregation to generate a new value to be used. The following is a simple example, in which we append `123` to all terms:

```
{ "aggs": {
    "tag": {
       "terms": { "field": "tag", "script": "_value + '123'"
       } } } }
```

In the previous `terms` aggregation examples, we have provided the `field` or `fields` parameter to select the field to be used to compute the aggregation. It's also possible to pass a `script` parameter, which replaces `field` and `fields`, to define the field to be used to extract the data. `script` can extract the term value from the `doc` script variable that's available in the script context.

In the case of `doc`, the previous example can be rewritten as follows:

```
POST /index-agg/_search?size=0
{ "aggs": {
    "tag": {
       "terms": {
          "script": {"source": """doc['tag'].value+ '123'"""}
       } } } }
```

See also

Refer to *Chapter 8*, *Scripting in Elasticsearch*, which covers how to use scripting languages in Elasticsearch.

Executing a significant terms aggregation

Significant terms aggregation is an evolution of the previous one, in that it's able to cover several scenarios, such as the following:

- Suggesting relevant terms related to current query text
- Discovering relations between terms
- Discovering common patterns in text

In these scenarios, the result must not be as simple as the previous terms aggregations; it must be computed as a variance between a foreground set (generally the query) and a background one (a large bulk of data).

Getting ready

You need an up and running Elasticsearch installation, as we described in the *Downloading and installing Elasticsearch* recipe in *Chapter 1*, *Getting Started*.

To execute the commands, any HTTP client can be used, such as cURL (`https://curl.haxx.se/`), Postman (`https://www.getpostman.com/`), or similar. Using Kibana Console is recommended, as it provides code completion and better character escaping for Elasticsearch.

To correctly execute the following commands, you will need an index populated with the `ch07/populate_aggregation.sh` commands available in the online code.

The index that's used in this recipe is `index-agg`.

How to do it...

To execute a significant terms aggregation, we will perform the following steps:

1. We want to calculate the significant terms given some tags. The REST call should be as follows:

```
POST /index-agg/_search?size=0
{"query": {
    "terms": { "tag": [ "ullam", "in", "ex" ] } },
```

```
    "aggs": {
      "significant_tags": {
        "significant_terms": { "field": "tag" } } } }
```

2. The result returned by Elasticsearch, if everything is okay, should be as follows:

```
    { ... truncated...
      "aggregations" : {
        "significant_tags" : {
          "doc_count" : 45, "bg_count" : 1000,
          "buckets" : [
            { "key" : "ullam", "doc_count" : 17,
              "score": 8.017283950617283,"bg_count" : 17},
    ... truncated...
            { "key" : "vitae", "doc_count" : 3,
              "score": 0.13535353535353536,
              "bg_count" : 22 } ] } } }
```

The aggregation result is composed of several buckets with the following:

- key: The term used to populate the bucket

- doc_count: The number of results with the key term

- score: The score for this bucket

- bg_count: The number of background documents that contain the key term

How it works...

The execution of the aggregation is similar to the previous ones. Internally, two terms aggregations are computed: one related to the documents matched with the query or parent aggregation, and the other based on all the documents on the knowledge base. Then, the two result datasets are scored to compute the significant result.

This kind of aggregation is very CPU-intensive due to the large cardinality of terms queries and the cost of significant relevance computation.

The significant aggregation returns terms that are evaluated as significant for the current query.

This aggregation works only on `Keyword` mapping fields that are indexed in Elasticsearch with the field data flags. To be able to use the same functionality on text mapping fields, there is `significant_text`. Because text fields have no field data (a special structure that keeps all the field values in memory), the distribution of the terms must be computed on the fly, and this requires a lot of resources (CPU and RAM).

To speed up the process, generally, you select a random subset of the documents to not waste resources. The best practice is to use a sampler to collect some documents and then extract the more significant terms from the text, as shown in the following example:

```
POST /index-agg/_search?size=0
{ "aggregations" : {
        "my_sample" : {
            "sampler" : { "shard_size" : 100   },
            "aggregations": {
                "keywords" : {
                    "significant_text" : { "field" :
"description" } } } } }
```

The sampler approach will speed up the processing, but it works only on a subset of the documents; for this reason, the drawbacks could be that the selected sample is not significative of the overall document context.

Executing a range aggregation

The previous recipe describes an aggregation type that can be very useful if buckets must be computed on fixed terms or on a limited number of items. Otherwise, it's often required to return the buckets aggregated in ranges; the **range aggregations** meet this requirement. Commons scenarios are as follows:

- Price range (used in shops)
- Size range
- Alphabetical range

Getting ready

You need an up and running Elasticsearch installation, as we described in the *Downloading and installing Elasticsearch* recipe in *Chapter 1, Getting Started*.

To execute the commands, any HTTP client can be used, such as cURL (https://curl.haxx.se/), Postman (https://www.getpostman.com/), or similar. Using Kibana Console is recommended, as it provides code completion and better character escaping for Elasticsearch.

To correctly execute the following commands, you will need an index populated with the ch07/populate_aggregation.sh commands available in the online code.

The index that's used in this recipe is index-agg.

How to do it...

We want to provide the following three types of aggregation ranges:

- **Price aggregation**: This aggregates the price of items in ranges.

- **Age aggregation**: This aggregates the age contained in a document in four ranges of 25 years.

- **Date aggregation**: The ranges of 6 months of the previous year and all this year.

To execute range aggregations, we will perform the following steps:

1. To execute the above three required aggregations, we need to execute a query similar to the following:

```
POST /index-agg/_search?size=0
{ "aggs": {
    "prices": {
      "range": { "field": "price",
        "ranges": [
          {"to": 10}, {"from": 10, "to": 20},
          {"from": 20, "to": 100}, {"from": 100} ]
    } },
    "ages": {
      "range": { "field": "age",
        "ranges": [
          {"to": 25}, {"from": 25, "to": 50},
          {"from": 50, "to": 75}, {"from": 75} ]
    } },
    "range": {
      "range": { "field": "date",
```

```
        "ranges" : [
            {"from" : "2020-01-01", "to": "2020-12-31"},
            {"from" : "2021-07-01", "to": "2021-12-31"},
            {"from" : "2022-01-01", "to": "2022-12-31"}
        ] } } } }
```

2. The result returned by Elasticsearch, if everything is okay, should be as follows:

```
{ … truncated…
  "aggregations" : {
    "range" : {
      "buckets" : [
        {
            "key" : "2020-01-01T00:00:00.000Z-2020-12-31T00:00:00.000Z",
            "from" : 1.5778368E12,
            "from_as_string" : "2020-01-01T00:00:00.000Z",
"to" : 1.6093728E12,
            "to_as_string" : "2020-12-31T00:00:00.000Z",
            "doc_count" : 166
        }
… truncated…
        { "key" : "20.0-100.0", "from" : 20.0, "to" :
100.0, "doc_count" : 788 },
        { "key" : "100.0-*", "from" : 100.0, "doc_count"
: 0 }
      ] } } }
```

All aggregation results have the following fields:

- to, to_as_string, from, and from_as_string: These define the original range of the aggregation.

- doc_count: This is the number of results in this range.

- key: This is a string representation of the range.

How it works...

This kind of aggregation is generally executed against numerical data types (`integer`, `float`, `long`, and `dates`). It can be considered as a list of range filters executed against the result of the query.

The `date` or `datetime` values, when used in a filter or query, must be expressed in string format; the valid string formats are `yyyy-MM-dd'T'HH:mm:ss` or `yyyy-MM-dd`.

Each range is computed independently, so in their definition, they can overlap.

There's more...

There are two special range aggregations used for targeting: date and IPv4 ranges. They are similar to the preceding range aggregation, but they provide special functionalities to control the range on the date and IP address.

The date range aggregation (`date_range`) allows for defining `from` and `to` in date math expressions. For example, to execute an aggregation of hits in the previous 6 months and after, the aggregation will be as follows:

```
POST /index-agg/_search?size=0
{ "aggs": {
    "range": {
      "date_range": {
        "field": "date", "format": "MM-yyyy",
        "ranges": [
          { "to": "now-6M/M" }, { "from": "now-6M/M" }]}}}}
```

The result will be the following:

```
{ ... truncated...
  "aggregations" : {
    "range" : {
      "buckets" : [
        { "key" : "*-09-2021", "to" : 1.6304544E12,
          "to_as_string" : "09-2021", "doc_count" : 896 },
        { "key" : "09-2021-*", "from" : 1.6304544E12,
          "from_as_string" : "09-2018", "doc_count" : 104 }
      ] } } }
```

In this sample, the buckets will be formatted in the form month-year (MM-YYYY), in two ranges. now means the actual DateTime, -6M means minus 6 months, and /M is a shortcut for dividing the months.

A complete reference of date math expressions and codes is available at https://www.elastic.co/guide/en/elasticsearch/reference/current/search-aggregations-bucket-daterange-aggregation.html.

The IPv4 range aggregation (ip_range) allows for defining the ranges as follows:

- IP range form:

```
{"aggs": {
    "ip_ranges": {
        "ip_range": { "field": "ip",
            "ranges": [
                { "to": "192.168.1.1" },
                { "from": "192.168.2.255" } ] } } } }
```

- **Classless Inter-Domain Routing (CIDR)** masks:

```
{ "aggs": {
    "ip_ranges": {
        "ip_range": { "field": "ip",
            "ranges": [
                { "mask": "192.168.1.0/25" },
                { "mask": "192.168.1.127/25" } ] } } } }
```

See also

Refer to the *Using range query* recipe in *Chapter 5*, *Text and Numeric Queries*, for details on using range queries, and the official documentation for IP aggregation at https://www.elastic.co/guide/en/elasticsearch/reference/master/search-aggregations-bucket-iprange-aggregation.html.

Executing a histogram aggregation

Elasticsearch numerical values can be used to process histogram data.

The histogram representation is a very powerful way to show data to end users, mainly using bar charts.

Getting ready

You need an up and running Elasticsearch installation, as we described in the *Downloading and installing Elasticsearch* recipe in *Chapter 1, Getting Started*.

To execute the commands, any HTTP client can be used, such as cURL (https://curl.haxx.se/), Postman (https://www.getpostman.com/), or similar. Using Kibana Console is recommended, as it provides code completion and better character escaping for Elasticsearch.

To correctly execute the following commands, you will need an index populated with the ch07/populate_aggregation.sh commands available in the online code.

The index that's used in this recipe is index-agg.

How to do it...

Using the items populated with the script, we will calculate the following histogram aggregations:

- Age with an interval of 5 years
- Price with an interval of $10
- Date with an interval of 6 months

To execute histogram aggregations, we will perform the following steps:

1. The query will be as follows:

```
POST /index-agg/_search?size=0
{ "aggregations": {
    "age": {
      "histogram": { "field": "age", "interval": 5 }},
    "price": {
      "histogram": {"field": "price","interval": 10} }
  } }
```

2. The result returned by Elasticsearch, if everything is okay, should be as follows:

```
{ ... truncated...
    "aggregations" : {
      "price" : {
        "buckets" : [
          { "key" : 0.0, "doc_count" : 105 },
```

```
            { "key" : 10.0, "doc_count" : 107 },
            { "key" : 20.0, "doc_count" : 79 },
            ... truncated...
        ]
    },
    "age" : {
      "buckets" : [
            { "key" : 0.0, "doc_count" : 34 },
            { "key" : 5.0, "doc_count" : 41 },
            { "key" : 10.0, "doc_count" : 42 },
            { "key" : 15.0, "doc_count" : 43 },
            { "key" : 20.0, "doc_count" : 50 },
            ... truncated...
            { "key" : 100.0, "doc_count" : 9 }
      ] } } }
```

The aggregation result is composed of buckets as a list of aggregation results. These results are composed of the following:

- key: The value that is always on the *x* axis in the histogram graph

- doc_count: The document bucket size

How it works...

This kind of aggregation is calculated in a distributed manner in each shard with search results, and then the aggregation results are aggregated in the search node server (arbiter) and returned to the user. The histogram aggregation works only on numerical fields (boolean, integer, long integer, and float) and date or datetime fields (these are internally represented as long).

To control the histogram generation on a defined field, the interval parameter is required, which is used to generate an interval to aggregate the hits.

The following are special parameters to control the histogram creation:

- min_doc_count: This allows you to set the minimum number of documents that there must be in the bucket for emitting it in the aggregation (default 1).

- order: This is used to change the key sorting. For a full reference, see the *Executing a terms aggregation* recipe.

- offset: This is used to shift the keys by the defined offset (default 0).

- missing: This allows you to provide a default value if it's not available in the document. It can be very handy when you have a document with a missing field value and you need to provide a default one for your analytics.

- extended_bounds: This is used to set the minimum or maximum value to be considered in the histogram. Because Elasticsearch doesn't emit buckets if they don't contain documents, the histogram could contain holes in the *x* axis. This option also allows Elasticsearch to emit empty buckets to prevent holes in the distribution:

```
POST /index-agg/_search?size=0
{ "aggs" : {
    "prices" : {
      "histogram" : {
        "field" : "price", "interval" : 5,
        "extended_bounds" : { "min" : 0, "max" : 150 }
      } } } }
```

In this case, we have prices up to 100, but using a maximum of 150 in extended_ bounds, we will generate empty buckets up to 150.

- The general representation of a histogram could be a bar chart, similar to the following:

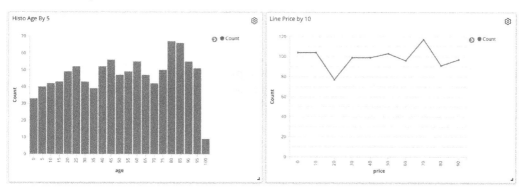

Figure 7.3 – Examples of graphs generated by histogram aggregation results

There's more...

The histogram aggregation can be also improved using Elasticsearch scripting functionalities. It is possible to script both by using `_value` if a field is stored, or by using the `doc` variable.

- An example of a scripted aggregation histogram using `_value` is as follows:

```
POST /index-agg/_search?size=0
{ "aggs": {
    "age": {
      "histogram": {
        "field": "age", "script": "_value*3",
        "interval": 5 } } } }
```

- An example of a scripted aggregation histogram using `_doc` is as follows:

```
{ ... truncated...
  "aggregations" : {
    "age" : {
      "buckets" : [
        { "key" : 0.0, "doc_count" : 8 },
        { "key" : 5.0, "doc_count" : 17 },
        { "key" : 10.0, "doc_count" : 9 },
        { "key" : 15.0, "doc_count" : 19 },
... truncated...
        { "key" : 295.0, "doc_count" : 11 },
        { "key" : 300.0, "doc_count" : 9 }
      ] } } }
```

See also

Refer to the *Executing a date histogram aggregation* recipe in this chapter for histogram aggregations based on date or time values.

Executing a date histogram aggregation

The previous recipe used mainly numeric fields. Elasticsearch provides special functionalities to compute the date histogram aggregation, which operates on `date` or `datetime` values.

This aggregation is required because date values need more customization to solve problems, such as timezone conversion and special time intervals.

Getting ready

You need an up and running Elasticsearch installation, as we described in the *Downloading and installing Elasticsearch* recipe in *Chapter 1, Getting Started*.

To execute these commands, any HTTP client can be used, such as cURL (`https://curl.haxx.se/`), Postman (`https://www.getpostman.com/`), or similar. Using Kibana Console is recommended, as it provides code completion and better character escaping for Elasticsearch.

To correctly execute the following commands, you will need an index populated with the `ch07/populate_aggregation.sh` commands available in the online code.

The index that's used in this recipe is `index-agg`.

How to do it...

We need two different date or time aggregations, which are as follows:

- An annual aggregation
- A quarter aggregation, but with a time zone +1:00

To execute these date histogram aggregations, we will perform the following steps:

1. The query will be as follows:

```
POST /index-agg/_search?size=0
{ "aggs": {
    "date_year": {
      "date_histogram": {
        "field": "date","calendar_interval": "year"}},
    "date_quarter": {
      "date_histogram": {
        "field":"date","calendar_interval": "quarter",
        "time_zone": "+01:00" } } } }
```

2. The result returned by Elasticsearch, if everything is okay, should be as follows:

```
{ ... truncated...
  "aggregations" : {
    "date_year" : {
      "buckets" : [
        {"key_as_string" : "2012-01-01T00:00:00.000Z",
          "key" : 1325376000000, "doc_count" : 190 },
        {"key_as_string" : "2013-01-01T00:00:00.000Z",
          "key" : 1356998400000, "doc_count" : 180 },
      ... truncated...
      ]
    },
    "date_quarter" : {
      "buckets" : [
  {"key_as_string":"2012-01-01T00:00:00.000+01:00",
          "key" : 1325372400000, "doc_count" : 48 },
  {"key_as_string":"2012-04-01T00:00:00.000+01:00",
          "key" : 1333234800000, "doc_count" : 52 },
      ... truncated... ] } } }
```

The aggregation result is composed by buckets as a list of aggregation results. These results are composed by the following:

- key: The value that is always on the *x* axis in the histogram graph

- key_as_string: A string representation of the key value

- doc_count: The document bucket size

How it works...

The main difference from the previous recipe histogram is that the interval is not numerical, as generally, date intervals are defined time constants. All the parameters that we have used in the histogram aggregation can be used in the DateTime.

The calendar_interval parameter allows you to use several values; the most commonly used ones are as follows:

- year

- quarter

- month
- week
- day
- hour
- minute

Like `calendar_interval`, there is `fixes_interval`, which is used when you want to specify a fixed value interval such as `10d` (10 days) or `5h` (5 hours).

The postfix suffixes used for `fixed_interval` are the following:

- `ms` for milliseconds
- `s` for seconds
- `m` for minutes
- `h` for hours
- `d` for days

When working with date values, it's important to use the correct timezone to prevent query time offset errors. By default, Elasticsearch uses the UTC milliseconds as the epoch to store DateTime values. To better handle the correct timestamp, there are some parameters that can be used, such as the following:

- `time_zone` (or `pre_zone`): This allows you to define a timezone offset to be used in the value calculation (optional). This value is used to preprocess the DateTime value for the aggregation. The value can be expressed in numeric form (that is, `-3`) if specifying hours, or, if minutes must be defined in the timezone, a string representation can be used (that is, `+07:30`).
- `post_zone`: This takes the result and applies the timezone offset (optional).
- `pre_zone_adjust_large_interval`: This applies the `hour` interval for `day` or larger intervals (default `false`) (optional).

There's more...

Estimating the number of buckets that a date histogram aggregation will produce is very hard. Often, it's very common to generate a large number of buckets and have memory and performances issues, not only on the Elasticsearch side but also at the application level.

To prevent this problem, Elasticsearch introduced a date histogram aggregation that auto-sizes the interval to generate the desired number of buckets: `auto_date_histogram`.

The extra parameter that it requires is `bucket`, along with the number of buckets to be generated. To test this, we can execute the following query:

```
POST /index-agg/_search?size=0
{ "aggs": {
    "10_buckets_date": {
        "auto_date_histogram": {
            "field": "date", "buckets": 10,
            "format": "yyyy-MM-dd" } } } }
```

The result will be as follows:

```
{ ... truncated...
    "aggregations" : {
        "10_buckets_date" : {
            "buckets" : [
                { "key_as_string" : "2012-01-01",
                    "key" : 1325376000000, "doc_count" : 190 },
                ... truncated...
                { "key_as_string" : "2022-04-04",
                    "key" : 1649030400000, "doc_count" : 1 } ] } } }
```

As you can see, this aggregation returns the interval that was used for this aggregation.

See also

You can refer to the following documentation for further reference to what we covered in this recipe:

- The official Elasticsearch documentation on date histogram aggregation is available at `https://www.elastic.co/guide/en/elasticsearch/reference/current/search-aggregations-bucket-datehistogram-aggregation.html`, for more details on managing time zone issues.

- The official Elasticsearch documentation on auto date histogram aggregation is available at `https://www.elastic.co/guide/en/elasticsearch/reference/master/search-aggregations-bucket-autodatehistogram-aggregation.html`.

Executing a filter aggregation

Sometimes, we need to reduce the number of hits in our aggregation to satisfy a particular filter. To obtain this result, filter aggregation is used.

The filter is one of the simpler ways to manipulate the bucket when filtering out values.

Getting ready

You need an up and running Elasticsearch installation, as we described in the *Downloading and installing Elasticsearch* recipe in *Chapter 1, Getting Started*.

To execute these commands, any HTTP client can be used, such as cURL (https://curl.haxx.se/), Postman (https://www.getpostman.com/), or similar. Using Kibana Console is recommended, as it provides code completion and better character escaping for Elasticsearch.

To correctly execute the following commands, you will need an index populated with the ch07/populate_aggregation.sh commands available in the online code.

The index that's used in this recipe is index-agg.

How to do it...

We need to compute two different filter aggregations, as follows:

- The count of documents that have "ullam" as a tag
- The count of documents that have an age equal to 37

To execute these filter aggregations, we will perform the following steps:

1. The query to execute these aggregations is as follows:

```
POST /index-agg/_search?size=0
{"aggregations": {
    "ullam_docs": {
        "filter": { "term": { "tag": "ullam" } } },
    "age37_docs": {
        "filter": { "term": { "age": 37 } } } } }
```

In this case, we have used simple filters, but they can be more complex if needed.

2. The result returned by Elasticsearch, if everything is okay, should be as follows:

```
{ ... truncated...
  "aggregations" : {
    "age37_docs" : { "doc_count" : 6 },
    "ullam_docs" : { "doc_count" : 17 } } }
```

How it works...

The filter aggregation is very straightforward; it executes a count on a filter on a matched element. You can consider this aggregation as a count query on the results. As we can see from the preceding result, the aggregation contains one value, doc_count, the count result.

It could be a very simple aggregation; generally, users tend not to use it as they prefer the statistic one, which also provides a count, or in the worst cases, they execute another search, generating more server workload.

The big advantage of this kind of aggregation is that the count, if possible, is executed using a filter; that is by far faster than iterating all the results. Another important advantage is that the filter can be composed of every possible valid Query DSL element.

There's more...

It's often required to have the count of the document that doesn't match a filter or generally doesn't have a particular field (or is null). For this kind of scenario, there is a special aggregation type: missing.

For example, to count the number of documents that have a missing code field, the query will be as follows:

```
POST /index-agg/_search?size=0
{ "aggs": {
    "missing_code": { "missing": { "field": "code" } } } }
```

The result will be as follows:

```
{ ... truncated...
  "aggregations":{"missing_code" : { "doc_count" : 1000 }}}
```

See also

Refer to the *Counting matched results* recipe in *Chapter 4, Exploring Search Capabilities,* for a standard count query.

Executing a filters aggregation

The filters aggregation answers the common requirement to split bucket documents using custom filters, which can be every kind of query supported by Elasticsearch.

Getting ready

You need an up and running Elasticsearch installation, as we described in the *Downloading and installing Elasticsearch* recipe in *Chapter 1, Getting Started.*

To execute the commands, any HTTP client can be used, such as cURL (https://curl.haxx.se/), Postman (https://www.getpostman.com/), or similar. Using Kibana Console is recommended, as it provides code completion and better character escaping for Elasticsearch.

To correctly execute the following commands, you will need an index populated with the ch07/populate_aggregation.sh commands available in the online code.

The index used in this recipe is index-agg.

How to do it...

We need to compute a filters aggregation composed of the following queries:

- Date greater than 2022/01/01 and price greater or equal to 50
- Date lower than 2022/01/01 and price greater or equal to 50
- All the documents that are not matched

To execute these filters aggregations, we will perform the following steps:

1. The query to execute these aggregations is as follows:

```
POST /index-agg/_search?size=0
{ "aggs": {
    "expensive_docs": {
        "filters": {
            "other_bucket": true,
            "other_bucket_key": "other_documents",
            "filters": {
                "2022_over_50": {
                    "bool": {
                        "must": [
```

```
                    { "range": {
                          "date": { "gte": "2022-01-01" }}},
                    { "range": {
                          "price": { "gte": 50 }}}]}
        },
        "previous_2022_over_50": {
          "bool": {
            "must": [
              { "range": {
                    "date": { "lt": "2022-01-01"}}},
              { "range": {
                    "price": { "gte": 50 } } }
          ]}}}}}}}
```

2. The result returned by Elasticsearch, if everything is okay, should be as follows:

```
{ ... truncated...
  "aggregations" : {
    "expensive_docs" : {
      "buckets" : {
        "2022_over_50" : { "doc_count" : 24 },
        "previous_2022_over_50" : {"doc_count" : 487},
        "other_documents" : { "doc_count" : 489 }}}}}
```

How it works...

The filters aggregation is a very handy one because it provides a convenient way to generate data buckets. The filters that compose the aggregation can be every kind of query that Elasticsearch supports. For this reason, this aggregation can be used to achieve complex relation management using children and nested queries.

Every query in the `filters` object generates a new bucket. Because the queries can be overlapped, the generated buckets can have overlapping documents. To collect all the documents that are not matched in filters, it's possible to use the `other_bucket` and `other_bucket_key` parameters. `other_bucket` is a `boolean` parameter: if it's `true`, it will return all the unmatched documents in `_other_` bucket.

The `other_bucket_key` is a string parameter that contains the label name of the other bucket that is used to control the name of the residual document bucket. If `other_bucket_key` is defined, it automatically implies that `other_bucket` is equal to `true`.

Executing a global aggregation

The aggregations are generally executed on query search results; Elasticsearch provides a special aggregation, `global`, that is executed globally on all the documents without being influenced by the query.

Getting ready

You need an up and running Elasticsearch installation, as we described in the *Downloading and installing Elasticsearch* recipe in *Chapter 1*, *Getting Started*.

To execute the commands, any HTTP client can be used, such as cURL (`https://curl.haxx.se/`), Postman (`https://www.getpostman.com/`), or similar. Using Kibana Console is recommended, as it provides code completion and better character escaping for Elasticsearch.

To correctly execute the following commands, you will need an index populated with the `ch07/populate_aggregation.sh` commands available in the online code.

The index used in this recipe is `index-agg`.

How to do it...

To execute global aggregations, we will perform the following steps:

1. Compare a global average with a query. The call will be something similar to the following:

```
POST /index-agg/_search?size=0
{ "query": { "term": { "tag": "ullam" } },
   "aggregations": {
      "query_age_avg": { "avg": { "field": "age" } },
     "all_persons": {
        "global": {},
        "aggs": {
          "age_global_avg": { "avg": { "field": "age" }
          } } } } }
```

2. The result returned by Elasticsearch, if everything is okay, should be as follows:

```
{ ... truncated...
  "aggregations" : {
    "all_persons" : { "doc_count" : 1000,
      "age_global_avg" : { "value" : 53.243  } },
      "query_age_avg":{ "value" : 53.470588235294116 }}}
```

In the preceding example, `query_age_avg` is computed on the query and `age_global_avg` is computed on all the documents.

How it works...

This kind of aggregation is mainly used as top aggregation as a starting point for other sub-aggregations. The JSON body of the global aggregations is empty: it doesn't have any optional parameters.

The most frequent use cases are comparing aggregations that are executed on filters with the ones without them, as we did in the preceding example.

Executing a geo distance aggregation

Among the other standard types that we have seen in the previous aggregations, Elasticsearch allows you to execute aggregations against a GeoPoint: the geo distance aggregations. This is an evolution of the previously discussed range aggregations that have been built to work on geo locations.

Getting ready

You need an up and running Elasticsearch installation, as we described in the *Downloading and installing Elasticsearch* recipe in *Chapter 1, Getting Started.*

To execute the commands, any HTTP client can be used, such as cURL (`https://curl.haxx.se/`), Postman (`https://www.getpostman.com/`), or similar. Using Kibana Console is recommended, as it provides code completion and better character escaping for Elasticsearch.

To correctly execute the following commands, you will need an index populated with the `ch07/populate_aggregation.sh` commands available in the online code.

The index used in this recipe is `index-agg`.

How to do it...

Using the `position` field that's available in the documents, we'll aggregate the other documents in the following five ranges:

- Less than 10 kilometers

- From 10 kilometers to 20

- From 20 kilometers to 50

- From 50 kilometers to 100

- Above 100 kilometers

To execute these geo distance aggregations, we will perform the following steps:

1. To achieve these goals, we will create a geo distance aggregation with a code similar to the following:

```
POST /index-agg/_search?size=0
{ "aggs": {
    "position": {
        "geo_distance": { "field": "position",
            "origin": { "lat": 83.76, "lon": -81.2 },
            "ranges": [
                { "to": 10 }, { "from": 10, "to": 20 },
                { "from": 20, "to": 50 },
                { "from": 50, "to": 100 }, { "from": 100 }
            ] } } } }
```

2. The result returned by Elasticsearch, if everything is okay, should be as follows:

```
{ ... truncated...
    "aggregations" : {
        "position" : {
            "buckets" : [
                { "key" : "*-10.0", "from" : 0.0,
                    "to" : 10.0, "doc_count" : 0 },
                { "key" : "10.0-20.0", "from" : 10.0,
                    "to" : 20.0, "doc_count" : 0 },
                { "key" : "20.0-50.0", "from" : 20.0,
                    "to" : 50.0, "doc_count" : 0 },
```

```
{ "key" : "50.0-100.0", "from" : 50.0,
  "to" : 100.0, "doc_count" : 0 },
{ "key" : "100.0-*", "from" : 100.0,
  "doc_count" : 1000 } ] } } }
```

How it works...

The geo range aggregation is an extension of the range aggregations that work on geo-localization. It works only if a field is mapped as geo_point. The field can contain single or multi-value geo points.

The aggregation requires at least the following three parameters:

- field: The field of the geo point to work on

- origin: The geo point to be used for computing the distances

- ranges: A list of ranges to collect documents based on their distance from the target point

The geo point can be defined in one of the following accepted formats:

- Latitude and longitude as properties, that is, {"lat": 83.76, "lon": -81.20}

- Longitude and latitude as an array, that is, [-81.20, 83.76]

- Latitude and longitude as a string, that is, 83.76, -81.20

- Geohash, that is, fnyk80 (see the *See also* section for more information)

The ranges are defined as a couple of from/to values. If one of them is missing, they are considered unbound.

The values that are used for the range are set to kilometers by default, but by using the unit property, it's possible to set them as follows:

- mi or miles

- in or inches

- yd or yards

- km or kilometers

- m or meters

- cm or centimeters
- mm or millimeters

It's also possible to set how the distance is computed with the distance_type parameter. Valid values for this parameter are as follows:

- arc: This uses the arc length formula. It is the most precise (see http://en.wikipedia.org/wiki/Arc_length for more details on the arc length algorithm).
- plane: This is used for the plane distance formula. It is the fastest and most CPU-intensive, but it's also the least precise.

As for the range filter, the range values are treated independently, so overlapping ranges are allowed. When the results are returned, this aggregation provides a lot of information in its fields as follows:

- from or to: Defines the analyzed range
- key: Defines the string representation of the range
- doc_count: Defines the number of documents in the bucket that match the range

See also

You can refer to the following recipes for further reference to what we covered in this recipe:

- The *Executing a range aggregation* recipe in this chapter for common functionalities of range aggregations
- The *Mapping a GeoPoint field* recipe in *Chapter 2, Managing Mapping*, to correctly define a GeoPoint field for executing geo aggregations
- Geohash grid aggregation at https://www.elastic.co/guide/en/elasticsearch/reference/current/search-aggregations-bucket-geohashgrid-aggregation.html to learn how to aggregate on a geohash

Executing a children aggregation

Children aggregation allows you to execute analytics based on parent documents and child documents. When working with complex structures, nested objects are very common.

Getting ready

You need an up and running Elasticsearch installation, as we described in the *Downloading and installing Elasticsearch* recipe in *Chapter 1, Getting Started*.

To execute these commands, any HTTP client can be used, such as cURL (https://curl.haxx.se/), Postman (https://www.getpostman.com/), or similar. Using Kibana Console is recommended, as it provides code completion and better character escaping for Elasticsearch.

To correctly execute the following commands, you will need an index populated with the ch04/populate_kibana.sh commands available in the online code.

The index used in this recipe is mybook-join.

How to do it...

To execute children aggregations, we will perform the following steps:

1. Index documents with child or parent relations, as discussed in the *Managing a child document* recipe in *Chapter 2, Managing Mapping*. For this example, we will use the same dataset as the child query.

2. Execute a terms aggregation on uuid of the parent, and for every uuid collecting the terms of the children value, we create a children aggregation with code similar to the following:

```
POST /mybooks-join/_search?size=0
{ "aggs": {
    "uuid": {
        "terms": { "field": "uuid", "size": 10 },
        "aggs": {
            "to-children": {
                "children": { "type": "author" },
                "aggs": {
                    "top-values": {
                        "terms": {
```

```
                    "field": "name.keyword", "size": 10   }
          } } } } } } }
```

3. The result returned by Elasticsearch, if everything is okay, should be as follows:

```
{ ... truncated...
  "aggregations" : {
    "uuid" : {
      "doc_count_error_upper_bound" : 0,
      "sum_other_doc_count" : 0,
      "buckets" : [
        { "key" : "11111", "doc_count" : 1,
          "to-children" : {
            "doc_count" : 2,
            "top-values" : {
              "doc_count_error_upper_bound" : 0,
              "sum_other_doc_count" : 0,
              "buckets" : [
                { "key" : "Mark", "doc_count" : 1 },
                { "key" : "Peter", "doc_count" : 1 }
              ] } } },
      ... truncated...
```

How it works...

The children aggregation works by following these steps:

1. All the parent IDs are collected by the matched query or by previous bucket aggregations.

2. The parent IDs are used to filter the children, and the matching document results are used to compute the children aggregation.

This type of aggregation, similar to the nested one, allows us to aggregate on different documents over searched ones. Because children documents are stored in the same shard as the parents, they are very fast.

Executing a nested aggregation

Nested aggregation allows you to execute analytics on nested documents. When working with complex structures, nested objects are very common.

Getting ready

You need an up and running Elasticsearch installation, as we described in the *Downloading and installing Elasticsearch* recipe in *Chapter 1, Getting Started*.

To execute these commands, any HTTP client can be used, such as cURL (https://curl.haxx.se/), Postman (https://www.getpostman.com/), or similar. Using Kibana Console is recommended, as it provides code completion and better character escaping for Elasticsearch.

To correctly execute the following commands, you will need an index populated with the ch04/populate_kibana.sh commands available in the online code.

The index used in this recipe is mybooks-join.

How to do it...

To execute nested aggregations, we will perform the following steps:

1. Create a nested aggregation to return the minimum size of the product version that can be purchased using the following code:

```
POST /mybooks-join/_search?size=0
{ "aggs": {
    "versions": {
        "nested": { "path": "versions" },
        "aggs": {
            "min_size": {
                "min": { "field": "versions.size" }}}}}}
```

2. The result returned by Elasticsearch, if everything is okay, should be as follows:

```
{ ... truncated...
    "aggregations" : {
        "versions" : {
            "doc_count" : 5,
            "min_size" : { "value" : 2.0 } }}}
```

In this case, the result aggregation is a simple `min` metric that we have already seen in the second recipe of this chapter.

How it works...

The nested aggregation requires only the path of the field, relative to the parent, which contains the nested documents.

After having defined the nested aggregation, all the other kinds of aggregations can be used in the sub-aggregations.

There's more...

Elasticsearch provides a way to aggregate values from nested documents to their parent; this aggregation is called `reverse_nested`.

For the preceding example, we can aggregate the top tags for the reseller with a similar query, as follows:

```
POST /mybooks-join/_search?size=0
{ "aggs": {
    "versions": {
       "nested": { "path": "versions" },
     "aggs": {
        "top_colors": {
           "terms": { "field": "versions.color" },
             "aggs": {
               "version_to_book": {
                  "reverse_nested": {},
                 "aggs": {
                    "top_uuid_per_version": {
                       "terms": { "field": "uuid" }}}}}}}}}
```

In this example, there are several steps:

1. We aggregate initially for nested versions.

2. Having activated the nested versions documents, we can term the aggregate by the `color` field (`versions.color`).

3. From the top versions aggregation, we go back to aggregate on the parent using `"reverse_nested"`.

4. Now, we can aggregate `uuid` of the parent document.

The response will be similar to this one:

```
{ ... truncated...
  "aggregations" : {
   "versions" : {
    "doc_count" : 5,
     "top_colors" : {
      "doc_count_error_upper_bound" : 0,
      "sum_other_doc_count" : 0,
      "buckets" : [
        { "key" : "blue", "doc_count" : 2,
         "version_to_book" : {
          "doc_count" : 2,
          "top_uuid_per_version" : {
           "doc_count_error_upper_bound" : 0,
           "sum_other_doc_count" : 0,
           "buckets" : [
             { "key" : "11111", "doc_count" : 1 },
             { "key" : "22222", "doc_count" : 1 }
           ] } } },
       ... truncated...
```

Executing a top hit aggregation

The top hit aggregation is different from the other aggregation types. All the previous aggregations have metric (simple) values or bucket values; the top hit aggregation returns buckets of search hits (documents).

Generally, the top hit aggregation is used as a sub-aggregation, so that the top matching documents can be aggregated in buckets. The most common scenario for this aggregation is to have, for example, the top n documents grouped by category (very common in search results in e-commerce websites).

Getting ready

You need an up and running Elasticsearch installation, as we described in the *Downloading and installing Elasticsearch* recipe in *Chapter 1, Getting Started*.

To execute these commands, any HTTP client can be used, such as cURL (`https://curl.haxx.se/`), Postman (`https://www.getpostman.com/`), or similar. Using Kibana Console is recommended, as it provides code completion and better character escaping for Elasticsearch.

To correctly execute the following commands, you will need an index populated with the `ch07/populate_aggregation.sh` commands available in the online code.

The index used in this recipe is `index-agg`.

How to do it...

To execute a top hit aggregation, we will perform the following steps:

1. Aggregate the document hits by the `tags` tag and return only the `name` field of the document with the maximum age (`top_tag_hits`). The search and aggregation will be executed with the following command:

```
POST /index-agg/_search?size=0
{ "aggs": {
    "tags": {
      "terms": { "field": "tag", "size": 2 },
      "aggs": {
        "top_tag_hits": {
          "top_hits": {
            "sort": [{ "age": { "order": "desc" } } ],
            "_source": {
              "includes": [ "name" ]
            }, "size": 1 } } } } } }
```

2. The result returned by Elasticsearch, if everything is okay, should be as follows:

```
{ ... truncated...
  "aggregations" : {
    "tags" : {
      "doc_count_error_upper_bound" : 0,
      "sum_other_doc_count" : 2828,
      "buckets" : [
        { "key" : "laborum", "doc_count" : 31,
          "top_tag_hits" : {
            "hits" : {
```

```
            "total" : 31, "max_score" : null,
        "hits" : [
            {
                "_index" : "index-agg",
                "_id" : "730", "_score" : null,
                "_source" : { "name" : "Gladiator"},
                "sort" : [ 90] } ] } } },
    ... truncated...
```

How it works...

The top hit aggregation allows you to collect buckets of hits of another aggregation. It provides optional parameters to control the results slicing. They are as follows:

- from: This is the starting position of the hits in the bucket (default: 0).

- size: This is the hit bucket size (default: the parent bucket size).

- sort: This allows us to sort for different values (default: score). Its definition is similar to the search sort of *Chapter 5, Text and Numeric Queries.*

To control the returned hits, it's possible to use the same parameters that were used for the search, as follows:

- _source: This allows us to control the returned source. It can be disabled (false), partially returned (obj.*), or have multiple exclude or include rules. In the preceding example, we have returned only the name field, as follows:

```
"_source": {
    "include": [
        "name"
    ]
},
```

- highlighting: This allows us to define the fields and settings to be used for calculating a query abstract.

- stored_fields: This allows us to return stored fields.

- explain: This returns information on how the score is calculated for a particular document.

- version: This adds the version of a document in the results (default: false).

The top hit aggregation can be used for implementing a field collapsing feature, first using a `terms` aggregation on the field that we want to collapse, and then collecting the documents with a top hit aggregation.

See also

Refer to the *Executing a search* recipe in *Chapter 4*, *Exploring Search Capabilities*, for common parameters that can be used during a search.

Executing a matrix stats aggregation

Elasticsearch 5.x or above provided a special module called `aggs-matrix-stats` that automatically computes advanced statistics on several fields.

Getting ready

You need an up and running Elasticsearch installation, as we described in the *Downloading and installing Elasticsearch* recipe in *Chapter 1*, *Getting Started*.

To execute these commands, any HTTP client can be used, such as cURL (https://curl.haxx.se/), Postman (https://www.getpostman.com/), or similar. Using Kibana Console is recommended, as it provides code completion and better character escaping for Elasticsearch.

To correctly execute the following commands, you will need an index populated with the ch07/populate_aggregation.sh commands available in the online code.

The index used in this recipe is `index-agg`.

How to do it...

To execute a matrix stats aggregation, we will perform the following steps:

1. First, we will evaluate statistics related to price and age in our knowledge base. The search and aggregation will be executed with the following command:

```
POST /index-agg/_search?size=0
{ "aggs": {
    "matrixstats": {
      "matrix_stats": {
        "fields": [ "age", "price" ] } } } }
```

2. The result returned by Elasticsearch, if everything is okay, should be as follows:

```
{ ... truncated...
  "aggregations" : {
    "matrixstats" : {
      "doc_count" : 1000,
      "fields" : [
        { "name" : "price", "count" : 1000,
          "mean" : 50.29545117592628,
          "variance" : 834.2714234338575,
          "skewness" : -0.04757692114597182,
          "kurtosis" : 1.808483274482735,
          "covariance" : {
            "price" : 834.2714234338575,
            "age" : 2.523682208250993 },
          "correlation" : {
            "price" : 1.0,
            "age" : 0.003051775248782358 } },
    ... truncated...
```

How it works...

The matrix stats aggregation allows us to compute different metrics on numeric fields, as follows:

- count: This is the number of per field samples included in the calculation.

- mean: This is the average value for each field.

- variance: This is the per field measurement for how spread out the samples are from the mean.

- skewness: This is the per field measurement quantifying the asymmetric distribution around the mean.

- kurtosis: This is the per field measurement quantifying the shape of the distribution.

- covariance: This is a matrix that quantitatively describes how changes in one field are associated with another.

- correlation: This is the covariance matrix scaled to a range of -1 to 1, inclusive. It describes the relationship between field distributions. The higher the value of the correlation, the more the numeric fields are correlated.

The matrix stats aggregation is also a good code sample for developing custom aggregation plugins to extend the power of the aggregation framework of Elasticsearch.

Executing a geo bounds aggregation

It's a very common scenario to have a set of documents that match a query, and you need to know the box that contains them; the solution to this scenario is the geo bounds metric aggregation.

Getting ready

You need an up and running Elasticsearch installation, as we described in the *Downloading and installing Elasticsearch* recipe in *Chapter 1*, *Getting Started*.

To execute these commands, any HTTP client can be used, such as cURL (`https://curl.haxx.se/`), Postman (`https://www.getpostman.com/`), or similar. Using Kibana Console is recommended, as it provides code completion and better character escaping for Elasticsearch.

To correctly execute the following commands, you will need an index populated with the `ch07/populate_aggregation.sh` commands available in the online code.

The index used in this recipe is `index-agg`.

How to do it...

To execute geo bounds aggregations, we will perform the following steps:

1. Execute a query and calculate the geo bounds on the results using the following code:

```
POST /index-agg/_search?size=0
{ "aggs": {
    "box": {
      "geo_bounds": {
        "field": "position", "wrap_longitude": true
      } } } }
```

2.   The result returned by Elasticsearch, if everything is okay, should be as follows:

```
{ ... truncated...
  "aggregations" : {
    "box" : {
      "bounds" : {
        "top_left" : {
          "lat" : 89.97587876860052,
          "lon" : 0.7563168089836836 },
        "bottom_right" : {
          "lat" : -89.8060692474246,
          "lon" : -0.2987125888466835 } } } } }
```

How it works...

The geo bounds aggregation is a metric aggregation that is able to compute the box of all the documents that are in a bucket.

It allows you to use the following parameters:

- `field`: This is the field that contains the geo point of the document.

- `wrap_longitude`: This is an optional parameter that specifies whether the bounding box should be allowed to overlap the International Date Line (default `true`).

The returned box (square) is given by two geo points: the top-left and the bottom-right.

See also

Refer to the *Mapping a geopoint field* recipe in *Chapter 2, Managing Mapping*, to correctly define a GeoPoint field for executing geo aggregations.

Executing a geo centroid aggregation

If you have a lot of geo-localized events and you need to know the center of these events, the geo centroid aggregation allows you to compute this geopoint.

Common scenarios could be as follows:

- During Twitter monitoring for events (earthquakes or tsunamis, for example): to detect the center of the event by monitoring the first top *n* events tweets.

- Having documents that have coordinates: to find the common center of these documents.

Getting ready

You need an up and running Elasticsearch installation, as we described in the *Downloading and installing Elasticsearch* recipe in *Chapter 1, Getting Started*.

To execute the commands, any HTTP client can be used, such as cURL (https://curl.haxx.se/), Postman (https://www.getpostman.com/), or similar. Using Kibana Console is recommended, as it provides code completion and better character escaping for Elasticsearch.

To correctly execute the following commands, you will need an index populated with the ch07/populate_aggregation.sh commands available in the online code.

The index used in this recipe is index-agg.

How to do it...

To execute geo centroid aggregations, we will perform the following steps:

1. Execute a query and calculate geo_centroid on the results using the following code:

```
POST /index-agg/_search?size=0
{ "aggs": {
    "centroid": {
       "geo_centroid": { "field": "position" } } } }
```

2. The result returned by Elasticsearch, if everything is okay, should be as follows:

```
{ ... truncated...
  "aggregations" : {
    "centroid" : {
      "location" : {
        "lat" : 3.0941622890532017,
        "lon" : 0.5758556071668863 },
        "count" : 1000 } } }
```

How it works...

The geo centroid aggregation is a metric aggregation that can compute the geo point centroid of a bucket of documents. It allows you to define only a single parameter in `field` that contains the geo point of the document.

The returned result is a geo point that is the centroid of the document distribution. For example, if your document contains earthquake events, by using the geo centroid aggregation, you are able to compute the epicenter of the earthquake.

See also

Refer to the *Mapping a GeoPoint field* recipe in *Chapter 2*, *Managing Mapping*, to correctly define a GeoPoint field for executing geo aggregations.

Executing a geotile grid aggregation

Using Elasticsearch to show data on maps is a very common pattern between Elasticsearch users. One of the most commonly used map formats is the tile one, in which a map is split into several small square parts and when the render of a location is required, the tiles near the location are fetched by a server.

Apart from commercial solutions, *OpenStreetMap* (`https://www.openstreetmap.org/`) maps are the most used, and a lot of Kibana maps are based on their tile servers. OpenStreetMap is open source and you can easily provide your own tile server via a Docker (`https://switch2osm.org/serving-tiles/using-a-docker-container/`).

The geotile grid aggregation allows to return buckets of documents with geopoints or geoshapes in the standard map `tile` format used for cells "`{zoom}/{x}/{y}`".

Getting ready

You need an up and running Elasticsearch installation, as we described in the *Downloading and installing Elasticsearch* recipe in *Chapter 1*, *Getting Started*.

To execute the commands, any HTTP client can be used, such as cURL (`https://curl.haxx.se/`), Postman (`https://www.getpostman.com/`), or similar. Using Kibana Console is recommended, as it provides code completion and better character escaping for Elasticsearch.

To correctly execute the following commands, you will need an index populated with the ch07/populate_aggregation.sh commands available in the online code.

The index used in this recipe is index-agg.

How to do it...

To execute a geotile grid aggregation on the position field available in index-agg, we will perform the following steps:

1. Execute a query and calculate geotile_grid on the results with a zoom precision of 5 using the following code:

```
POST /index-agg/_search?size=0
{ "aggs": {
    "tiles": {
      "geotile_grid": {
        "field": "position", "precision": 5  } } } }
```

2. The result returned by Elasticsearch, if everything is okay, should be as follows:

```
{ ... truncated...
  "aggregations" : {
    "tiles" : {
      "buckets" : [
          { "key" : "5/2/11", "doc_count" : 7 },
          { "key" : "5/12/17", "doc_count" : 6 },
        ... truncated ...
```

3.   If we want to search the first value, "5/2/11", on OpenStreetMap maps, the address will be the following one: https://www.openstreetmap.org/#map=5/2/11.

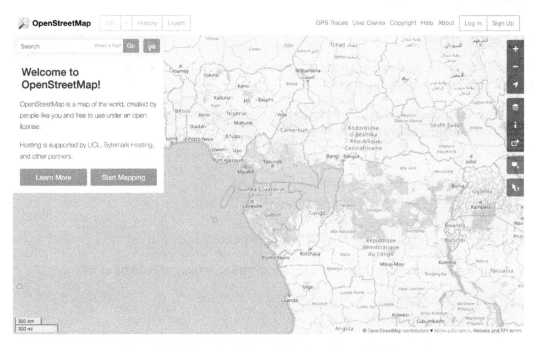

Figure 7.4 – Example of tile position from OpenStreetMap

How it works...

The geotile aggregation works on buckets of documents and returns all the tiles that contain matched documents.

The only mandatory parameter that must be provided is field, which is the geopoint mapped field to be used to compute the data.

For controlling the generated tiles, the other available parameters are as follows:

- precision (default 6): That is the level of the zoom from 0 to 29. The 29 value is a high precision value and covers about 10 cm x 10 cm of land. Generally, more commonly used values are between 6 and 12.

- size (default 10,000): That is the maximum number of tiles to return.

- bounds: A bounding box query that is used to filter the buckets to reduce the number of returned tiles. This is very useful for reducing the memory consumption of the aggregation.

See also

You can refer to the following links for further reference to what we covered in this recipe:

- The *Mapping a GeoPoint field* recipe in *Chapter 2*, *Managing Mapping*, to correctly define a GeoPoint field for executing geo aggregations

- *OpenStreetMap* maps project and documentation for a deep dive on managing maps (https://www.openstreetmap.org/)

- The official Elasticsearch documentation about the geotile grid aggregation at https://www.elastic.co/guide/en/elasticsearch/reference/current/search-aggregations-bucket-geotilegrid-aggregation.html

Executing a sampler aggregation

It's quite common to use a significant sample of documents to return the most important analytics/**Key Performance Indicators** (**KPI**) without computing the analytics on all the datasets.

This aggregation works with a query that can score the documents and, based on the score, it takes the *n* ranked first document to compute its sub-aggregations; from this behavior, it's called *sampler*.

Getting ready

You need an up and running Elasticsearch installation, as we described in the *Downloading and installing Elasticsearch* recipe in *Chapter 1*, *Getting Started*.

To execute these commands, any HTTP client can be used, such as curl (https://curl.haxx.se/), Postman (https://www.getpostman.com/), or similar. Using Kibana Console is recommended, as it provides code completion and better character escaping for Elasticsearch.

To correctly execute the following commands, you will need an index populated with the ch07/populate_aggregation.sh commands available in the online code.

The index used in this recipe is index-agg.

How to do it...

To execute a sampler aggregation, we will perform the following steps:

1. We execute a query that is mandatory to be able to sort the document on the score in the sampler, a sampler aggregation to keep 10 documents, and a sub-aggregation to compute the first three tags of the sampler, using the following code:

```
POST /index-agg/_search?size=0
{ "query": { "terms": { "tag": ["ex", "ullam"] } },
  "aggs": {
    "sample": { "sampler": { "shard_size": 10 },
      "aggs": {
        "keywords": {
          "terms": { "field": "tag",
            "exclude": [ "ex", "ullam" ], "size":3 }
        } } } } }
```

2. The result returned by Elasticsearch, if everything is okay, should be as follows:

```
{ ... truncated...
  "aggregations" : {
    "sample" : {
      "doc_count" : 10,
      "keywords" : {
        "doc_count_error_upper_bound" : 0,
        "sum_other_doc_count" : 26,
        "buckets" : [
          { "key" : "architecto", "doc_count" : 1 },
          { "key" : "atque", "doc_count" : 1 },
          { "key" : "deleniti", "doc_count" : 1 }
        ] } } } }
```

How it works...

The sampler aggregation works in the same way as the filter aggregation: it takes a bucket and filters out maximum *n* documents based on their score. The documents that pass this aggregation and that are available at its sub-aggregations are the most pertinent (high scores).

The only parameter that can be used to change the sampler is `shard_size` (default `100`), which defines the maximum number of documents to collect for a single shard.

Executing a pipeline aggregation

Elasticsearch allows you to define aggregations that are a mix of the results of other aggregations (for example, by comparing the results of two metric aggregations); these are pipeline aggregations.

They are very common when you need to compute results from different aggregations, such as statistics on results.

Getting ready

You need an up and running Elasticsearch installation, as we described in the *Downloading and installing Elasticsearch* recipe in *Chapter 1, Getting Started*.

To execute the commands, any HTTP client can be used, such as cURL (https://curl.haxx.se/), Postman (https://www.getpostman.com/), or similar. Using Kibana Console is recommended, as it provides code completion and better character escaping for Elasticsearch.

To correctly execute the following commands, you will need to create `index-pipagg` with the following command:

```
PUT /index-pipagg
{ "mappings": {
    "properties": {
       "type": { "type": "keyword" },
       "date": { "type": "date" } } } }
```

Then, populate it with some documents, as follows:

```
PUT /_bulk
{"index":{"_index":"index-pipagg"}}
{"date": "2022-01-01", "price": 200, "promoted": true,
"rating": 1, "type": "hat"}
{"index":{"_index":"index-pipagg"}}
{"date": "2022-01-01", "price": 200, "promoted": true,
"rating": 1, "type": "t-shirt"}
{"index":{"_index":"index-pipagg"}}
{"date": "2022-01-01", "price": 150, "promoted": true,
```

```
"rating": 5, "type": "bag"}
{"index":{"_index":"index-pipagg"}}
{"date": "2022-02-01", "price": 50, "promoted": false,
"rating": 1, "type": "hat"}
{"index":{"_index":"index-pipagg"}}
{"date": "2022-02-01", "price": 10, "promoted": true, "rating":
4, "type": "t-shirt"}
{"index":{"_index":"index-pipagg"}}
{"date": "2022-03-01", "price": 200, "promoted": true,
"rating": 1, "type": "hat"}
{"index":{"_index":"index-pipagg"}}
{"date": "2022-03-01", "price": 175, "promoted": false,
"rating":2, "type": "t-shirt"}
```

How to do it...

To execute a pipeline aggregation, we will perform the following steps:

1. Execute a query and calculate a composed aggregation that will divide the sales across the month, and for every month, we will compute the incoming `price`. To get the extended aggregation on these sales, we will execute the following code:

```
POST /index-pipagg/_search?size=0
{ "aggs" : {
        "sales_per_month" : {
            "date_histogram" : {
                "field" : "date",
                "calendar_interval" : "month"
            },
            "aggs": {
                "sales": {
                    "sum": { "field": "price" }}}},
        "stats_monthly_sales": {
            "stats_bucket": {
                "buckets_path": "sales_per_month>sales" }
} } }
```

The result returned by Elasticsearch, if everything is okay, should be as follows:

```
{ ... truncated...
  "aggregations" : {
    "sales_per_month" : {
      "buckets" : [
        {"key_as_string" : "2022-01-01T00:00:00.000Z",
          "key" : 1640995200000, "doc_count" : 3,
          "sales" : { "value" : 550.0 } },
        {"key_as_string" : "2022-02-01T00:00:00.000Z",
          "key" : 1643673600000, "doc_count" : 2,
          "sales" : { "value" : 60.0 } },
        {"key_as_string" : "2022-03-01T00:00:00.000Z",
          "key" : 1646092800000, "doc_count" : 2,
          "sales" : { "value" : 375.0 } } ] },
    "stats_monthly_sales" : {
      "count" : 3, "min" : 60.0, "max" : 550.0,
      "avg" : 328.3333333333333, "sum" : 985.0 } } }
```

How it works...

The pipeline aggregation can compute an aggregation based on another one. You can consider the pipeline aggregation similar to a metric (see the *Executing a stats aggregation* recipe in this chapter) that is working on the results of other aggregations.

The most commonly used types of pipeline aggregation are as follows:

- `avg_bucket`: Used to compute the average of parent aggregations.
- `derivative`: Used to compute the derivative of parent aggregations.
- `max_bucket`: Used to compute the maximum of related aggregations.
- `min_bucket`: Used to compute the minimum of related aggregations.
- `sum_bucket`: Used to compute the sum of related aggregations.
- `stats_bucket`: Used to compute the statistics of related aggregations.
- `extended_stats_bucket`: Used to compute the statistics of related aggregations.

- `percentile_bucket`: Used to compute the percentile of related aggregations.
- `moving_fn`: A moving function, which is used to compute the percentile of related aggregations.
- `cumulative_sum`: Used to compute the derivative of parent aggregations.
- `bucket_script`: Used to define the operation between related aggregations. This is the most powerful one if you need to customize complex value computations between aggregation metrics.
- `bucket_select`: Used to filter out parent bucket aggregation.
- `bucket_sort`: Used to sort parent bucket aggregation.

Every pipeline type aggregation has additional parameters that relate to how the metric is computed; the online official documentation covers all the corner cases of these usages.

See also

The official Elasticsearch documentation on pipeline aggregations at `https://www.elastic.co/guide/en/elasticsearch/reference/master/search-aggregations-pipeline.html`

8
Scripting in Elasticsearch

Elasticsearch has a powerful way of extending its capabilities by using custom scripts, which can be written in several programming languages. The most common ones are Painless, Express, and Mustache. In this chapter, we will explore how it's possible to create custom scoring algorithms, specially processed return fields, custom sorting, complex update operations on records, and ingest processors. The scripting concept of Elasticsearch is an advanced stored-procedure system in the NoSQL world; due to this, every advanced user of Elasticsearch should learn how to master it.

Elasticsearch natively provides scripting in Java (that is, Java code compiled in JAR files), Painless, Express, and Mustache; however, a lot of other interesting languages are also available as plugins, such as Kotlin and Velocity. In older Elasticsearch releases, prior to version 5.0, the official scripting language was **Groovy**. But, for better sandboxing and performance, the official language is now **Painless**, which is provided in Elasticsearch by default.

In this chapter, we will cover the following recipes:

- Painless scripting
- Installing additional scripting languages
- Managing scripts

- Sorting data using scripts

- Computing return fields with scripting

- Filtering a search using scripting

- Using scripting in aggregations

- Updating a document using scripts

- Reindexing with a script

- Scripting in ingest processors

Painless scripting

Painless is a simple, secure scripting language that is available in Elasticsearch by default. It was designed by the Elasticsearch team to be used specifically with Elasticsearch and can safely be used with inline and stored scripting. Its syntax is similar to Groovy, from which it was originally born.

In this recipe, we will see how to create a custom score function in Painless.

Getting ready

You will need an up-and-running Elasticsearch installation, similar to the one that we described in the *Downloading and installing Elasticsearch* recipe in *Chapter 1, Getting Started*.

In order to execute the commands, any HTTP client can be used, such as cURL (https://curl.haxx.se/) or Postman (https://www.getpostman.com/). You can use Kibana Console as it provides code completion and better character escaping for Elasticsearch.

To correctly execute the following commands, you will need an index that is populated with the ch07/populate_aggregation.txt commands – these are available in the online code.

To be able to use regular expressions in Painless scripting, you will need to activate them in elasticsearch.yml by adding the following code:

```
script.painless.regex.enabled: true
```

The index used in this recipe is the index-agg index.

How to do it...

We'll use Painless scripting to compute the scoring by performing the following steps:

1. We can execute a search with a scripting function in Kibana using the following code:

```
POST /index-agg/_search
{ "query": {
    "function_score": {
      "script_score": {
        "script": {
          "lang": "painless",
          "source": "doc['price'].value * 1.2" }}}}}
```

If everything works correctly, the result will be as follows:

```
{ ... truncated...
  "hits" : {
    "total" : 1000, "max_score" : 119.97963,
    "hits" : [
        { "_index" : "index-agg", "_id" : "857",
          "_score" : 119.97963,
          "_ignored" : ["description.keyword"],
          "_source" : { ... truncated... },
        { "_index" : "index-agg", "_id" : "136",
          "_score" : 119.90164,
          "_ignored" : ["description.keyword"],
          "_source" : { ... truncated...
```

You can see that, in this case, using the scripting on only the price field has the same meaning as sorting by price!

How it works...

Painless is a scripting language that was developed for Elasticsearch for rapid data processing and security (it is sandboxed in order to prevent a malicious code injection). The syntax is based on Groovy, and it's provided by default in every installation.

Elasticsearch processes the scripting language in two steps:

1. The script code is compiled in an object to be used in a script call; if the scripting code is invalid, then an exception is raised.

2. For each element, the script is called, and the result is collected; if the script fails on some elements, then the search or computation may fail.

> **Note**
> Using scripting is a powerful Elasticsearch functionality, but it costs a lot in terms of memory and CPU cycles. The best practice, if possible, is to optimize the indexing of data to search or aggregate – and to avoid using scripting completely.

The method that is used to define a script in Elasticsearch is always the same; the script is contained in a `script` object using the following format:

```
"script": {
    "lang":     "...",
    "source" | "id": "...",
    "params": { ... }
}"
```

It accepts several parameters, as follows:

- `source`/`id`: These are the references for the script, which can be defined as follows:

  - `source`: If the string with your script code is provided with the call.

  - `id`: If the script is stored in the cluster, then it refers to the `id` parameter, which is used to save the script in the cluster.

- `params` (an optional JSON object): This defines the parameters to be passed, which are, in the context of scripting, available using the `params` variable.

- `lang` (the default is `painless`): This defines the scripting language that is to be used.

For complex scripts that contain the " special character in the text, I suggest using Kibana and a triple " to escape the script text (similar to the Python and Scala special """"). In this way, you can improve the readability of your script code. Otherwise, if you want to use curl or other tools, you need to escape the " with \".

There's more...

Painless is the preferred choice if the script is not too complex; otherwise, a native plugin provides a better environment in order to implement complex logic and data management.

For accessing document properties in Painless scripts, the same approach works as with other scripting languages:

- `doc._score`: This stores the document score. It's generally available in searching, sorting, and aggregations.

- `doc._source`: This allows access to the source of the document. Use it wisely because it requires the entire source to be fetched and it's very CPU- and memory-intensive.

- `_fields['field_name'].value`: This allows you to load the value from the stored field (in mapping, the field has the `stored:true` parameter).

- `doc['field_name']`: This extracts the document field from the `doc` values of the field. In Elasticsearch, the `doc` values are automatically stored for every field that is not of the `text` type.

- `doc['field_name'].value`: This extracts the value of the `field_name` field from the document. If the value is an array, or if you want to extract the value as an array, then you can use `doc['field_name'].values`.

- `doc['field_name'].empty`: This returns `true` if the `field_name` field has no value in the document.

- `doc['field_name'].multivalue`: This returns `true` if the `field_name` field contains multiple values.

For performance, the fastest access method for a field value is through the `doc` value, then the stored field, and, finally, from the source.

If the field contains a geopoint value, then additional methods are available, such as the following:

- `doc['field_name'].lat`: This returns the latitude of a geopoint value. If you need the value as an array, then you can use `doc['field_name'].lats`.

- `doc['field_name'].lon`: This returns the longitude of a geopoint value. If you need the value as an array, then you can use `doc['field_name'].lons`.

- `doc['field_name'].distance(lat,lon)`: This returns the plane distance in miles from a latitude/longitude point.

- `doc['field_name'].arcDistance(lat,lon)`: This returns the arc distance in miles, which is given as a latitude/longitude point.

- `doc['field_name'].geohashDistance(geohash)`: This returns the distance in miles, which is given as a geohash value.

By using these helper methods, it is possible to create advanced scripts in order to boost a document by a distance, which can be very handy for developing geospatial-centered applications.

See also

You can refer to the following URLs for further reference, which are related to this recipe:

- The official announcement by Jack Conradson about the Painless development at `https://www.elastic.co/blog/painless-a-new-scripting-language`

- The Elasticsearch official page of Painless at `https://www.elastic.co/guide/en/elasticsearch/painless/master/painless-guide.html` to learn about its main functionalities

- The Painless syntax reference at `https://www.elastic.co/guide/en/elasticsearch/painless/master/painless-lang-spec.html` to learn about Painless powerful syntax

Installing additional scripting languages

Elasticsearch provides native scripting (that is, Java code compiled in JAR files) and Painless, but a lot of other interesting languages are also available, such as Kotlin.

> **Note**
>
> At the time of writing this book, there are no available language plugins as part of Elasticsearch's official ones. Usually, plugin authors will take a week or up to a month to update their plugins to the new version after a major release. This section will be a reference for this use case based on Elasticsearch 7.x. As previously stated, the official language is now Painless, and this is provided by default in Elasticsearch for better sandboxing and performance.

Getting ready

You will need an up-and-running Elasticsearch installation, similar to the one that we described in the *Downloading and installing Elasticsearch* recipe in *Chapter 1, Getting Started*.

How to do it...

In order to install the Kotlin language support for Elasticsearch, we will perform the following steps:

1. From the command line, simply call the following command:

```
bin/elasticsearch-plugin install lang-kotlin
```

It will print the following output:

```
-> Downloading lang-kotlin from elastic
[=================================================]
100%??
@@@@@@@@@@@@@@@@@@@@@@@@@@@@@@@@@@@@@@@@@@@@@@@@@@@@@@@@@@@@
@@
@ WARNING: plugin requires additional permissions @
@@@@@@@@@@@@@@@@@@@@@@@@@@@@@@@@@@@@@@@@@@@@@@@@@@@@@@@@@@@@
@@
* java.lang.RuntimePermission createClassLoader
* org.elasticsearch.script.ClassPermission <<STANDARD>>
* org.elasticsearch.script.ClassPermission
org.mozilla.javascript.ContextFactory
* org.elasticsearch.script.ClassPermission
org.mozilla.javascript.Callable
* org.elasticsearch.script.ClassPermission
org.mozilla.javascript.NativeFunction
* org.elasticsearch.script.ClassPermission
org.mozilla.javascript.Script
* org.elasticsearch.script.ClassPermission
org.mozilla.javascript.ScriptRuntime
* org.elasticsearch.script.ClassPermission
org.mozilla.javascript.Undefined
org.mozilla.javascript.optimizer.OptRuntime
See http://docs.oracle.com/javase/8/docs/technotes/
guides/security/permissions.html
for descriptions of what these permissions allow and the
associated risks.
Continue with installation? [y/N]y
-> Installed lang-javascript
```

If the installation is successful, the output will end with `Installed`; otherwise, an error is returned. You can refer to http://docs.oracle.com/javase/8/ docs/technotes/guides/security/permissions.html for descriptions of what these permissions allow and their associated risks.

2. Restart your Elasticsearch server to check that the scripting plugins are loaded:

```
[...][INFO ][o.e.n.Node ] [] initializing ...
```

How it works...

Language plugins allow an extension for the number of supported languages that can be used in scripting. During installation, they require special permissions in order to access classes and methods that are banned by the Elasticsearch security layer, such as access to `ClassLoader` or class permissions.

During the Elasticsearch startup, an internal Elasticsearch service, known as `PluginService`, loads all the installed language plugins.

Installing or upgrading a plugin requires the restarting of a node.

From version 7.x, all the plugins have the same version of Elasticsearch.

The Elasticsearch community provides a number of common scripting languages (a full list is available on the Elasticsearch site plugin page at http://www.elastic.co/ guide/en/elasticsearch/reference/current/modules-plugins.html), while other languages are available in the GitHub repository (a simple search on GitHub will allow you to find them).

There's more...

The plugin manager that we used to install the plugins also provides the following commands:

*   `list`: This command is used to list all the installed plugins.

    For example, you can execute the following command:

    ```
    bin/elasticsearch-plugin list
    ```

    The result of the preceding command will be as follows:

    ```
    lang-kotlin@7.0.0
    ```

- `remove`: This command is used to remove an installed plugin.

  For example, you can execute the following command:

  ```
  bin/elasticsearch-plugin remove lang-kotlin
  ```

  The result of the preceding command will be as follows:

  ```
  -> Removing lang-kotlin...
  ```

Managing scripts

Depending on your scripting usage, there are several ways of customizing Elasticsearch in order to use your script extensions.

In this recipe, we will demonstrate how you can manage scripts by storing them in Elasticsearch or providing them inline in API calls.

Getting ready

You will need an up-and-running Elasticsearch installation, similar to the one that we described in the *Downloading and installing Elasticsearch* recipe in *Chapter 1, Getting Started*.

In order to execute the commands, any HTTP client can be used, such as cURL (`https://curl.haxx.se/`) or Postman (`https://www.getpostman.com/`). You can use Kibana Console, as it provides code completion and better character escaping for Elasticsearch.

To correctly execute the following commands, you will need an index populated with the `ch08/populate_aggregation.txt` commands – these are available in the online code.

In order to be able to use regular expressions in Painless scripting, you will need to activate them in `elasticsearch.yml` by adding the following code:

```
script.painless.regex.enabled: true
```

The index used in this recipe is the `index-agg` index.

How to do it...

To store a script in Elasticsearch, we will perform the following steps:

1. Dynamic scripting (except Painless) is disabled by default for security reasons. We need to activate it in order to use dynamic scripting languages, such as JavaScript and Python. To do this, we need to enable the scripting flags in the Elasticsearch configuration file (`config/elasticseach.yml`), and then restart the cluster:

    ```
    script.inline: true
    ```

2. If the dynamic script is enabled (as done in the first step), Elasticsearch lets us store the scripts in a special part of the cluster state, "`_scripts`". In order to put `my_script` in the cluster state, execute the following code:

    ```
    POST /_scripts/my_script
    { "script": {
        "source": """doc['price'].value * params.factor""",
    "lang":"painless" } }
    ```

3. The script can be used by simply referencing it in the `script/id` field:

    ```
    POST /index-agg/_search?size=2
    { "sort": {
        "_script": {
          "script": {
            "id": "my_script",
            "params": { "factor": 1.2 } },
          "type": "number",
          "order": "desc" } },
      "_source": {
        "includes": [ "price" ] } }
    ```

How it works...

Elasticsearch permits different ways in which to load your script, and each approach has its pros and cons. The scripts can also be available in the special `_script` cluster state. The REST endpoints are as follows:

```
POST http://<server>/_scripts/<id> (to retrieve a script)
PUT http://<server>/_scripts/<id> (to store a script)
DELETE http://<server>/_scripts/<id> (to delete a script)
```

The stored script can be referenced in the code using `"script":{"id": "id_of_the_script"}`.

The following sections of the book will use inline scripting because it's easier to use during the development and testing phases.

Generally, a good workflow is to do the development using inline dynamic scripting on request – this is because it's faster to prototype. Once the script is ready and no more changes are required, then it should be stored in the index, so that it will be simpler to call and manage in all the cluster nodes. In production, the best practice is to disable dynamic scripting and store the script on a disk (by dumping the indexed script to the disk) to improve security.

When you are storing files on a disk, pay attention to the file extension. The following table summarizes the status of the plugins:

Language	Provided as	File extension	Status
Painless	Built-in/module	painless	Default
Expression	Built-in/module	expression	Deprecated
Mustache	Built-in/module	mustache	Default

Table 8.1 – Default scripting languages in Elasticsearch

The other scripting parameters that can be set in `config/elasticsearch.yml` are as follows:

- `script.max_compilations_per_minute` (the default is `25`): This default scripting engine has a global limit for how many compilations can be done per minute. You can change this to a higher value, for example, `script.max_compilations_per_minute: 1000`.

- `script.cache.max_size` (the default is `100`): This defines how many scripts are cached; it depends on context, but, in general, it's better to increase this value.

- `script.max_size_in_bytes` (the default is `65535`): This defines the maximum text size for a script; for large scripts, the best practice is to develop native plugins.

- `script.cache.expire` (the default is disabled): This defines a time-based expiration for the cached scripts.

There's more...

In the preceding example, we activated Elasticsearch scripting for all the engines, but Elasticsearch provides fine-grained settings in order to control them.

In Elasticsearch, the scripting can be used in the following different contexts:

Context	Description
Aggs	Aggregations
Search	The search API, percolator API, and suggest API
Update	The update API
Ingester	In an ingest processor
Plugin	The special scripts under the generic plugin category

Table 8.2 – Contexts that can be used in scripting

Here, scripting is enabled by default for all the contexts.

You can disable all the scripting by setting the `script.allowed_contexts: none` value in `elasticsearch.yml`.

To activate the scripting only for `update` and `search`, you can use `script.allowed_contexts: search, update`.

For more fine-grained control, scripting can be controlled by the active type of scripting using the `elasticsearch.yml` entry, `script.allowed_types`.

In order to only enable inline scripting, we can use the following command:

```
script.allowed_types: inline
```

See also

You can refer to the following URLs for further reference, which are related to this recipe:

- The scripting page on the Elasticsearch website at `https://www.elastic.co/guide/en/elasticsearch/reference/current/modules-scripting.html` for general information about scripting

- The scripting security page at `https://www.elastic.co/guide/en/elasticsearch/reference/current/modules-scripting-security.html` for other borderline cases on security management

Sorting data using scripts

Elasticsearch provides scripting support for sorting functionality. In real-world applications, there is often a need to modify the default sorting using an algorithm that is dependent on the context and some external variables. Some common scenarios are as follows:

- Sorting places near a point
- Sorting by most read articles
- Sorting items by custom user logic
- Sorting items by revenue

Because the computing of scores on a large dataset is very CPU-intensive, if you use scripting, then it's better to execute it on a small dataset using standard score queries for detecting the top documents, and then execute a rescoring on the top subset.

Getting ready

You will need an up-and-running Elasticsearch installation, similar to the one that we described in the *Downloading and installing Elasticsearch* recipe in *Chapter 1, Getting Started*.

To execute the commands, any HTTP client can be used, such as cURL (`https://curl.haxx.se/`) or Postman (`https://www.getpostman.com/`). You can use Kibana Console, as it provides code completion and better character escaping for Elasticsearch.

To correctly execute the following commands, you will need an index that is populated with the `ch07/populate_aggregation.txt` commands – these are available in the online code.

To be able to use regular expressions in Painless scripting, you will need to activate them in `elasticsearch.yml` by adding `script.painless.regex.enabled: true`.

The index used in this recipe is the `index-agg` index.

How to do it...

For sorting using a custom script, we will perform the following steps:

1. If we want to order our documents by the `price` field multiplied by a `factor` parameter (that is, sales tax), then the search will be as follows:

```
POST /index-agg/_search?size=3
{ "sort": {
    "_script": {
        "script": {
            "source": """
Math.sqrt(doc["price"].value *
params.factor)
""",
            "params": { "factor": 1.2  } },
        "type": "number", "order": "desc"  } },
    "_source": { "includes": [ "price" ] } }
```

Here, we have used a `sort` script; in real-world applications, the documents that are to be sorted should not have a high cardinality.

2. If everything's correct, then the result that is returned by Elasticsearch will be as follows:

```
{ ... truncated...
  "hits" : {
    "total" : 1000, "max_score" : null,
    "hits" : [
      { "_index" : "index-agg", "_id" : "857",
        "_ignored" : ["description.keyword"],
        "_score" : null,
        "_source" : { "price" : 99.98302508488757 },
        "sort" : [ 10.953521329536066 ] },
... truncated...
```

How it works...

The `sort` parameter, which we discussed in *Chapter 4*, *Exploring Search Capabilities*, can be extended with the help of scripting.

The `sort` scripting allows you to define several parameters, such as the following:

- `order (default "asc") ("asc" or "desc")`: This determines whether the order must be ascending or descending.

- `type`: This defines the type in order to convert the value.

- `script`: This contains the script object that is to be executed.

Extending the `sort` parameter using scripting allows you to use a broader approach to score your hits.

Elasticsearch scripting permits the use of any code that you want to use; for instance, you can create custom complex algorithms for scoring your documents.

There's more...

Painless provides a lot of built-in functions (mainly taken from the Java `Math` class) that can be used in scripts, such as the following:

Function	Description
`time()`	This is the current time in milliseconds
`sin(a)`	This returns the trigonometric sine of an angle
`cos(a)`	This returns the trigonometric cosine of an angle
`tan(a)`	This returns the trigonometric tangent of an angle
`asin(a)`	This returns the arc sine of a value
`acos(a)`	This returns the arc cosine of a value
`atan(a)`	This returns the arctangent of a value
`toRadians(angdeg)`	This converts an angle that is measured in degrees to an approximately equivalent angle that is measured in radians
`toDegrees(angrad)`	This converts an angle that is measured in radians to an approximately equivalent angle that is measured in degrees
`exp(a)`	This returns Euler's number raised to the power of a value
`log(a)`	This returns the natural logarithm (base e) of a value
`log10(a)`	This returns the base 10 logarithm of a value
`sqrt(a)`	This returns the correctly rounded positive square root of a value
`cbrt(a)`	This returns the cube root of a double value
`IEEEremainder (f1, f2)`	This computes the remainder operation on two arguments as prescribed by the IEEE 754 standard

Function	Description
`ceil(a)`	This returns the smallest (closest to negative infinity) value that is greater than or equal to the argument and is equal to a mathematical integer
`floor(a)`	This returns the largest (closest to positive infinity) value that is less than or equal to the argument and is equal to a mathematical integer
`rint(a)`	This returns the value that is closest in value to the argument and is equal to a mathematical integer
`atan2(y, x)`	This returns the angle, theta, from the conversion of rectangular coordinates (x, y_) to polar coordinates (r, _theta)
`pow(a, b)`	This returns the value of the first argument raised to the power of the second argument
`round(a)`	This returns the closest integer to the argument
`random()`	This returns a random double value
`abs(a)`	This returns the absolute value of a value
`max(a, b)`	This returns the greater of two values
`min(a, b)`	This returns the smaller of two values
`ulp(d)`	This returns the size of the unit in the last place of the argument
`signum(d)`	This returns the signum function of the argument
`sinh(x)`	This returns the hyperbolic sine of a value
`cosh(x)`	This returns the hyperbolic cosine of a value
`tanh(x)`	This returns the hyperbolic tangent of a value
`hypot(x,y)`	This returns sqrt(x2+y2) without intermediate overflow or underflow

Table 8.3 – List of most common scripting functions

If you want to retrieve records in a random order, then you can use a script with a random method, as shown in the following code:

```
POST /index-agg/_search?&size=2
{ "sort": {
    "_script": {
      "script": { "source": "Math.random()" },
      "type": "number", "order": "asc" } } }
```

In this example, for every hit, the new sort value is computed by executing the `Math.random()` scripting function.

Computing return fields with scripting

Elasticsearch allows us to define custom complex expressions that can be used to return a newly calculated field value.

The most common scenarios for these use cases are as follows:

- Merge field values (that is, *first name + last name*)
- Compute values (that is, *total=quantity*price*)
- Apply transformations (that is, convert dollars to euros, string manipulation)

These special fields are called `script_fields`, and they can be expressed with a script in every available Elasticsearch scripting language.

Getting ready

You will need an up-and-running Elasticsearch installation, similar to the one that we described in the *Downloading and installing Elasticsearch* recipe in *Chapter 1, Getting Started*.

To execute the commands, any HTTP client can be used, such as cURL (`https://curl.haxx.se/`) or Postman (`https://www.getpostman.com/`). You can use Kibana Console, as it provides code completion and better character escaping for Elasticsearch.

In order to correctly execute the following commands, you will need an index that is populated with the `ch07/populate_aggregation.txt` commands – these are available in the online code.

To be able to use regular expressions in Painless scripting, you need to activate them in `elasticsearch.yml` by adding `script.painless.regex.enabled: true`.

The index used in this recipe is the `index-agg` index.

How to do it...

For computing return fields with scripting, we will perform the following steps:

1. Return the following script fields:

    *   `"my_calc_field"`: This concatenates the texts of the `"name"` and `"description"` fields.

    *   `"my_calc_field2"`: This multiplies the `"price"` value by the `"discount"` parameter.

2. From the command line, we will execute the following code:

```
POST /index-agg/_search?size=2
{ "script_fields": {
    "my_calc_field": {
      "script": {
        "source": """params._source.name + " -- " +
params._source.description""" } },
    "my_calc_field2": {
      "script": {
        "source": """doc["price"].value * params.
discount""",
        "params": { "discount": 0.8 } } } } }
```

If everything is all right, then the result that is returned by Elasticsearch will be as follows:

```
{ ... truncated...
  "hits" : {
    "total" : 1000,
    "max_score" : 1.0,
    "hits" : [
      { "_index" : "index-agg", "_id" : "1",
        "_score" : 1.0,
        "_ignored" : ["description.keyword"],
        "fields" : {
          "my_calc_field" : [
            "Valkyrie -- ducimus nobis harum doloribus
voluptatibus libero nisi omnis officiis exercitationem
amet odio odit dolor perspiciatis minima quae voluptas
```

```
        dignissimos facere ullam tempore temporibus laboriosam
        ad doloremque blanditiis numquam placeat accusantium
        at maxime consectetur esse earum velit officia dolorum
        corporis nemo consequatur perferendis cupiditate eum
        illum facilis sunt saepe"
               ],
                "my_calc_field2" : [
                    15.696847534179689
               ] } },
            { "_index" : "index-agg", "_id" : "2",
                "_score" : 1.0,
                "_ignored" : ["description.keyword"],
                "fields" : {
                    "my_calc_field" : [
                        "Omega Red -- quod provident sequi rem
        placeat deleniti exercitationem veritatis quasi
        accusantium accusamus autem repudiandae"
                   ],
                    "my_calc_field2" : [
                        56.201733398437504
               ] } } ] } }
```

How it works...

The script fields are similar to executing a SQL function on a field during a select. In Elasticsearch, after a search phase is executed and the hits to be returned are calculated, if some fields (standard or script) are defined, then they are calculated and returned.

The script field, which can be defined using all supported languages, is processed by passing a value to the source of the document and, if some other parameters are defined in the script (such as the discount factor), then they are passed to the script function.

The script function is a code snippet, so it can contain everything that the language allows to be written; however, it must be evaluated to a value (or a list of values).

See also

You can refer to the following recipes for further reference:

- The *Installing additional script plugins* recipe in this chapter to install additional languages for scripting

- The *Sorting data using scripts* recipe in this chapter for a reference to extra built-in functions for Painless scripts

Filtering a search using scripting

In *Chapter 4, Exploring Search Capabilities,* we explored many filters. Elasticsearch scripting allows the extension of a traditional filter by using custom scripts.

Using scripting to create a custom filter is a convenient way to write scripting rules that are not provided by Lucene or Elasticsearch, and to implement business logic that is not available in a DSL query.

Getting ready

You will need an up-and-running Elasticsearch installation, similar to the one that we described in the *Downloading and installing Elasticsearch* recipe in *Chapter 1, Getting Started.*

To execute the commands, any HTTP client can be used, such as cURL (`https://curl.haxx.se/`) or Postman (`https://www.getpostman.com/`). You can use Kibana Console, as it provides code completion and better character escaping for Elasticsearch.

To correctly execute the following commands, you will need an index that is populated with the `ch07/populate_aggregation.txt` commands – these are available in the online code.

In order to be able to use regular expressions in Painless scripting, you will need to activate them in `elasticsearch.yml` by adding `script.painless.regex.enabled: true`.

The index used in this recipe is the `index-agg` index.

How to do it...

For filtering results using a custom script, we will perform the following steps:

1. We'll write a search using a filter that filters out a document with a price value that is less than a parameter value:

```
POST /index-agg/_search?pretty&size=3
{ "query": {
    "bool": {
```

```
        "filter": {
            "script": {
                "script": {
                    "source": """doc['price'].value > params.
    param1""",
                    "params": { "param1": 80 } } } } },
        "_source": { "includes": [ "name", "price" ] } }
```

In this example, all the documents where the age value is greater than `param1` are taken as qualified for return.

2. This script filter is done for demonstration purposes; in real-world applications, it can be replaced with a `range` query, which is much faster.

If everything is correct, the result that is returned by Elasticsearch will be as follows:

```
{ ... truncated...
  "hits" : {
    "total" : 190, "max_score" : 0.0,
    "hits" : [
      { "_index" : "index-agg",  "_id" : "7",
        "_score" : 0.0,
        "_ignored" : ["description.keyword"],
        "_source" : {
          "price" : 86.65705393127125,
          "name" : "Bishop" } },
      { "_index" : "index-agg", "_id" : "14",
        "_score" : 0.0,
        "_ignored" : ["description.keyword"],
        "_source" : {
          "price" : 84.9516714617024,
          "name" : "Crusader" } },
      ... truncated...
```

How it works...

The script filter is a language script that returns a Boolean value (`true` or `false`). For every hit, the script is evaluated and, if it returns `true`, then the hit passes the filter. This type of scripting can only be used as Lucene filters and not as queries because it doesn't affect the score of the search.

The script code can be any code in your preferred supported scripting language that returns a Boolean value.

See also

You can refer to the following recipes for further reference:

- The *Installing additional script plugins* recipe in this chapter to install additional languages for scripting
- The *Sorting data using script* recipe for a reference to extra built-in functions that are available for Painless scripts

Using scripting in aggregations

Scripting can be used in aggregations for extending its analytics capabilities to manipulate and transform the values used in metric aggregations or to define new rules to create buckets.

Getting ready

You will need an up-and-running Elasticsearch installation, similar to the one that we described in the *Downloading and installing Elasticsearch* recipe in *Chapter 1, Getting Started*.

To execute the commands, any HTTP client can be used, such as cURL (`https://curl.haxx.se/`) or Postman (`https://www.getpostman.com/`). You can use Kibana Console, as it provides code completion and better character escaping for Elasticsearch.

To correctly execute the following commands, you will need an index that is populated with the `ch07/populate_aggregation.txt` commands – these are available in the online code.

In order to be able to use regular expressions in Painless scripting, you will need to activate them in `elasticsearch.yml` by adding `script.painless.regex.enabled: true`.

The index used in this recipe is the `index-agg` index.

How to do it...

For using a scripting language in an aggregation, we will perform the following steps:

1. Write a metric aggregation that selects the field using `script`:

```
POST /index-agg/_search?size=0
{ "aggs": {
    "my_value": {
        "sum": {
            "script": {
                "source": """doc["price"].value * doc["price"].
value""" } } } } }
```

If everything is correct, then the result that is returned by Elasticsearch will be as follows:

```
{ ... truncated...
    "hits" : {
        "total" : 1000, "max_score" : null, "hits" : [] },
    "aggregations" : {
        "my_value" : { "value" : 3363069.561000406 } } }
```

2. Then, write a metric aggregation that uses the value field using `script`:

```
POST /index-agg/_search?size=0
{ "aggs": {
    "my_value": {
        "sum": { "field": "price",
            "script": {"source": "_value * _value" }}}}}
```

If everything is correct, then the result that is returned by Elasticsearch will be as follows:

```
{ ... truncated...,
    "aggregations" : {
        "my_value" : { "value" : 3363069.561000406 } } }
```

3. Again, write a term bucket aggregation that changes the terms using `script`:

```
POST /index-agg/_search?size=0
{ "aggs": {
    "my_value": {
```

```
        "terms": { "field": "tag", "size": 5,
            "script": {
                "source": """
if(params.replace.containsKey(_value.toUpperCase())) {
    params.replace[_value.toUpperCase()]
} else {
    _value.toUpperCase()
}
""",
                "params": {
                    "replace": {
                        "LABORUM": "Result1",
                        "MAIORES": "Result2",
                        "FACILIS": "Result3" } } } } } } }
```

If everything is correct, then the result that is returned by Elasticsearch will be as follows:

```
{ ... truncated...
    "aggregations" : {
        "my_value" : {
            "doc_count_error_upper_bound" : 0,
            "sum_other_doc_count" : 2755,
            "buckets" : [
                { "key" : "Result1", "doc_count" : 31 },
                { "key" : "Result2", "doc_count" : 25 },
                { "key" : "Result3", "doc_count" : 25 },
                { "key" : "IPSAM", "doc_count" : 24 },
                { "key" : "SIT", "doc_count" : 24 }]}}}
```

How it works...

Elasticsearch provides two kinds of aggregation that we saw in *Chapter 7, Aggregations*, as follows:

- Metrics that compute some values
- Buckets that aggregate documents in a bucket

In both cases, you can use script or value script (if you define the field to be used in the aggregation). The object accepted in aggregation is the standard `script` object; the value that is returned by the script will be used for the aggregation.

If a `field` value is defined in the aggregation, then you can use the value script aggregation. In this case, in the context of the script, there is a special `_value` variable available that contains the value of the field.

Using scripting in aggregation is a very powerful feature; however, using it on large cardinality aggregation could be very CPU-intensive and could slow down query times.

Updating a document using scripts

Elasticsearch allows you to update a document in place. Updating a document using scripting reduces network traffic (otherwise, you need to fetch the document, change the field or fields, and then send them back) and improves performance when you need to process a large number of documents.

Getting ready

You will need an up-and-running Elasticsearch installation, similar to the one that we described in the *Downloading and installing Elasticsearch* recipe in *Chapter 1, Getting Started*.

To execute the commands, any HTTP client can be used, such as cURL (https:// curl.haxx.se/) or Postman (https://www.getpostman.com/). You can use Kibana Console, as it provides code completion and better character escaping for Elasticsearch.

To correctly execute the following commands, you will need an index that is populated with the ch07/populate_aggregation.txt commands – these are available in the online code.

In order to be able to use regular expressions in Painless scripting, you will need to activate them in elasticsearch.yml by adding the following code:

```
script.painless.regex.enabled: true
```

The index used in this recipe is the index-agg index.

How to do it...

For updating using scripting, we will perform the following steps:

1. Write an `update` action that adds a tag value to the list of tags that are available in the source of the document:

```
POST /index-agg/_doc/10/_update
{ "script": {
    "source": "ctx._source.age = ctx._source.age +
params.sum",
    "params": { "sum": 2 } } }
```

If everything is correct, then the result that is returned by Elasticsearch will be as follows:

```
{ "_index" : "index-agg", "_id" : "10",
  "_version" : 3, "result" : "updated",
  "_shards" : { "total" : 2, "successful" : 1,
    "failed" : 0 },
  "_seq_no" : 2002, "_primary_term" : 3 }
```

2. If we now retrieve the document, we will have the following code:

```
GET /index-agg/_doc/10
```

The result will be as follows:

```
{ "_index" : "index-agg",   "_id" : "10",
  "_version" : 3, "found" : true,
  "_source" : { ... truncated...
```

From the preceding result, we can see that the version number has increased by one.

How it works...

The REST HTTP method that is used to update a document is POST. The URL contains only the index name, the type, the document ID, and the action:

```
http://<server>/<index_name>/_doc/<document_id>/_update
```

The update action is composed of three different steps, as follows:

- **The Get API call**: This operation is very fast and works on real-time data (there is no need to refresh) and retrieves the record.

- **The script execution**: The script is executed in the document and, if required, it is updated.

- **Saving the document**: The document, if needed, is saved.

The script execution follows the workflow in the following manner:

- The script is compiled, and the result is cached to improve re-execution. The compilation depends on the scripting language; that is, it detects errors in the script, such as typographical errors, syntax errors, and language-related errors. The compilation step can also be CPU-bound so that Elasticsearch caches the compilation results for further execution.

- The document is executed in the script context; the document data is available in the ctx variable in the script.

The update script can set several parameters in the ctx variable; the most important parameters are as follows:

- ctx._source: This contains the source of the document.

- ctx._timestamp: If it's defined, this value is set to the document timestamp.

- ctx.op: This defines the main operation type to be executed. There are several available values, such as the following:

  - index: This is the default value; the record is reindexed with the update values.

  - delete: The document is deleted and not updated (that is, this can be used for updating a document or removing it if it exceeds a quota).

  - none: The document is skipped without reindexing the document.

If you need to execute a large number of update operations, then it's better to perform them in bulk in order to improve your application's performance.

There's more...

In the following example, we'll execute an update that adds new tags and labels values to an object. However, we will mark the document for indexing only if the tags or labels values are changed:

```
POST /index-agg/_doc/10/_update
{ "script": {
    "source": """
    ctx.op = "none";
```

```
        if(ctx._source.containsValue("tags")){
           for(def item : params.new_tags){
              if(!ctx._source.tags.contains(item)){
                 ctx._source.tags.add(item);
                 ctx.op = "index";
              }
           }
        }else{
           ctx._source.tags=params.new_tags;
           ctx.op = "index"
        }

        if(ctx._source.containsValue("labels")){
           for(def item : params.new_labels){
              if(!ctx._source.labels.contains(item)){
                 ctx._source.labels.add(item);
                 ctx.op = "index"
              }
           }
        }else{
           ctx._source.labels=params.new_labels;
           ctx.op = "index"
        }
    """,
        "params": {
        "new_tags": [ "cool",  "nice" ],
        "new_labels": [ "red", "blue", "green" ] } } }
```

The preceding script uses the following steps:

1. It marks the operation as none to prevent indexing if, in the following steps, the original source is not changed.

2. It checks whether the tags field is available in the source object.

3. If the tags field is available in the source object, then it iterates all the values of the new_tags list. If the value is not available in the current tags list, then it adds it and updates the operation to the index.

4. If the `tags` field doesn't exist in the source object, then it simply adds it to the source and marks the operation to the index.

5. *Steps 2* to *4* are repeated for the `labels` value. The repetition is present in this example to show the Elasticsearch user how it is possible to update multiple values in a single update operation.

You can consolidate different script operations into a single script. To do this, use the previously explained workflow of building the script, adding sections in the script, and changing `ctx.op` only if the record is changed.

This script can be quite complex, but it shows Elasticsearch's powerful scripting capabilities.

Reindexing with a script

Reindexing is a functionality for automatically copying your data into a new index. This action is often done to cover different scenarios, as follows:

*   Reindexing after a mapping change
*   Removing a field from an index
*   Adding new fields based on a function

Getting ready

You will need an up-and-running Elasticsearch installation, similar to the one that we described in the *Downloading and installing Elasticsearch* recipe in *Chapter 1*, *Getting Started*.

To execute `curl` using the command line, you will need to install `curl` for your operating system.

In order to correctly execute the following commands, you will need an index that is populated with the `ch07/populate_aggregation.txt` script (available in the online code), and the JavaScript or Python language scripting plugins installed.

How to do it...

For reindexing with a script, we will perform the following steps:

1. Create the destination index, as this is not created by the `reindex` API:

```
PUT /reindex-scripting
{ "mappings": {
    "properties": {
      "name":{"term_vector": "with_positions_offsets",
        "store": true, "type": "text" },
      "title":{"term_vector":"with_positions_offsets",
        "store": true, "type": "text"        },
      "parsedtext":{"term_vector":"with_positions_
offsets",
        "store": true, "type": "text" },
      "tag": { "type": "keyword", "store": true },
      "processed": { "type": "boolean" },
      "date": { "type": "date", "store": true },
      "position": {"type": "geo_point","store": true},
      "uuid": {"store": true, "type": "keyword"}}}}
```

2. Write a reindex action that adds a `processed` field (a Boolean field set to `true`).
    It should look as follows:

```
POST /_reindex
{ "source": { "index": "index-agg" },
  "dest": { "index": "reindex-scripting" },
  "script": {
    "source": """
if(!ctx._source.containsKey("processed")){
  ctx._source.processed=true
}
""" } }
```

If everything is correct, then the result that is returned by Elasticsearch should be
as follows:

```
{ "took" : 386, "timed_out" : false,
  "total" : 1000, "updated" : 0, "created" : 1000,
  "deleted": 0, "batches" : 1,"version_conflicts" : 0,
```

```
    "noops" : 0, "retries" : {"bulk" : 0,"search" : 0 },
    "throttled_millis" : 0,
    "requests_per_second" : -1.0,
    "throttled_until_millis" : 0, "failures" : [ ] }
```

3. Now, if we retrieve the same documents, we will have the following code:

```
GET /reindex-scripting/_doc/10
```

The result will be as follows:

```
{ "_index" : "reindex-scripting", "_id" : "10",
    "_version" : 1, "found" : true,
    "_source" : {
      "date" : "2012-06-21T16:46:01.689622" } }
```

From the preceding result, we can see that the script is applied.

How it works...

The scripting in `reindex` offers very powerful functionality because it allows the execution of a lot of useful actions, such as the following:

- Computing new fields
- Removing fields from a document
- Adding a new field with default values
- Modifying the field values

The scripting works as for `update`, but during reindexing, you can also change the following document metadata fields:

- `_id`: This is the ID of the document.
- `_index`: This is the destination index of the document.
- `_version`: This is the version of the document.
- `_routing`: This is the routing value to send the document in a specific shard.
- `_parent`: This is the parent of the document.

The possibility of changing these values provides a lot of options during reindexing; for example, splitting a type into two different indices, or partitioning an index into several indices and changing the `_index` value.

Scripting in ingest processors

In *Chapter 12*, *Using the Ingest Module*, we will see several types of ingest processors.

Ingest processors are the building blocks for an ingestion pipeline; they describe an action that can be executed on a document to modify it.

Scripting is the main functionality used in processors to provide the core functionalities for completeness. Their scripting functionalities are discussed in this chapter.

Getting ready

You will need an up-and-running Elasticsearch installation, similar to the one that we described in the *Downloading and installing Elasticsearch* recipe in *Chapter 1*, *Getting Started*.

How to do it...

We will simulate a pipeline with a set and a script processor. To modify our documents before ingesting them, we will perform the following steps:

1. Execute a pipeline simulation API call with the two processor steps and two documents as a sample:

```
POST /_ingest/pipeline/_simulate
{ "pipeline": {
    "processors": [
        { "set": { "if": "ctx.cost!=null",
            "field": "vat", "value": 20 } },
        { "script": {
            "lang": "painless",
            "if": "ctx.vat!=null && ctx.cost!=null ",
            "source": "ctx.total = ctx.cost*(ctx.vat -
params.discount)",
            "params": { "discount": 10 } } } ] },
    "docs": [
      { "_index": "index", "_id": "id",
        "_source": { "cost": 5 } },
      { "_index": "index", "_id": "id",
        "_source": { "foo": "no touch" } } ] }
```

The result of a simulation is the document has changed. The result that is returned by Elasticsearch should be as follows:

```
{ "docs" : [
    { "doc" : { "_index" : "index", "_id" : "id",
        "_source" : { "vat" : 20, "total" : 50,
          "cost" : 5 },
        "_ingest":{"timestamp":"2022-02-
12T20:44:54.070675554Z" } } },
    { "doc" : { "_index" : "index", "_id" : "id",
        "_source" : { "foo" : "rab" },
        "_ingest" : {
          "timestamp" : "2022-02-12T20:44:54.070689911Z"
} } } ] }
```

From the preceding result, we can see that the steps are applied to the first returned document.

How it works...

Inside ingester processor definition, scripting is widely used. One of the most commonly used operators is the if operator, which validates if the processor should be executed.

The if operator requires a script that returns a boolean value:

- In the preceding example we have defined a set processor as follows:

```
{ "set": {
    "if": "ctx.cost!=null",
    "field": "vat", "value": 20 } }
```

The set processor will set the value of field ("vat" in the example) to value. In this case, the painless script in the if statement is only executed if the cost field is defined in the document and it's not null.

- The second processor that we have used in the example is a script one, as follows:

```
{ "script": { "lang": "painless",
    "if": "ctx.vat!=null && ctx.cost!=null",
    "source": "ctx.total = ctx.cost*(ctx.vat - params.
discount)",
    "params": { "discount": 10 } } }
```

The `script` processor has similar behavior to scripting used in the `update` API. I take an optional `params` dictionary and `source` that is used to modify the values in the document.

As usual, inside `source`, you can put all the scripting code that you need.

In this case, we put an `if` statement to check whether `vat` and `cost` are null values.

The pipeline simulation API accepts an array of documents that need to be evaluated by the pipeline.

In this case, for the first document, we have the following scenario:

- It has the `cost` field, so the first processor will add the `vat` field.

- It has the `vat` and `cost` fields, so the second processor will evaluate the script.

The final document result is the application of both processors:

```
"_source" : { "vat" : 20, "total" : 50, "cost" : 5 },
```

The second document is unable to pass the `if` statements of the two processors so it's unchanged.

See also

You can refer to the following for further reference:

- *Chapter 12*, *Using the Ingest Module*, which describes the overall ingest functionalities

- The *Simulating an ingest pipeline* recipe in *Chapter 12*, *Using the Ingest Module*, for an in-depth description of the simulate pipeline API

- The *Updating a document using scripts* recipe in this chapter for a reference about document manipulation with Painless scripts

9
Managing Clusters

In the **Elasticsearch** ecosystem, it's important to monitor nodes and clusters in order to manage and improve their performance and state. Several issues can arise at the cluster level, such as the following:

- **Node overheads**: Some nodes can have too many shards allocated and become a bottleneck for the entire cluster.

- **Node shutdown**: This can happen due to a number of reasons, for example, full disks, hardware failures, and power problems.

- **Shard relocation problems or corruptions**: Some shards can't get an online status.

- **Shards that are too large**: If a shard is too big, then the index performance decreases due to the merging of massive Lucene segments.

- **Empty indices and shards**: These waste memory and resources; however, because each shard has a lot of active threads, if there are a large number of unused indices and shards, then the general cluster performance is degraded.

Detecting malfunctioning or poor performance at the cluster level can be done through an API or frontends (as we will see in *Chapter 11*, *User Interfaces*). These allow users to have a working web dashboard on their Elasticsearch data; it works by monitoring the cluster health, backing up or restoring data, and allowing the testing of queries before implementing them in code.

In this chapter, we will explore the following topics:

- Using the health API to check the health of the cluster
- Using the task API that controls jobs at the cluster level
- Using hot threads to check inside nodes for problems due to high CPU usage
- Learning how to monitor Lucene segments so as not to reduce the performance of a node due to there being too many of them

In this chapter, we will cover the following recipes:

- Controlling the cluster health using the health API
- Controlling the cluster state using the API
- Getting cluster node information using the API
- Getting node statistics using the API
- Using the task management API
- Using the hot threads API
- Managing the shard allocation
- Monitoring segments with the segment API
- Cleaning the cache

Controlling the cluster health using the health API

Elasticsearch provides a health check API to control the cluster state, and this is one of the first things to check whether any problems do occur.

Getting ready

You will need an up-and-running Elasticsearch installation, similar to the one that we described in the *Downloading and installing Elasticsearch* recipe in *Chapter 1, Getting Started*.

To execute the commands, any HTTP client can be used, such as **curl** (`https://curl.haxx.se/`) or **Postman** (`https://www.getpostman.com/`). You can use the **Kibana** console, as it provides code completion and better character escaping for Elasticsearch.

How to do it...

To control the cluster health, we will perform the following steps:

1. In order to view the cluster health, the HTTP method that we use is GET:

    ```
    GET /_cluster/health
    ```

    The result will be as follows:

    ```
    { "cluster_name":"elasticsearch", "status": "yellow",
      "timed_out" : false, "number_of_nodes" : 1,
      "number_of_data_nodes" : 1,
      "active_primary_shards" : 17,
      "active_shards" : 17, "relocating_shards" : 0,
      "initializing_shards" : 0, "unassigned_shards" : 15,
      "delayed_unassigned_shards" : 0,
      "number_of_pending_tasks" : 0,
      "number_of_in_flight_fetch" : 0,
      "task_max_waiting_in_queue_millis" : 0,
      "active_shards_percent_as_number" : 53.125 }
    ```

How it works...

Every Elasticsearch node keeps the cluster's status. The status value of the cluster can be one of three types, as follows:

-   green: This means that everything is okay.
-   yellow: This means that some nodes or shards are missing, but they don't compromise the cluster's functionality. For instance, some replicas could be missing (either a node is down or there are insufficient nodes for replicas), but there is at least one copy of each active shard; additionally, read and write functions are working. The yellow state is very common during the development stage when users typically start a single Elasticsearch server.
-   red: This indicates that some primary shards are missing, and these indices are in the red state. You cannot write to indices that are in the red state and, additionally, the results may not be complete, or only partial results may be returned. Usually, you'll need to restart the node that is down and possibly create some replicas.

The yellow or red states could be transient if some nodes are in recovery mode. In this case, just wait until the recovery completes.

The cluster health API contains an enormous amount of information, as follows:

- `cluster_name`: This is the name of the cluster.
- `timeout`: This is a Boolean value indicating whether the REST API hits the timeout set in the call.
- `number_of_nodes`: This indicates the number of nodes that are in the cluster.
- `number_of_data_nodes`: This indicates the number of nodes that can store data (you can refer to *Chapter 2, Managing Mapping*, and the *Downloading and setup* recipe, in order to set up different node types for different types of nodes).
- `active_primary_shards`: This shows the number of active primary shards; the primary shards are the masters of writing operations.
- `active_shards`: This shows the number of active shards; these shards can be used for searches.
- `relocating_shards`: This shows the number of shards that are relocating or migrating from one node to another node – this is mainly due to cluster-node balancing.
- `initializing_shards`: This shows the number of shards that are in the initializing status. The initializing process is done at shard startup. It's a transient state before becoming active and is composed of several steps; the most important steps are as follows:

  A. Copy the shard data from a primary one if its translation log is too old or a new replica is needed.

  B. Check the Lucene indices.

  C. Process the transaction log as needed.

- `unassigned_shards`: This shows the number of shards that are not assigned to a node. This is usually due to having set a replica number that is larger than the number of nodes. During startup, shards that are not already initialized or initializing will be counted here.
- `delayed_unassigned_shards`: This shows the number of shards that will be assigned, but their nodes are configured for a delayed assignment. You can find more information about delayed shard assignments at `https://www.elastic.co/guide/en/elasticsearch/reference/master/delayed-allocation.html`.

- `number_of_pending_tasks`: This is the number of pending tasks at the cluster level, such as updates to the cluster state, the creation of indices, and shard relocations. It should rarely be anything other than `0`.

- `number_of_in_flight_fetch`: This is the number of cluster updates that must be executed in the shards. As the cluster updates are asynchronous, this number tracks how many updates still have to be executed in the shards.

- `task_max_waiting_in_queue_millis`: This is the maximum amount of time that some cluster tasks have been waiting in the queue. It should rarely be anything other than `0`. If the value is different to `0`, then it means that there is some kind of cluster saturation of resources or a similar problem.

- `active_shards_percent_as_number`: This is the percentage of active shards that are required by the cluster. In a production environment, it should rarely differ from 100 percent – apart from some relocations and shard initializations.

Installed plugins play an important role in shard initialization; for example, if you use a mapping type that is provided by a native plugin and you remove the plugin (or if the plugin cannot be initialized due to API changes), then the shard initialization will fail. These issues are easily detected by reading the Elasticsearch log file.

When upgrading your cluster to a new Elasticsearch release, make sure that you upgrade your mapping plugins or, at the very least, check that they work with the new Elasticsearch release. If you don't do this, you risk your shards failing to initialize and giving a red status to your cluster.

There's more...

This API call is very useful; it's possible to execute it against one or more indices in order to obtain their health in the cluster. This approach allows the isolation of those indices that have problems. The API call to execute this is as follows:

```
GET /_cluster/health/index1,index2,indexN
```

The previous call also has additional request parameters in order to control the health of the cluster. These additional parameters are as follows:

- `level`: This controls the level of the health information that is returned. This parameter accepts only `cluster`, `index`, and `shards`.

- `timeout`: This is the wait time for a `wait_for_*` parameter (the default is `30s`).

- `wait_for_status`: This allows the server to wait for the provided status (green, yellow, or red) until timeout.

- `wait_for_relocating_shards`: This allows the server to wait until the provided number of relocating shards has been reached, or until the timeout period has been reached (the default is 0).

- `wait_for_nodes`: This waits until the defined number of nodes is available in the cluster. The value for this parameter can also be an expression, such as >N, >=N, <N, <=N, ge(N), gt(N), le(N), and lt(N).

If the number of pending tasks is different to zero, then it's good practice to investigate what these pending tasks are. They can be shown using the following API URL:

```
GET /_cluster/pending_tasks
```

The return value is a list of pending tasks; beware that Elasticsearch applies cluster changes very quickly, so many of these tasks have a lifespan of some milliseconds to apply those that show themselves to you.

See also

You can refer to the following recipe and web page for more information:

- The *Setting up different node types* recipe in *Chapter 1, Getting Started*, in order to set up nodes as masters

- The official documentation about pending cluster tasks at `https://www.elastic.co/guide/en/elasticsearch/reference/current/cluster-pending.html`, for example, the returned value from this call

Controlling the cluster state using the API

The previous recipe returns information only about the health of the cluster. If you need more details on your cluster, then you need to query its state.

The **cluster state** is the core of all functionalities and the metadata stored in Elasticsearch.

Getting ready

You will need an up-and-running Elasticsearch installation, similar to the one that we described in the *Downloading and installing Elasticsearch* recipe in *Chapter 1, Getting Started*.

In order to execute the commands, any HTTP client can be used, such as curl (`https://curl.haxx.se/`) or Postman (`https://www.getpostman.com/`). You can use the Kibana console, as it provides code completion and better character escaping for Elasticsearch.

How to do it...

To check the cluster state, we will perform the following steps:

1. In order to view the cluster state, the HTTP method that you can use is GET, and the `curl` command is as follows:

    ```
    GET /_cluster/state
    ```

2. The result is very long and contains a lot of data sections, so I'll present the most important ones in the following bullet points:

    -   The general cluster information is as follows:

        ```
        { "cluster_name" : "elastic-cookbook",
          "compressed_size_in_bytes" : 4714,
          "cluster_uuid" : "02UhFNltQXOqtz1JH6ec8w",
          "version" : 9,
          "state_uuid" : "LZcYMc3PRdKSJ9MMAyM-ew",
          "master_node" : "-IFjP29_TOGQF-1axtNMSg",
          "blocks" : { },
        ```

    -   The node address information is as follows:

        ```
        "nodes" : {
            "-IFjP29_TOGQF-1axtNMSg" : {
                "name" : "5863a2552d84",
                "ephemeral_id" : "o6xo1mowRIGVZ7ZfXkClww",
                "transport_address" : "172.18.0.2:9300",
                "attributes" : {
                    "ml.machine_memory" : "68719476736",
                    "xpack.installed" : "true",
                    "ml.max_jvm_size" : "33285996544"
                },
                "roles" : [
                    "data", "data_cold", "data_content", "data_
        ```

```
frozen",
        "data_hot", "data_warm", "ingest", "master",
"ml",
        "remote_cluster_client", "transform"
    ]}},
```

- The cluster metadata information (such as templates, indices with mappings, and the aliases) is as follows:

```
"metadata" : {
    "cluster_uuid" : "02UhFNltQXOqtz1JH6ec8w",
    "cluster_coordination" : { "term" : 0,
        "last_committed_config" : [ ],
        "last_accepted_config" : [ ],
        "voting_config_exclusions" : [ ] },
    "templates" : {
        "kibana_index_template:.kibana" : {
            "index_patterns" : [".kibana"],
            "order" : 0,
            "settings" : {
                "index" : {
                    "format" : "7",
                    "codec" : "best_compression",
                    "number_of_shards" : "1",
                    "auto_expand_replicas" : "0-1",
                    "number_of_replicas" : "0" } },
            "mappings" : {
                "_doc" : {
                    "dynamic" : "strict",
                    "properties" : {
                        "server" : {
                            "properties" : {"uuid" : {"type" :
"keyword"}}}},
        "index-graveyard" : {"tombstones" : [ ] } },
```

- You can route the tables in order to find the shards by using the following code:

```
"routing_table" : {
    "indices" : {
```

```
       ".kibana_1" : {
           "shards" : {
               "0" : [
                   {
                       "state" : "STARTED",
                       "primary" : true,
                       "node" : "-IFjP29_TOGQF-1axtNMSg",
                       "relocating_node" : null,
                       "shard": 0, "index": ".kibana_1",
                       "allocation_id" : {
                           "id" : "QjMusIOIRRqOIsL8kEdudQ"
                   } } ] } } } },
```

- You can route the nodes by using the following code:

```
"routing_nodes" : {
     "unassigned" : [ ],
     "nodes" : {
         "-IFjP29_TOGQF-1axtNMSg" : [
             {
                 "state" : "STARTED","primary" : true,
                 "node" : "-IFjP29_TOGQF-1axtNMSg",
                 "relocating_node" : null, "shard" : 0,
                 "index" : ".kibana_1",
                 "allocation_id" : {
                     "id" : "QjMusIOIRRqOIsL8kEdudQ"}}]}}
```

How it works...

The cluster state contains information about the whole cluster; the fact that its output is very large is normal.

The call output also contains common fields, and they are as follows:

- cluster_name: This is the name of the cluster.

- master_node: This is the identifier of the master node. The master node is the primary node that is used for cluster management.

- blocks: This section shows the active blocks in a cluster.

- nodes: This shows the list of nodes in the cluster. For each node, we have the following information:

  - id: This is the hash that is used to identify the node in Elasticsearch (for example, 7NwnFF1JTPOPhOYuP1AVN).

  - name: This is the name of the node.

  - transport_address: This is the IP address and port used to connect to this node.

  - attributes: These are additional node attributes.

  - metadata: This is the definition of the **indices** (including their settings and mappings), index/component templates, ingest pipelines, and stored_scripts.

  - routing_table: These are the indices or shards routing tables that are used to select primary and secondary shards and their nodes.

  - routing_nodes: This is the routing for the nodes.

The metadata section is the most used field because it contains all the information that is related to the indices and their mappings. This is a convenient way in which to gather all the indices mappings in one go; otherwise, you'll need to call the get mapping instance for every type.

The metadata section is composed of several sections, as follows:

- templates, index_templates, and component_templates: These are the templates that control the dynamic mapping for your created indices.

- indices: These are the indices that exist in the cluster.

- index-graveyard: This contains the history of deleted indices.

- * ingest: This stores all the ingest pipelines that are defined in the system.

- stored_scripts: This stores the scripts, which are usually in the form of language#script_name.

The indices subsection returns a full representation of all the metadata descriptions for each index, and contains the following:

- state (open or closed): This describes whether an index is open (that is, it can be searched and can index data) or closed (you can refer to the *Opening/closing an index* recipe in *Chapter 3, Basic Operations*).

- `settings`: These are the index settings. The most important ones are as follows:

  - `index.number_of_replicas`: This is the number of replicas of this index; it can be changed using an update index settings call.

  - `index.number_of_shards`: This is the number of shards in this index. This value cannot be changed in an index.

  - `index.codec`: This is the codec that is used to store index data; `default` is not shown, but the **LZ4** algorithm is used. If you want a high compression rate, then use `best_compression` and the **DEFLATE** algorithm (this will slow down the writing performances slightly).

  - `index.version.created`: This is the index version.

  - `mappings`: These are defined in the index. This section is similar to the `get` mapping response (you can refer to the *Getting a mapping* recipe in *Chapter 3, Basic Operations*).

  - `alias`: This is a list of index aliases, which allows the aggregation of indices in a single name or the definition of alternative names for an index.

The routing records for the index and shards have similar fields, and they are as follows:

- `state` (`UNASSIGNED`, `INITIALIZING`, `STARTED`, or `RELOCATING`): This shows the state of the shard or the index.

- `primary` (`true` or `false`): This shows whether the shard or node is primary.

- `node`: This shows the ID of the node.

- `relocating_node`: This field, if validated, shows the `id` node in which the shard is relocated.

- `shard`: This shows the number of the shard.

- `index`: This shows the name of the index in which the shard is contained.

There's more...

The cluster state call returns a lot of information, and it's possible to filter out the different section parts through the URL.

The complete URL of the cluster state API is as follows:

```
http://{elasticsearch_server}/_cluster/state/{metrics}/
{indices}
```

The `metrics` value could be used to return only parts of the response. It consists of a comma-separated list and includes the following values:

- `* version`: This is used to show the version part of the response.
- `blocks`: This is used to show the blocks part of the response.
- `master_node`: This is used to show the master node part of the response.
- `nodes`: This is used to show the node part of the response.
- `metadata`: This is used to show the metadata part of the response.
- `routing_table`: This is used to show the routing table part of the response.

The `indices` value is a comma-separated list of index names to include in the metadata.

See also

You can refer to the following recipes for more information:

- The *Opening or closing an index* recipe in *Chapter 3, Basic Operations*, for APIs on opening and closing indices—remember that closed indices cannot be searched
- The *Getting a mapping* recipe in *Chapter 3, Basic Operations*, for returning single mappings

Getting cluster node information using the API

The previous recipe allows information to be returned to the cluster level; Elasticsearch provides calls to gather information at the node level. In production clusters, it's very important to monitor nodes using this API in order to detect misconfiguration and any problems relating to different plugins and modules.

Getting ready

You will need an up-and-running Elasticsearch installation, similar to the one that we described in the *Downloading and installing Elasticsearch* recipe in *Chapter 1, Getting Started*.

In order to execute the commands, any HTTP client can be used, such as curl (https://curl.haxx.se/) or Postman (https://www.getpostman.com/). You can use the Kibana console, as it provides code completion and better character escaping for Elasticsearch.

How to do it...

To get the cluster node information, we will perform the following steps:

1. To retrieve the node information, the HTTP method you can use is GET, and the curl command is as follows:

```
GET /_nodes
GET /_nodes/<nodeId1>,<nodeId2>
```

2. The result will contain a lot of information about the node; it's huge, so the repetitive parts have been truncated:

```
{
    ... truncated ...,
    "cluster_name" : "elastic-cookbook",
    "nodes" : {
        "-IFjP29_TOGQF-1axtNMSg" : {
            "name" : "5863a2552d84",
            "transport_address" : "172.18.0.2:9300",
            "host" : "172.18.0.2","ip" : "172.18.0.2",
            ... truncated...
            "settings" : {
                "cluster":{"name" : "elastic-cookbook"},
                "node" : {
                    "attr" : {
                        "xpack": { "installed" : "true"} ,
                        "ml" : {
                            "max_jvm_size" : "33285996544",
                            "machine_memory": "68719476736"}},
                    "name" : "5863a2552d84"
                },
                "path" : {
                    "logs": "/usr/share/elasticsearch/logs",
                    "home": "/usr/share/elasticsearch"},
                ... truncated... },
            "os" : { ... truncated... },
            ... truncated...
```

```
          "plugins" : [
              { "name" : "analysis-icu",
                  ... truncated...
```

The full response is very long, so I keep only the most important part for further discussion.

How it works...

The cluster node information call provides an overview of the node configuration. It covers a lot of information. The most important sections are as follows:

- `hostname`: This is the name of the host.

- `ip`: This is the IP address of the host.

- `version`: This is the Elasticsearch version. It is best practice for all the nodes of a cluster to have the same Elasticsearch version.

- `roles`: This is a list of roles that this node can cover. The developer nodes usually support three roles: `master`, `data`, and `ingest`.

- `settings`: This section contains information about the current cluster and the path of the Elasticsearch node. The most important fields are as follows:

  - `cluster_name`: This is the name of the cluster.

  - `node.name`: This is the name of the node.

  - `path.*`: This is the configured path of this Elasticsearch instance.

  - `script`: This section is useful to check the `script` configuration of the node.

- `os`: This section provides the **operating system (OS)** information about the node that is running Elasticsearch, including the processors that are available or allocated, and the OS version.

- `process`: This section contains information about the currently-running Elasticsearch process:

  - `id`: This is the process identifier ID of the process.

  - `mlockall`: This flag defines whether Elasticsearch can use direct memory access; in production, this must be set to `active`.

- `max_file_descriptors`: This is the maximum file descriptor number.

- jvm: This section contains information about the **Java Virtual Machine (JVM)** node; this includes the version, vendor, name, PID, and memory (heaps and non-heaps).

> **Note**
>
> It's highly recommended to run all the nodes on the same JVM version and type.

- thread_pool: This section contains information about several types of thread pools running in a node.

- transport: This section contains information about the transport protocol. The transport protocol is used for intra-cluster communication, or by the native client in order to communicate with a cluster. The response format is similar to the HTTP one, as follows:

  - bound_address: If a specific IP is not set in the configuration, then Elasticsearch binds all the interfaces together.

  - publish_address: This is the address that is used for publishing the native transport protocol.

- http: This section gives information about the HTTP configuration:

  - max_content_length_in_bytes (the default is 104857600 of 100 MB): This is the maximum size of HTTP content that Elasticsearch will allow to be received. HTTP payloads that are larger than this size are rejected.

> **Note**
>
> The default 100 MB HTTP limit, which can be changed in elasticsearch.yml, can result in a malfunction due to a large payload, so it's important to bear this limit in mind when doing bulk actions.

  - publish_address: This is the address that is used to publish the Elasticsearch node.

- plugins: This section lists every plugin installed in the node and provides information about the following:

  - name: This is the plugin name.

  - description: This is the plugin description.

- version: This is the plugin version.

- classname: This is the Java class used to load the plugin.

> **Warning!**
> All the nodes must have the same plugin version; different plugin versions in a node bring unexpected failures.

- modules: This section lists every module installed in the node. The structure is the same as the plugin section.

- ingest: This section contains the list of active processors in the ingest node.

There's more...

The API call allows you to filter the section that must be returned. In this example, we've returned the whole section. Alternatively, we could select one or more of the following sections:

- http
- thread_pool
- transport
- jvm
- os
- process
- plugins
- modules
- ingest
- settings

For example, if you need only the os and plugins information, the call will be as follows:

```
GET /_nodes/os,plugins
```

See also

You can refer to the following for more information:

- The *Networking setup* recipe in *Chapter 1*, *Getting Started*, about how to configure networking for Elasticsearch

- *Chapter 13*, *Using the Ingest Module*, for more information about Elasticsearch ingestion

Getting node statistics using the API

The node statistics call API is used to collect real-time metrics of your node, such as memory usage, threads usage, the number of indices, and searches.

Getting ready

You will need an up-and-running Elasticsearch installation, similar to the one that we described in the *Downloading and installing Elasticsearch* recipe in *Chapter 1*, *Getting Started*.

In order to execute the commands, any HTTP client can be used, such as curl (https://curl.haxx.se/) or Postman (https://www.getpostman.com/). You can use the Kibana console, as it provides code completion and better character escaping for Elasticsearch.

How to do it...

To get the node statistics, we will perform the following steps:

1. To retrieve the node statistics, the HTTP method that we will use is GET, and the command is as follows:

   ```
   GET /_nodes/<nodeId1>,<nodeId2>/stats
   ```

2. The result will be a long list of all the node statistics. The most significant parts of the results can be broken up. First, a header that describes the cluster name and the nodes section, as follows:

   ```
   { "_nodes" : {"total" : 1, "successful" : 1,
       "failed" : 0 },
     "cluster_name" : "elastic-cookbook",
     "nodes" : {
       "-IFjP29_TOGQF-1axtNMSg" : {
   ```

```
         "timestamp" : 1545580226575,
         "name" : "5863a2552d84",
         "transport_address" : "172.18.0.2:9300",
         "host": "172.18.0.2","ip": "172.18.0.2:9300",
         "roles": ["data", "data_cold", "data_content",
             "data_frozen", "data_hot", "data_warm",
             "ingest", "master", "ml",
             "remote_cluster_client", "transform" ],
         "attributes" : {
             "ml.machine_memory" : "68719476736",
             "xpack.installed" : "true",
             "ml.max_jvm_size" : "33285996544" },
```

Here are the statistics that are related to the indices:

```
"indices" : {
              "docs": { "count" : 3, "deleted" : 0},
              "store":{"size_in_bytes" : 12311 }},
```

Here are the statistics that are related to the OS:

```
"os" : {
    "timestamp" : 1545580226579,
    "cpu" : { "percent" : 0,
        "load_average" : {
            "1m" : 0.0, "5m" : 0.02, "15m" : 0.0 } },
    "mem" : {
        "total_in_bytes" : 2095869952,
        "free_in_bytes" : 87678976,
        "used_in_bytes" : 2008190976,
        "free_percent" : 4, "used_percent" : 96 },
... truncated...
    "memory" : {
      "control_group" : "/",
      "limit_in_bytes" : "9223372036854771712",
      "usage_in_bytes" : "1360773120" } } },
```

Here are the statistics that are related to the current Elasticsearch process:

```
"process" : {
    "timestamp" : 1545580226580,
```

```
        "open_file_descriptors" : 257,
        "max_file_descriptors" : 1048576,
        "cpu" : {"percent" : 0,"total_in_millis" : 50380},
        "mem" : {"total_virtual_in_bytes" : 4881367040 }},
```

Here are the statistics that are related to the current JVM:

```
  "jvm" : { "timestamp" : 1545580226581,
      "uptime_in_millis" : 3224543,
      "mem" : { "heap_used_in_bytes" : 245981600,
          "heap_used_percent" : 23,
          "heap_committed_in_bytes" : 1038876672,
          "heap_max_in_bytes" : 1038876672,
          "non_heap_used_in_bytes" : 109403072,
          "non_heap_committed_in_bytes" : 119635968,
  … truncated… },
```

Here are the statistics related to thread pools:

```
  "thread_pool" : {
      "analyze" : {
          "threads" : 0, "queue" : 0, "active" : 0,
          "rejected" : 0, "largest" : 0, "completed" : 0},
  … truncated… },
```

Here are the node filesystem statistics:

```
  "fs" : {
      "timestamp" : 1545580226582,
      "total" : {"total_in_bytes" : 62725623808,
          "free_in_bytes" : 59856470016,
          "available_in_bytes" : 56639754240 },
      … truncated…
```

Here are the statistics relating to the communications between the nodes:

```
  "transport" : { "server_open" : 0,
      "rx_count" : 0, "rx_size_in_bytes" : 0,
      "tx_count" : 0, "tx_size_in_bytes" : 0 },
```

Here are the statistics related to the HTTP connections:

```
  "http" : {"current_open" : 4,   "total_opened" : 175 },
```

Here are the statistics related to the `breaker` caches:

```
"breakers" : {
   "request" : { "limit_size_in_bytes" : 623326003,
     "limit_size" : "594.4mb",
     "estimated_size_in_bytes" : 0,
     "estimated_size" : "0b",   "overhead" : 1.0,
     "tripped" : 0 }, ... truncated... },
```

Here are the statistics related to the script:

```
"script" : { "compilations" : 0,"cache_evictions" : 0,
   "compilation_limit_triggered" : 0},
```

Here is the cluster state queue:

```
"discovery" : { },
```

Here are the ingest statistics:

```
"ingest" : {
   "total" : { "count" : 0, "time_in_millis" : 0,
     "current" : 0, "failed" : 0  },
   "pipelines" : { }
},
```

Here are the `adaptive_selection` statistics:

```
"adaptive_selection" : {
   "-IFjP29_TOGQF-1axtNMSg" : {
     "outgoing_searches" : 0,
     "avg_queue_size" : 0,
     "avg_service_time_ns" : 7479391,
     "avg_response_time_ns" : 13218805,
     "rank" : "13.2"  } }
```

How it works...

During execution, each Elasticsearch node collects statistics about several aspects of node management; these statistics are accessible using the statistics API call. In the next recipe, we will see an example of a monitoring application that uses this information to provide the real-time status of a node or a cluster.

The main statistics collected by this API are as follows:

- `fs`: This section contains statistics about the filesystem. This includes the free space that is on devices, the mount points, and reads and writes. It can also be used to remotely control the disk usage of your nodes.

- `http`: This gives the number of current open sockets and their maximum number.

- `indices`: This section contains statistics about several indexing aspects:

  - The use of fields and caches

  - Statistics about operations such as `get`, `index`, `flush`, `merge`, `refresh`, and `warmer`

- `jvm`: This section provides statistics about buffers, pools, garbage collectors (this refers to the creation or destruction of objects and their memory management), memory (such as used memory, heaps, and pools), threads, and uptime. You should check to see whether the node is running out of memory.

- `network`: This section provides statistics about **Transmission Control Protocol (TCP)** traffic, such as open connections, closed connections, and data I/O.

- `os`: This section collects statistics about the OS, such as the following:

  - CPU usage

  - Node load

  - Virtual and swap memory

  - Uptime

- `process`: This section contains statistics about the CPU that are used by Elasticsearch, memory, and open file descriptors.

> **Note**
>
> It's very important to monitor the open file descriptors. This is because if you run out of them, then the indices may be corrupted.

- `thread_pool`: This section monitors all the thread pools that are available in Elasticsearch. It's important, in the case of low performance, to control whether there are pools that have an excessive overhead. Some of them can be configured to a new maximum value.

- `transport`: This section contains statistics about the transport layer and, in particular, the bytes that are read and transmitted.
- `breakers`: This section monitors the circuit breakers. This must be checked to see whether it's necessary to optimize resources, queries, or aggregations to prevent them from being called.
- `adaptive_selection`: This section contains information about the adaptive node selection that is used for executing searches. Adaptive selection allows you to choose the best replica node from a coordinator node to execute searches.

There's more...

The API response is very large. It's possible to limit it by requesting only the parts that are required. In order to do this, you will need to pass a `query` parameter to the API call specifying the following desired sections:

- `fs`
- `http`
- `indices`
- `jvm`
- `network`
- `os`
- `process`
- `thread_pool`
- `transport`
- `breaker`
- `discovery`
- `script`
- `ingest`
- `breakers`
- `adaptive_selection`

For example, to request only `os` and `http` statistics, then the call will be as follows:

```
GET /_nodes/stats/os,http
```

Using the task management API

Elasticsearch, from version 5.x, allows you to define actions that are executed on the server side. These actions can take some time to complete, and they can consume huge cluster resources. The most common ones are as follows:

- `delete_by_query`
- `update_by_query`
- `reindex`

When these actions are called, they create a server-side task that executes the job; the task management API allows you to control these jobs.

Getting ready

You will need an up-and-running Elasticsearch installation, similar to the one that we described in the *Downloading and installing Elasticsearch* recipe in *Chapter 1, Getting Started*.

In order to execute the commands, any HTTP client can be used, such as curl (`https://curl.haxx.se/`) or Postman (`https://www.getpostman.com/`). You can use the Kibana console, as it provides code completion and better character escaping for Elasticsearch.

How to do it...

To get task information, we will perform the following steps:

1. Retrieve the node information using the GET HTTP method. The command is as follows:

```
GET /_tasks
GET /_tasks?nodes=nodeId1,nodeId2
GET /_tasks?nodes=nodeId1,nodeId2&actions=cluster
```

The result will be as follows:

```
{ "nodes" : {
    "-IFjP29_TOGQF-1axtNMSg" : {
        "name" : "5863a2552d84",
        "transport_address" : "172.18.0.2:9300",
        "host" : "172.18.0.2",
        … truncated…
```

```
        "tasks" : {
            "-IFjP29_TOGQF-1axtNMSg:92797" : {
                "node" : "-IFjP29_TOGQF-1axtNMSg",
                "id" : 92797,
                "type" : "transport",
                "action" : "cluster:monitor/tasks/lists",
                "start_time_in_millis" : 1545642518460,
                "running_time_in_nanos" : 7937700,
                "cancellable" : false, "headers" : { }
            },
        ... truncated...
```

How it works...

Every task that is executed in Elasticsearch is available in the task list.

The most important properties for the tasks are as follows:

- node: This defines the node that is executing the task.
- id: This defines the unique ID of the task.
- action: This is the name of the action. It is generally composed of an action type, the : separator, and the detailed action.
- cancellable: This defines whether the task can be canceled. Some tasks, such as delete/update by query or reindex, can be canceled; however, other tasks are mainly management tasks and so they cannot be canceled.
- parent_task_id: This defines the group of tasks. Some tasks can be split and executed in several subtasks. This value can be used to group these tasks by the parent.

The id property of the task can be used to filter the response through the node_id parameter of the API call, as follows:

```
GET /_tasks/-IFjP29_TOGQF-1axtNMSg:92797
```

If you need to monitor a group of tasks, you can filter according to their parent_task_id property using an API call, as follows:

```
GET /_tasks?parent_task_id=-IFjP29_TOGQF-1axtNMSg:92797
```

There's more...

In general, canceling a task could produce some data inconsistency in Elasticsearch due to the partial updating or deleting of documents; however, when reindexing, it can make good sense. When you are reindexing a large amount of data, it's common to change the mapping or to reindex a script in the middle of it. So, in order to not waste time and CPU usage, canceling the reindexing is a sensible solution.

To cancel a task, the API URL is as follows:

```
POST /_tasks/task_id:1/_cancel
```

In the case of a group of tasks, then they can be stopped with a single `cancel` call using `query` arguments to select them, as follows:

```
POST /_tasks/_cancel?nodes=nodeId1,nodeId2&actions=*reindex
```

See also

For further details, I suggest visiting the following links:

- The official documentation regarding task management for some more borderline cases is available at `https://www.elastic.co/guide/en/elasticsearch/reference/current/tasks.html`.

- A QBOX blog post, *Introduction to Elasticsearch Task and Cancel API*, is available at `https://qbox.io/blog/elasticsearch-task-api-track-reindex-update-by-query-tasks-tutorial`.

Using the hot threads API

Sometimes, your cluster will slow down due to high levels of CPU usage, and you will need to understand why. Elasticsearch provides the ability to monitor hot threads in order to be able to understand where the problem is.

In Java, hot threads are threads that use a lot of CPU and take a long time to execute.

Getting ready

You will need an up-and-running Elasticsearch installation, similar to the one that we described in the *Downloading and installing Elasticsearch* recipe in *Chapter 1, Getting Started*.

In order to execute the commands, any HTTP client can be used, such as curl (`https://curl.haxx.se/`) or Postman (`https://www.getpostman.com/`). You can use the Kibana console, as it provides code completion and better character escaping for Elasticsearch.

How to do it...

To get the task information, we will perform the following steps:

1. To retrieve the node information, the HTTP method that we use is GET, and the `curl` command is as follows:

```
GET /_nodes/hot_threads
GET /_nodes/{nodesIds}/hot_threads'
```

The result will be as follows:

```
::: {5863a2552d84}{-IFjP29_TOGQF-1axtNMSg}
{o6xo1mowRIGVZ7ZfXkClww}{172.18.0.2}{172.18.0.2:9300}
{xpack.installed=true}
   Hot threads at 2018-12-24T09:22:30.481,
interval=500ms, busiestThreads=3, ignoreIdleThreads=true:

   16.1% (80.6ms out of 500ms) cpu usage by thread
'elasticsearch[5863a2552d84][write][T#2]'
     10/10 snapshots sharing following 2 elements
       java.base@11.0.1/java.util.concurrent.
ThreadPoolExecutor$Worker.run(ThreadPoolExecutor.
java:628)
       java.base@11.0.1/java.lang.Thread.run(Thread.
java:834)

   8.7% (43.3ms out of 500ms) cpu usage by thread
'elasticsearch[5863a2552d84][write][T#3]'
     2/10 snapshots sharing following 35 elements
       app//org.elasticsearch.index.mapper.
DocumentMapper.parse(DocumentMapper.java:264)
       app//org.elasticsearch.index.shard.IndexShard.
prepareIndex(IndexShard.java:733)
```

```
        app//org.elasticsearch.index.shard.IndexShard.
applyIndexOperation(IndexShard.java:710)
        app//org.elasticsearch.index.shard.IndexShard.
applyIndexOperationOnPrimary(IndexShard.java:691)
        app//org.elasticsearch.action.bulk.Transport
ShardBulkAction.lambda$executeIndexRequestOnPrimary$3
(TransportShardBulkAction.java:462)
  ... truncated ...
```

How it works...

The hot threads API is quite particular. It works by returning a text representation of the currently-running hot threads so that it's possible to check the causes of the slowdown of every single thread by using the stack trace.

To control the returned values, there are additional parameters that can be provided as query arguments:

- threads: This is the number of hot threads to provide (the default is 3).

- interval: This is the interval for the sampling of threads (the default is 500ms).

- type: This allows the control of different types of hot threads, for example, to check, wait, and block states (the default is cpu; the possible values are cpu, wait, and block).

- ignore_idle_threads: This is used to filter out any known idle threads (the default is true).

The hot threads API is an advanced monitor feature that is provided by Elasticsearch. It is very useful in helping you to debug the slow speed of a production cluster as it can be used as a runtime debugger. If your nodes or clusters have performance problems, then the hot threads API is the only call that can help you to understand how the CPU is being used.

> **Tip**
>
> It is common to have a high overhead in computation due to the wrong regex usage, or due to scripting problems. The hot threads API is the only way to track these issues down.

Managing the shard allocation

During normal Elasticsearch usage, it is not generally necessary to change the shard allocation, because the default settings work very well with all standard scenarios. Sometimes, however, due to massive relocation, nodes restarting, or some other cluster issues, it's necessary to monitor or define a custom shard allocation.

Getting ready

You will need an up-and-running Elasticsearch installation, similar to the one that we described in the *Downloading and installing Elasticsearch* recipe in *Chapter 1, Getting Started*.

In order to execute the commands, any HTTP client can be used, such as curl (https://curl.haxx.se/) or Postman (https://www.getpostman.com/). You can use the Kibana console, as it provides code completion and better character escaping for Elasticsearch.

To correctly execute the following commands, you will need an index that is populated with the ch04/populate_kibana.sh commands. These are available in the online code.

How to do it...

To get information about the current state of unassigned shard allocations, we will perform the following steps:

1. To retrieve the cluster allocation information, the HTTP method that we use is GET, and the command is as follows:

```
GET /_cluster/allocation/explain
```

The result will be as follows:

```
{ "index" : "mybooks", "shard" : 0,
  "primary" : false, "current_state" : "unassigned",
  "unassigned_info" : {
    "reason" : "INDEX_CREATED",
    "at" : "2018-12-24T09:47:23.192Z",
    "last_allocation_status" : "no_attempt" },
  "can_allocate" : "no",
  "allocate_explanation" : "cannot allocate because
allocation is not permitted to any of the nodes",
  "node_allocation_decisions" : [
```

```
{ "node_id" : "-IFjP29_TOGQF-1axtNMSg",
    "node_name" : "5863a2552d84",
    "transport_address" : "172.18.0.2:9300",
    "node_attributes" : {
        "xpack.installed" : "true" },
    "node_decision" : "no", "weight_ranking" : 1,
    "deciders" : [
        { "decider" : "same_shard",
            "decision" : "NO",
            "explanation" : "the shard cannot be
allocated to the same node on which a copy of the
shard already exists [[mybooks][0], node[-IFjP29_TOGQF-
1axtNMSg], [P], s[STARTED], a[id=4IEkiR-JS7adyFCHN_
GGTw]]" } ] } ] }
```

How it works...

Elasticsearch allows for different shard allocation mechanisms. Sometimes, your shards are not assigned to nodes, and it's useful to investigate why Elasticsearch has not allocated them by querying **the cluster allocation explain API**.

The call returns a lot of information about the unassigned shard, but the most important one is `decisions`. This is a list of objects that explain why the shard cannot be allocated in the node. In the preceding example, the result was `the shard cannot be allocated on the same node id [-IFjP29_TOGQF-1axtNMSg] on which it already exists`, which is returned because the shard needs a replica. However, in this case, the cluster is composed of only one node, so it's not possible to initialize the replicated shard in the cluster.

There's more...

The cluster allocation explain API provides capabilities to filter the result for searching for a particular shard; this is very useful if your cluster has a lot of shards. This can be done by adding parameters to be used as a filter in the `get` body. These parameters are as follows:

- `index`: This is the index that the shard belongs to.

- `shard`: This is the number of the shard; shard numbers start from `0`.

- `primary` (`true` or `false`): This indicates whether the shard to be checked is the primary one or not.

The preceding example shard can be filtered using a similar call as follows:

```
GET /_cluster/allocation/explain
{ "index": "mybooks", "shard": 0, "primary": false }
```

To manually relocate shards, Elasticsearch provides a cluster reroute API that allows the migration of shards between nodes. The following is an example of this API:

```
POST /_cluster/reroute
{ "commands": [
    { "move": {
        "index": "test-index", "shard": 0,
        "from_node": "node1", "to_node": "node2" } } ] }
```

In this case, the 0 shard of the test-index index is migrated from node1 to node2. If you force a shard migration, the cluster starts moving the other shard in order to rebalance itself.

See also

You can refer to the following web pages for more information related to this recipe:

- The official documentation about shard allocations and the settings that control them can be found at https://www.elastic.co/guide/en/elasticsearch/reference/current/shards-allocation.html.

- The cluster reroute API official documentation can be found at https://www.elastic.co/guide/en/elasticsearch/reference/current/cluster-reroute.html. It describes the complexity of the manual relocation of shards in depth.

Monitoring segments with the segment API

Monitoring the index segments means monitoring the health of an index. It contains information about the number of segments and the data that is stored in them.

Getting ready

You will need an up-and-running Elasticsearch installation, similar to the one that we described in the *Downloading and installing Elasticsearch* recipe in *Chapter 1, Getting Started*.

In order to execute the commands, any HTTP client can be used, such as curl (`https://curl.haxx.se/`) or Postman (`https://www.getpostman.com/`). You can use the Kibana console, as it provides code completion and better character escaping for Elasticsearch.

To correctly execute the following commands, you will need an index that is populated with the `ch04/populate_kibana.sh` commands. These are available in the online code.

How to do it...

To get information about index segments, we will perform the following steps:

1. To retrieve the index segments, the HTTP method that we use is GET, and the `curl` command is as follows:

```
GET /mybooks/_segments
```

The result will be as follows:

```
{ ... truncated...
   "indices" : {
      "mybooks" : {
         "shards" : {
            "0" : [
               { "routing" : {
                     "state" : "STARTED",
                     "primary" : true,
                     "node": "eNBAg76DT2-Ul3jNFZf_tA"},
                  "num_committed_segments" : 1,
                  "num_search_segments" : 1,
                  "segments" : {
                     "_0" : {
                        "generation" : 0,
                        "num_docs" : 3,
                        "deleted_docs" : 0,
                        "size_in_bytes" : 6683,
                        "committed" : true,
                        "search" : true,
                        "version" : "9.0.0",
                        "compound" : true,
```

```
                                    "attributes" : {
                               "Lucene90StoredFieldsFormat.
    mode" : "BEST_SPEED" } } } } ] } } }
```

In Elasticsearch, there is the special `alias` `_all` value, which defines all the indices. This can be used in all the APIs that require a list of index names.

How it works...

The indices segment API returns statistics about the segments in an index. This is an important indicator of the health of an index. It returns the following information:

- `num_docs`: The number of documents that are stored in the index.
- `deleted_docs`: The number of deleted documents in the index. If this value is high, then a lot of space is wasted to tombstone documents in the index.
- `size_in_bytes`: The size of the segments in bytes. If this value is too high, the writing speed will be very low.
- `memory_in_bytes`: The memory taken up, in bytes, by the segment.
- `committed`: This indicates whether the segment is committed to the disk.
- `search`: This indicates whether the segment is used for searching. During force merge or index optimization, new segments are created and returned by the API, but they are not available for searching until the end of the optimization.
- `version`: The Lucene version that is used for creating the index.
- `compound`: This indicates whether the index is a compound one.
- `attributes`: This is a key-value list of attributes about the current segment.

The most important elements that are needed to monitor the segments are `deleted_docs` and `size_in_bytes`. This is because they either mean a waste of disk space or that the shard is too large. If the shard is too large (that is, it is over 10 GB), then for improved performances in writing, the best solution is to reindex the index with a large number of shards.

Having large shards also creates a problem in relocating, due to massive data moving between nodes.

It is impossible to define the perfect size for a shard. In general, a good size for a shard that doesn't need to be frequently updated is between 10 GB and 25 GB.

See also

You can refer to the following recipes for more information related to this recipe:

- The *ForceMerge an index* recipe in *Chapter 3, Basic Operations*, is about how to optimize an index with a small number of fragments in order to improve search performances.

- The *Shrinking an index* recipe in *Chapter 3, Basic Operations*, is about how to reduce the number of shards if too large a number of shards are defined for an index.

Cleaning the cache

During its execution, Elasticsearch caches data in order to speed up aspects of searching, such as results, items, and filter results.

Elasticsearch frees up memory automatically by following internal indicators, such as the percentage size of the cache regarding the memory (that is, 20%). If you want to start some performance tests or free up memory manually, then it's necessary to call the cache API.

Getting ready

You will need an up-and-running Elasticsearch installation, similar to the one that we described in the *Downloading and installing Elasticsearch* recipe in *Chapter 1, Getting Started*.

In order to execute the commands, any HTTP client can be used, such as curl (https://curl.haxx.se/) or Postman (https://www.getpostman.com/). You can use the Kibana console, as it provides code completion and better character escaping for Elasticsearch.

To correctly execute the following commands, you will need an index that is populated with the ch04/populate_kibana.sh commands. These are available in the online code.

How to do it...

For cleaning the cache, we will perform the following steps:

1. We call the _cache/clean API on an index as follows:

```
POST /mybooks/_cache/clear
```

If everything is okay, then the result that is returned by Elasticsearch will be as follows:

```
{ "_shards" : { "total" : 2, "successful" : 1,
    "failed" : 0 } }
```

How it works...

The cache clean API frees the memory used to cache values in Elasticsearch – both queries and aggregations.

Generally, it is not a good idea to clean the cache because Elasticsearch manages the cache internally by itself and cleans obsolete values. However, it can be very handy if your node is running out of memory or if you want to force a complete cache clean-up.

If you need to execute a performance test, before firing a query you can execute a clean cache API to have a real-time sample of query execution, without the boost due to the caching.

10
Backups and Restoring Data

Elasticsearch is commonly used as a data store for logs and other kinds of data. Therefore, if you store valuable data, you will also need tools to back up and restore that data to support disaster recovery.

In earlier versions of Elasticsearch, the only viable solution was to dump your data with a complete scan and then reindex it. As Elasticsearch has matured as a complete product, it now supports native functionalities to back the data up and restore it.

In this chapter, we'll explore how you can configure shared storage, using a **Network File System** (**NFS**), to store your backups and how to execute and restore a backup. In the case of cloud and managed deployments, you can use other supported repositories such as Google Cloud Storage, Azure Blob Storage, and Amazon S3.

In the last recipe of the chapter, we will demonstrate how you can use the reindex functionality to clone data between different Elasticsearch clusters. This approach is very useful if you cannot use standard backup or restore functionalities because you are moving from an old Elasticsearch version to a newer one.

In this chapter, we will cover the following recipes:

- Managing repositories
- Executing a snapshot
- Restoring a snapshot
- Setting up an NFS share for backups
- Reindexing from a remote cluster

Managing repositories

Elasticsearch provides a built-in system that rapidly backs up and restores your data. When working with live data, keeping a backup is complex. This is due to a large number of concurrency problems (such as record writing during the backup phases).

An Elasticsearch snapshot allows you to create snapshots of individual indices (or aliases), or an entire cluster, in a remote repository.

Before starting to execute a snapshot, a **repository** must be created – this is where your backups or snapshots will be stored.

Getting ready

You will need an up-and-running Elasticsearch installation – similar to the one that we described in the *Downloading and installing Elasticsearch* recipe of *Chapter 1, Getting Started*.

In order to execute the commands, any HTTP client can be used, such as curl (`https://curl.haxx.se/`) or Postman (`https://www.getpostman.com/`). Additionally, you can use the Kibana console as it provides code completion and better character escaping for Elasticsearch.

We need to edit `config/elasticsearch.yml` and add the directory of our backup repository – `path.repo: /backup/`.

Generally, in a production cluster, the `/backup` directory should be a shared repository.

If you are using the Docker Compose file that was provided in *Chapter 1, Getting Started*, then everything has already been configured for you.

How to do it...

To manage a repository, we will perform the following steps:

1. Create a repository called `my_repository` to store the file inside `/backup/my_repository` (you should choose every path that Elasticsearch can write on). The HTTP method that we use is `PUT`, and the command will be as follows:

    ```
    PUT /_snapshot/my_repository
    { "type": "fs",
      "settings": {
          "location": "/backup/my_repository",
          "compress": true } }
    ```

    The result will be as follows:

    ```
    { "acknowledged" : true }
    ```

    If you check on your filesystem, the `/backup/my_repository` directory will have been created.

2. To retrieve the repository information, the HTTP method that we use is `GET`, and the `curl` command is as follows:

    ```
    GET /_snapshot/my_repository
    ```

    The result will be as follows:

    ```
    {"my_repository" : {
         "type" : "fs",
         "settings" : { "compress" : "true",
             "location" : "/backup/my_repository" } } }
    ```

3. To delete a repository, the HTTP method that we use is `DELETE`, and the `curl` command is as follows:

    ```
    DELETE /_snapshot/my_repository
    ```

    The result will be as follows:

    ```
    { "acknowledged" : true }
    ```

How it works...

Before taking a snapshot of our data, we must create a **repository**, that is, a place where we can store our backup data. The parameters that can be used to create a repository are as follows:

- `type`: This is used to define the type of shared filesystem repository (generally, it is `fs`).

- `settings`: These are options that we can use to set up the shared filesystem repository.

In the case of the `fs` type usage, the settings are as follows:

- `location`: This is the location on the filesystem for storing snapshots.

- `compress`: This turns on the compression for the snapshot files. Compression is only applied to metadata files (that is, the index mapping and settings); data files are not compressed (the default is `true`).

- `chunk_size`: This defines the size of the file chunks during snapshotting. The chunk size can be specified in bytes or by using size value notation (that is, 1GB, 10MB, or 5KB; the default is `disabled`).

- `max_restore_bytes_per_sec`: This controls the throttle per node restore rate (the default is `40mb`).

- `max_snapshot_bytes_per_sec`: This controls the throttle per node snapshot rate (the default is `40mb`).

- `readonly`: This flag defines the repository as read-only (the default is `false`). It is possible to return all of the defined repositories by executing `GET` without providing the repository name:

```
GET /_snapshot
```

The default values for `max_restore_bytes_per_sec` and `max_snapshot_bytes_per_sec` are too low for production environments. Usually, production systems use SSDs or more efficient solutions, so it's better to configure values that are related to your real network and storage performance.

There's more...

The most common type of repository backend is a filesystem, but there are other official repository backends, such as the following:

- **S3 repository**: `https://www.elastic.co/guide/en/elasticsearch/plugins/master/repository-s3.html`

- **HDFS**: `https://www.elastic.co/guide/en/elasticsearch/plugins/master/repository-hdfs.html` for Hadoop environments

- **Azure cloud**: `https://www.elastic.co/guide/en/elasticsearch/plugins/master/repository-azure.html` for Azure storage repositories

- **Google Cloud**: `https://www.elastic.co/guide/en/elasticsearch/plugins/master/repository-gcs.html` for Google Cloud Storage repositories

When a repository is created, it's immediately verified on all of the data nodes to ensure it's functional.

Elasticsearch also provides a manual way to verify the node status repository, which is very useful in order to check the status of the cloud repository storage. The command to manually verify a repository is as follows:

```
POST /_snapshot/my_repository/_verify
```

See also

The official Elasticsearch documentation, which can be viewed at `https://www.elastic.co/guide/en/elasticsearch/reference/master/modules-snapshots.html`, provides a lot of information about borderline cases for repository usage.

Executing a snapshot

In the previous recipe, we defined a repository, that is, the place where we will store the backups. Now we can create snapshots of indices (using the full backup of an index) in the exact instant that the command is called.

For each repository, it's possible to define multiple snapshots.

Getting ready

You will need an up-and-running Elasticsearch installation – similar to the one that we described in the *Downloading and installing Elasticsearch* recipe of *Chapter 1, Getting Started*.

To execute the commands, any HTTP client can be used, such as Curl (https://curl. haxx.se/) or Postman (https://www.getpostman.com/). Additionally, you can use the Kibana console as it provides code completion and better character escaping for Elasticsearch.

To correctly execute the following commands, the repository that was created in the *Managing repositories* recipe is required.

To have a mybooks-1 index with a record, we will execute the following command:

```
POST mybooks-1/_doc/1
{
    "name":"alberto"
}
```

How to do it...

To manage a snapshot, we will perform the following steps:

1. To create a snapshot called snap_1 for the index*, mybooks*, and mygeo* indices, the HTTP method that we use is PUT. The command is as follows:

```
PUT /_snapshot/my_repository/snap_1?wait_for_
completion=true
{ "indices": "index*,mybooks*,mygeo*",
    "ignore_unavailable": "true",
    "include_global_state": false }
```

The result will be as follows:

```
{ "snapshot" : {
    "snapshot" : "snap_1",
    "uuid" : "OhOpA-8sRYOdrstwgY2sFA",
    "repository" : "my_repository",
    "version_id" : 8000099, "version" : "8.0.0",
    "indices" : [ "index-agg", "mybooks",
    "mygeo-index", "mybooks-join" ],
    "data_streams" : [ ],
    "include_global_state": false,"state": "SUCCESS",
    "start_time" : "2022-02-19T18:40:13.571Z",
    "start_time_in_millis" : 1645296013571,
    "end_time" : "2022-02-19T18:40:14.995Z",
    "end_time_in_millis" : 1645296014995,
    "duration_in_millis" : 1424, "failures" : [ ],
    "shards" : { "total" : 4, "failed" : 0,
        "successful" : 4 },
    "feature_states" : [ ] } }
```

2. If you check your filesystem, you will see that the /backup/my_repository directory has been populated with several files, such as index (this is a directory that contains our data), metadata-snap_1, and snapshot-snap_1.

3. To retrieve the snapshot information, the HTTP method that we use is GET. The command is as follows:

```
GET /_snapshot/my_repository/snap_1
```

The result will be the same as the previous step.

4. To delete a snapshot, the HTTP method that we use is DELETE. The command is as follows:

```
DELETE /_snapshot/my_repository/snap_1
```

The result will be as follows:

```
{ "acknowledged" : true }
```

How it works...

The minimum configuration that is required to create a snapshot is the name of the repository and the name of the snapshot (that is, `snap_1`).

If no other parameters are set, then the snapshot command will dump all of the cluster data. To control the snapshot process, the following parameters are available:

- `indices` (a comma-delimited list of indices; wildcards are accepted): This controls the indices that must be dumped.

- `ignore_unavailable` (the default is `false`): This prevents the snapshot from failing if some indices are missing.

- `include_global_state` (this defaults to `true`; the available values are `true`, `false`, and `partial`): This controls the storing of the global state in the snapshot. If a primary shard is not available, then the snapshot fails.

The `wait_for_completion` query argument allows you to wait for the snapshot to end before returning the call. It's very useful if you want to automate your snapshot script to sequentially back up indices.

If the `wait_for_completion` argument is not set, then in order to check the snapshot status, a user must monitor it using the snapshot's `GET` call.

The snapshots are incremental; this means that only changed files are copied between two snapshots of the same index. This approach reduces both the time and disk usage during snapshots.

The snapshot process is designed to be as fast as possible, so it implements a direct copy of the Lucene index segments in the repository. To prevent any changes and index corruption that might occur during the copy, all the segments that need to be copied are blocked from changing until the end of the snapshot.

Lucene's segment copy is at the shard level. So, if you have a cluster of several nodes and you have a local repository, the snapshot is spread through all the nodes. For this reason, in a production cluster, the repository must be shared to easily collect all the backup fragments.

Elasticsearch takes care of everything during a snapshot, including preventing data being written to files that are in the snapshot process and managing the cluster events (such as shard relocating, failures, and more).

To retrieve all the available snapshots for a repository, the command to use is as follows:

```
GET /_snapshot/my_repository/_all
```

There's more...

The snapshot process can be monitored using the `_status` endpoint, which provides a complete overview of the snapshot status.

For the current example, the snapshot's `_status` API call will be as follows:

```
GET /_snapshot/my_repository/snap_1/_status
```

The result is very long and consists of the following sections:

- Here is the information about the snapshot:

```
"snapshots" : [{
        "snapshot" : "snap_1",
        "repository" : "my_repository",
        "uuid" : "_uWUZ_n3RAmAM5NWszS0hg",
        "state" : "SUCCESS",
        "include_global_state" : false,
```

- Here are the global shard's statistics:

```
"shards_stats" : {
    "initializing" : 0, "started" : 0,
    "finalizing" : 0, "done" : 4, "failed" : 0,
    "total" : 4 },
```

- Here are the snapshot's global statistics:

```
"stats" : {
    "incremental" : { "file_count" : 16,
        "size_in_bytes" : 837344   },
    "total" : { "file_count" : 16,
        "size_in_bytes" : 837344 },
    "start_time_in_millis" : 1546779914447,
    "time_in_millis" : 52 },
```

- Here is a list of the snapshot index statistics:

```
"indices" : {
    "mybooks-join" : {
      "shards_stats" : {
          "initializing" : 0, "started" : 0,
```

```
      "finalizing" : 0, "done" : 1, "failed" : 0,
      "total" : 1 },
   "stats" : {
      "incremental" : { "file_count" : 4,
         "size_in_bytes" : 10409   },
         "total" : { "file_count" : 4,
            "size_in_bytes" : 10409 },
         "start_time_in_millis" : 1546779914449,
         "time_in_millis" : 15   },
```

- Here are the statistics for each index and shard:

```
  "shards" : {
    "0" : {
       "stage" : "DONE",
       "stats" : {
          "incremental" : { "file_count" : 4,
             "size_in_bytes" : 10409 },
          "total" : { "file_count" : 4,
             "size_in_bytes" : 10409 },
          "start_time_in_millis" : 1546779914449,
          "time_in_millis" : 15 }}}}, … truncated…
```

The status response is very rich. It can also be used to estimate the performance of the snapshot and the size that is required in time for the incremental backups.

Restoring a snapshot

Once you have snapshots of your data, it can be restored. The restoration process is very fast – the indexed shard data is simply copied onto the nodes and activated.

Getting ready

You will need an up-and-running Elasticsearch installation—similar to the one that we described in the *Downloading and installing Elasticsearch* recipe of *Chapter 1, Getting Started*.

To execute the commands, any HTTP client can be used, such as Curl (`https://curl.haxx.se/`) or Postman (`https://www.getpostman.com/`). Additionally, you can use the Kibana console as it provides code completion and better character escaping for Elasticsearch.

To correctly execute the following commands, the backup that was created in the *Executing a snapshot* recipe is required.

How to do it...

To restore a snapshot, we will perform the following steps:

1. To restore a snapshot called `snap_1` for the `mybooks-*` indices, the HTTP method that we use is POST. The command is as follows:

```
POST /_snapshot/my_repository/snap_1/_restore
{ "indices": "mybooks-*",
  "ignore_unavailable": "true",
  "include_global_state": false,
  "rename_pattern": "mybooks-(.+)",
  "rename_replacement": "copy_$1" }
```

    The result will be as follows:

```
{ "accepted" : true }
```

2. The restore is finished when the cluster state changes from `red` to `yellow` or `green`.

    In this example, the `"mybooks-*"` index pattern matches `mybooks-join`. The `rename_pattern` parameter captures `"join"`, and the new index will be generated by `rename_placement` and `"copy-join"`.

How it works...

The restoration process is very fast; the process comprises the following steps:

1. The data is copied on the primary shard of the restored index (during this step, the cluster is in the `red` state).
2. The primary shards are recovered (during this step, the cluster turns from `red` to `yellow` or `green`).
3. If a replica is set, then the primary shards are copied onto other nodes.

It's possible to control the restoration process using some parameters, as follows:

- `indices`: This controls the indices that must be restored. If they are not defined, then all indices in the snapshot will be restored (a comma-delimited list of indices; wildcards are accepted).

- `ignore_unavailable`: This stops the restoration process from failing if some indices are missing (the default is `false`).

- `include_global_state`: This allows the restoration of the global state from the snapshot (this defaults to `true`; the available values are `true` and `false`).

- `rename_pattern` and `rename_replacement`: The first one is a pattern that must be matched, and the second one uses regular expression replacement to define a new index name.

- `partial`: If set to `true`, this allows the restoration of indices with missing shards (the default is `false`).

Setting up an NFS share for backups

Managing the repository (where the data is stored) is the most crucial part of Elasticsearch backup management. Due to its native distributed architecture, the snapshot and the restoration process are designed in a cluster style.

During a snapshot, the shards are copied to the defined repository. If this repository is local to the nodes, then the backup data is spread across all the nodes. For this reason, it's necessary to have shared repository storage if you have a multi-node cluster.

A common approach is to use an NFS, as it's very easy to set up, and it's a very quick solution (additionally, standard Windows Samba shares can be used).

Getting ready

We have a network with the following nodes:

- Host server: `192.168.1.30` (where we will store the backup data)
- Elasticsearch master node 1: `192.168.1.40`
- Elasticsearch data node 1: `192.168.1.50`
- Elasticsearch data node 2: `192.168.1.51`

You will need an up-and-running Elasticsearch installation – similar to the one that we described in the *Downloading and installing Elasticsearch* recipe of *Chapter 1, Getting Started*.

In order to execute the commands, any HTTP client can be used, such as Curl (`https://curl.haxx.se/`) or Postman (`https://www.getpostman.com/`). Additionally, you can use the Kibana console as it provides code completion and better character escaping for Elasticsearch.

The following instructions are for standard Debian or Ubuntu distributions; note that they can be easily changed for another Linux distribution.

How to do it...

To create an NFS shared repository, we need to execute the following steps onto the NFS server:

1. Install the NFS server (using the `nfs-kernel-server` package) on the host server. On the `192.168.1.30` host server, we will execute the following commands:

```
sudo apt-get update
sudo apt-get install nfs-kernel-server
```

2. Once the package has been installed, create a directory to be shared among all the clients:

```
sudo mkdir /mnt/shared-directory
```

3. Give access permissions for this directory to the `nobody` user and the `nogroup` group. `nobody` and `nogroup` are special user/group values that are used to allow share read/write permissions. To apply them, you need root access. Execute the following command:

```
sudo chown -R nobody:nogroup /mnt/shared-directory
```

4. Then, we need to configure the NFS exports, where we can specify that this directory will be shared with certain machines. Edit the /etc/exports file (sudo nano /etc/exports), and add the following lines containing the directory that is to be shared along with a list of client IPs that are allowed to access the exported directory:

```
/mnt/shared-directory 192.168.1.40(rw,sync,no_subtree_
check)
192.168.1.50(rw,sync,no_subtree_check)
192.168.1.51(rw,sync,no_subtree_check)
```

5. To refresh the NFS table that holds the export of the share, the following command must be executed:

```
sudo exportfs -a
```

6. Finally, we can start the NFS service by running the following command:

```
sudo service nfs-kernel-server start
```

After the NFS server is up and running, we need to configure the clients. We'll repeat the following steps inside each Elasticsearch node:

1. Install the NFS client inside our Elasticsearch node:

```
sudo apt-get update
sudo apt-get install nfs-common
```

2. Now, create a directory on the client machine, and we'll try to mount the remote shared directory:

```
sudo mkdir /mnt/nfs
sudo mount 192.168.1.30:/mnt/shared-directory /mnt/nfs
```

3. If everything is fine, we can add the mount directory to our /etc/fstab node file. This is so that it will be mounted during the next boot:

```
sudo nano /etc/fstab
```

4. Then, add the following lines to this file:

```
192.168.1.30:/mnt/shared-directory /mnt/nfs/ nfs
auto,noatime,nolock,bg,nfsvers=4,sec=krb5p,intr,tcp,ac-
timeo=1800 0 0
```

5. We can update the Elasticsearch node configuration (`config/elasticsearch.yml`) of `path.repo` as follows:

```
path.repo: /mnt/nfs/
```

6. After having restarted all the Elasticsearch nodes, we can create our shared repository on the cluster using a single standard repository creation call:

```
PUT /_snapshot/my_repository
{ "type": "fs",
  "settings": {
      "location": "/mnt/nfs/my_repository",
      "compress": true } }
```

How it works...

NFS is a distributed filesystem protocol that is very common in the Unix world – it allows you to mount remote directories onto your server. The mounted directories look like the local directory of the server. Therefore, by using NFS, multiple servers can write to the same directory.

This is very handy if you need to do a shared backup; this is because all the nodes will read/write from the same shared directory.

If you need to snapshot an index that will be rarely updated, such as an old time-based index, the best practice is to optimize it before backing it up, cleaning up any deleted documents, and reducing the Lucene segments.

Reindexing from a remote cluster

The snapshot and restore APIs are very fast and are the preferred way to back up data. However, they do have some limitations:

- The backup is a safe Lucene index copy, so it depends on the Elasticsearch version that has been used. If you are switching from a version of Elasticsearch that is earlier than version 5.x, then it's not possible to restore the old indices.

- It's not possible to restore the backups of a newer Elasticsearch version in an older version; the restore is only forward compatible.

- It's not possible to restore partial data from a backup.

To be able to copy data in this scenario, the solution is to use the reindex API using a remote server.

Getting ready

You will need an up-and-running Elasticsearch installation – similar to the one that we described in the *Downloading and installing Elasticsearch* recipe of *Chapter 1, Getting Started*.

In order to execute the commands, any HTTP client can be used, such as Curl (https://curl.haxx.se/) or Postman (https://www.getpostman.com/). Additionally, you can use the Kibana console as it provides code completion and better character escaping for Elasticsearch.

How to do it...

To copy an index from a remote server, we need to execute the following steps:

1. We need to add the remote server address in the config/elasticsearch.yml section using reindex.remote.whitelist, as follows:

    ```
    reindex.remote.whitelist: ["192.168.1.227:9200"]
    ```

2. After having restarted the Elasticsearch node to take the new configuration, we can call the reindex API to copy test-source index data in test-dest using the remote REST endpoint:

    ```
    POST /_reindex
    { "source": {
        "remote": { "host": "http://192.168.1.227:9200"},
        "index": "test-source" },
      "dest": { "index": "test-dest" } }
    ```

    The result will be similar to a local reindex, as we saw in the *Reindexing an index* recipe of *Chapter 3, Basic Operations*.

How it works...

The reindex API allows you to call a remote cluster. Note that every version of the Elasticsearch server is supported.

The reindex API executes a scan query on the remote index cluster and puts the data in the current cluster. This process can take a lot of time, depending on the amount of data that needs to be copied and the time that is required to index the data.

The source section contains important parameters to control the fetched data, such as the following:

- `remote`: This is a section that contains information regarding the remote cluster connection.

- `index`: This is the remote index that has to be used to fetch the data; it can also be an alias or multiple indices via glob patterns.

- `query`: This parameter is optional; it's a standard query that can be used to select the document that must be copied.

- `size`: This parameter is optional and the buffer is up to 200 MB – that is, the number of the documents to be used for the bulk read and write operations.

The `remote` section of the configuration is composed of the following parameters:

- `host`: The remote REST endpoint of the cluster

- `username`: The username to be used for copying the data (this is an optional parameter)

- `password`: The password for the user to access the remote cluster (this is an optional parameter)

There are a lot of advantages to using this approach on a standard snapshot and restore, including the following:

- The ability to copy data from older clusters (that is, version 1.x or later).

- The ability to use a query to copy from a selection of documents. This is very handy for copying data from a production cluster to a development or test one.

See also

For further details, I suggest that you take a look at the following references:

- The *Reindexing an index* recipe of *Chapter 3, Basic Operations*.

- The official Elasticsearch documentation can be viewed at `https://www.elastic.co/guide/en/elasticsearch/reference/master/docs-reindex.html`. This provides more detailed information about the reindex API and some borderline cases for using this API.

11
User Interfaces

In an Elasticsearch ecosystem, it can be immensely useful to monitor nodes and clusters in order to manage and improve their performance and state.

Detecting malfunction or bad performance can be done through the API or through some frontends that are designed to be used in Elasticsearch.

Some of the frontends introduced in this chapter will allow you to have a working web dashboard in your Elasticsearch data; these work by monitoring cluster health, backing up or restoring your data, and allowing test queries before implementing them in the code. In this chapter, we will only briefly examine these frontends; this is due to their complexity and the large number of features, which are beyond the scope of this book. For an in-depth description, I suggest that you have a look at the official documentation of **Kibana**, which is available at `https://www.elastic.co/guide/en/kibana/current/index.html`.

In this chapter, we will explore some aspects of Kibana (covering all of Kibana's functionalities is beyond the scope of this book).

Grafana (`https://grafana.com/`) is another open source solution, which is used to visualize Elasticsearch data and monitor **Public Key Infrastructure** (**PKI**); however, it is not covered in this book.

In this chapter, we will cover the following recipes:

- Installing Kibana
- Managing Kibana Discover
- Visualizing data with Kibana
- Using Kibana Dev Tools

Installing Kibana

The most well-known Elasticsearch interface is Kibana, and it is always released with Elasticsearch.

Kibana is a pluggable interface and is free to use with Elasticsearch. It provides data visualization, and it can be extended with a commercial product called **X-Pack** that provides security, graph capabilities, cluster monitoring, and many other features.

In this chapter, we will mainly cover the Kibana core components. Kibana with X-Pack offers a lot of functionalities and, as these are beyond the scope of this book, I suggest that you look for books related to Kibana for a full description of all Kibana's capabilities.

Getting ready

You will need an up-and-running Elasticsearch installation, similar to the one that we described in the *Downloading and installing Elasticsearch* recipe in *Chapter 1, Getting Started*.

If you are installing using Docker Compose, which is available in the ch01 directory, then you don't need to manually install it.

The Kibana version must be the same version of Elasticsearch, so if you update your Elasticsearch cluster, then it is best practice to update the Kibana nodes as well.

How to do it...

To install Kibana, we will perform the following steps:

1. Download a binary version of the Elasticsearch website and unpack it. For Linux, the commands are as follows:

```
wget https://artifacts.elastic.co/downloads/kibana/
kibana-7.13.2-linux-x86_64.tar.gz
tar xfvz kibana-7.13.2-linux-x86_64.tar.gz
```

2. For macOS, you can install Kibana using the following command:

```
brew install kibana
```

How it works...

Kibana is the official Elasticsearch frontend. It's an analytics and visualization platform based on ReactJS that works with Elasticsearch. It is served by a Node.js backend web server.

Kibana allows you to navigate data in Elasticsearch and organize it in dashboards that are created, shared, and updated in real time.

After setting up and starting Elasticsearch and Kibana, you can navigate to Kibana using `http://localhost:5601`, as shown in the following screenshot:

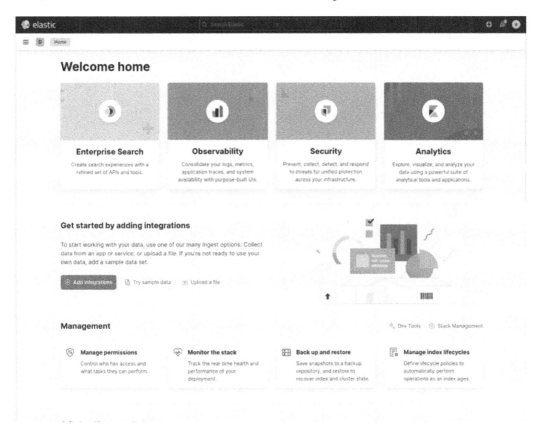

Figure 11.1 – Kibana home page

Before playing with Kibana, I suggest that you load some datasets, which are provided in the installation. Just click on **Try sample data** (the button in the middle next to the blue **Add integrations** button) and add a Kibana dashboard in the **More ways to add data** title:

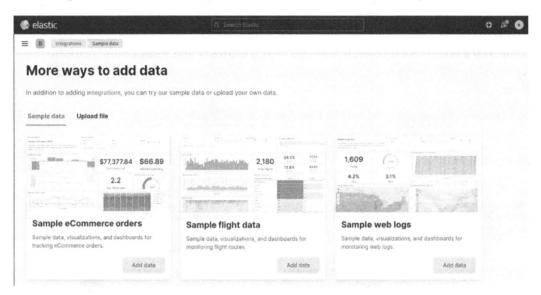

Figure 11.2 – Kibana sample dataset

The Kibana installation provides some data samples that you can start to use. These datasets are very handy because they show many of the advanced Kibana dashboard capabilities; just click on **Add data** to initialize all the demonstrations that you need.

If you select the first data sample, you will gain access to a full feature dashboard, as follows:

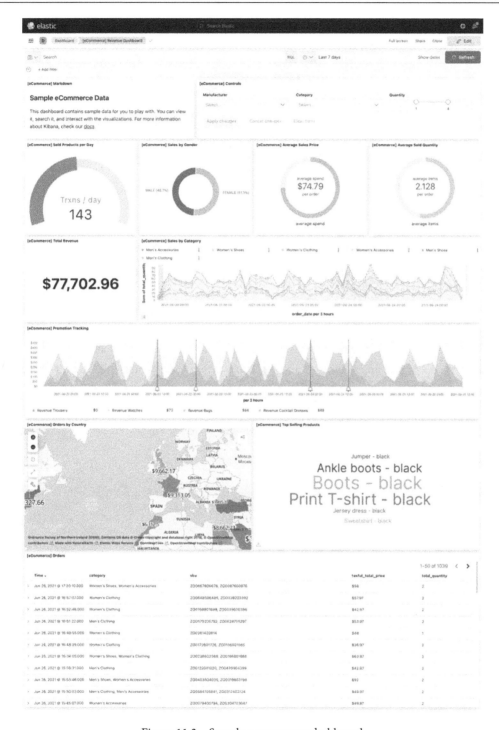

Figure 11.3 – Sample e-commerce dashboard

On the left side of the dashboard page, you have the navigation bar (this will change from version to version because new features are always added with every release):

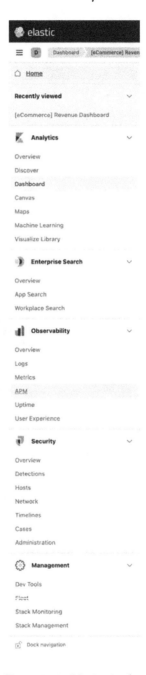

Figure 11.4 – Navigation bar

From the navigation bar, you have access to the following macro sections:

- **Spaces** (the green **D**): This is a way to group your interface in its own defined spaces.

- The **Analytics** section with the following parts:

  - **Overview**: This is used to present an overview of Kibana capabilities.

  - **Discover**: This is used to navigate your data in order to discover information.

  - **Dashboard**: This section hosts your **Space** dashboards.

  - **Canvas**: This section allows you to create pixel art dashboards that are similar to infographics.

  - **Maps**: This section allows you to manage maps.

  - **Machine Learning**: This section allows you to manage machine learning capabilities.

  - **Visualize Library**: This is used to create visualizations that can be used to populate pages.

- **Enterprise Search**: This section allows you to manage integration with enterprise search solutions by Elastic, such as **App Search** and **Workplace Search**.

- **Observability**: This section allows you to manage and monitor an infrastructure with the following parts:

  - **Logs**: This section is used to manage logs.

  - **Metrics**: This section is used to collect different application metrics.

  - **APM** (**Application Performance Monitoring**): This section is used to manage the Elastic APM application.

  - **Uptime**: This section is used to manage the uptime of applications.

- **Security**: This section collects all the stack components that allow you to manage security aspects. It's part of the **security information and event management** (**SIEM**) components of Elastic.

  - **Detections**: This section allows you to define detection engines.

  - **Hosts**: This section allows you to manage the hosts in the SIEM solution.

  - **Network**: This section allows you to manage the hosts in the SIEM solution.

  - **Timelines**: This section allows you to investigate events in timelines.

- **Cases**: This section allows you to manage security issues and send the cases to external systems, such as ServiceNow, Jira, or IBM Resilient.

- **Administration**: This section allows you to manage security endpoints.

- **Management**: This section lets you manage all the **Elasticsearch, Logstash, and Kibana (ELK)** stack components and manually execute REST commands to Elasticsearch. It consists of the following parts:

  - **Dev Tools**: This section allows you to manually send REST commands to Elasticsearch, profile queries, debug Grok statements, and test Painless scripts.

  - **Fleet**: This section enables you to add and manage integrations for popular services and platforms, as well as manage Elastic Agent installations.

  - **Stack Monitoring**: This section is used to monitor your node functionalities and the cluster overall.

  - **Stack Management**: This section allows you to configure the ingest pipelines, indices life cycle, alerts, and Kibana and stack licenses.

The preceding menu can change depending on Kibana configurations that can disable some sections, or on new functionalities added on every Kibana release.

See also

For further material not covered in this recipe, I suggest visiting the following links:

- An overview of Kibana is available at `https://www.elastic.co/products/kibana`.

- The Elastic Enterprise Search overview is at `https://www.elastic.co/enterprise-search`.

- The Elastic Observability overview is at `https://www.elastic.co/observability`.

- The Elastic Security overview is at `https://www.elastic.co/security`.

Managing Kibana Discover

One of the most popular aspects of Kibana is the Discover dashboard. This is because it allows you to dynamically navigate your data. With the evolution of Kibana, a lot of new features have been added to the Discover dashboard in order to allow you to easily filter and analyze your data.

Getting ready

You will need an up-and-running Elasticsearch installation, similar to the one that we described in the *Downloading and installing Elasticsearch* recipe in *Chapter 1, Getting Started*. Additionally, a working Kibana instance is required, as described in the *Installing Kibana* recipe of this chapter.

If you have used Docker Compose, which is available in the ch01 directory, then everything should be correctly installed.

How to do it...

For managing Kibana dashboards, we will perform the following steps:

1. Access the **Analytics | Discover** section of Kibana, as shown in the following screenshot:

Figure 11.5 – Kibana Discover

2. Now, you can play with, and analyze, your indexed data.

The **Discover** section is one of the most used Kibana parts to analyze data. In the case of logs, this is the main entry point to start analyzing issues with your servers and services.

How it works...

The **Discover** section is designed to allow you to explore your data.

You can save and share your created Discover dashboard, which can then be reused to build other dashboards. In the middle of the screen, you should be able to view your documents, which are available in tabular and JSON formats:

Figure 11.6 – View of a record on the Discover dashboard

As you can see from the preceding screenshot, when you hover over a field, special actions are enabled:

- **Filter using this value**
- **Filter not using this value**
- **Toggle column in table**
- **Filter for field present** (exists query)

Sometimes, your data is not shown; this is mainly because you have selected the wrong date range.

You can change it easily from the calendar drop-down menu, as follows:

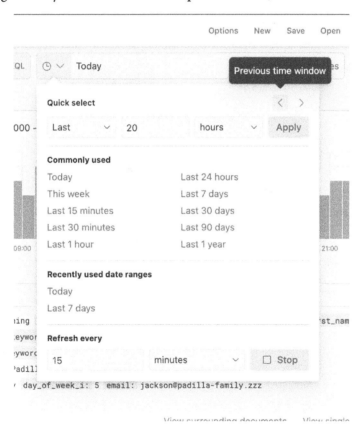

Figure 11.7 – Time management in Kibana

The majority of the filtering is done from the search box. Here, you can provide Google-like syntax in order to search for your data quickly:

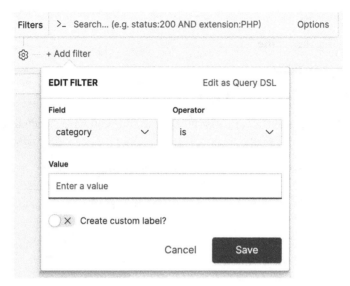

Figure 11.8 – Custom filter in Kibana

You can add a filter using the following two options:

- Using the web interface: This is for simple filters. All the fields are available in the drop-down menus. In this way, building the query should be very easy.

- Using **Edit as Query DSL**: This allows you to input your complex JSON.

You can also create filters using the facets created automatically when you select a field on the left side of the screen:

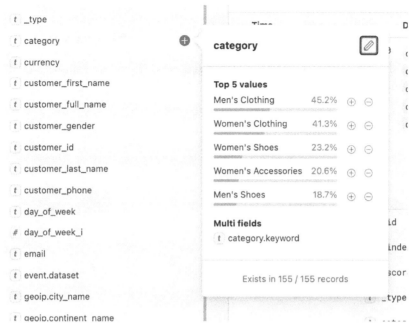

Figure 11.9 – Facet filtering in Kibana

For every field, the interface suggests the most used values in order to easily use them as a filter. Using the + and - symbols lets you choose whether to filter or unfilter values.

Visualizing data with Kibana

Kibana allows you to create reusable data representations called **visualizations**. These are representations of aggregations and can be used to power up the dashboard using custom graphs. In general, you can consider visualization as a building block for your dashboard.

Getting ready

You will need an up-and-running Elasticsearch installation, similar to the one that we described in the *Downloading and installing Elasticsearch* recipe in *Chapter 1, Getting Started*. Additionally, a working Kibana instance is required, as described in the *Installing Kibana* recipe of this chapter.

If you have used Docker Compose, which is available in the ch01 directory, then everything should be correctly installed.

How to do it...

To use Kibana to create custom widgets, we will perform the following steps:

1. Access the **Analytics | Visualize Library** section of Kibana, as shown in the following screenshot:

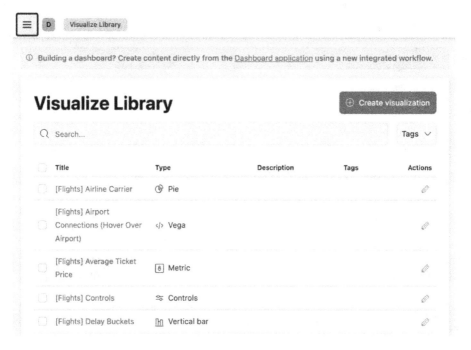

Figure 11.10 – Visualize Library page

Now, we can choose the visualization that we want to create, as shown in the following screenshot:

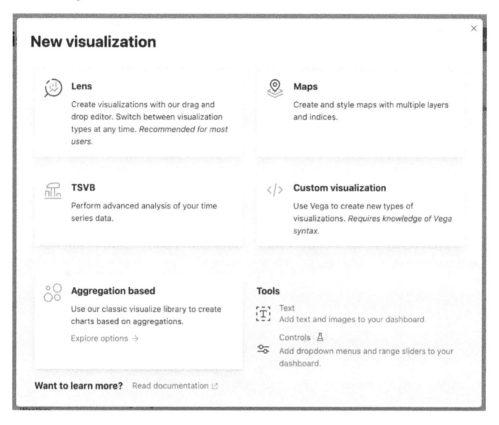

Figure 11.11 – Visualization selection page

2. If we want to create a **Tag cloud** visualization under the **Aggregation based** tile, then we select it and populate the required fields, as shown in the following screenshot:

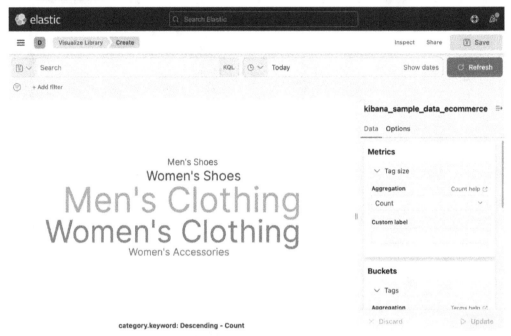

Figure 11.12 – Custom tag cloud visualization

Mastering the use of visualization is the main part of mastering the dashboards, as a dashboard is a collection of visualizations.

How it works...

Kibana allows you to define five main types of visualizations (they can be extended with plugins), as follows:

- **Lens**: This is an interface similar to Power BI, Qlik, or other tools, in which you drag and drop fields to build your visualization:

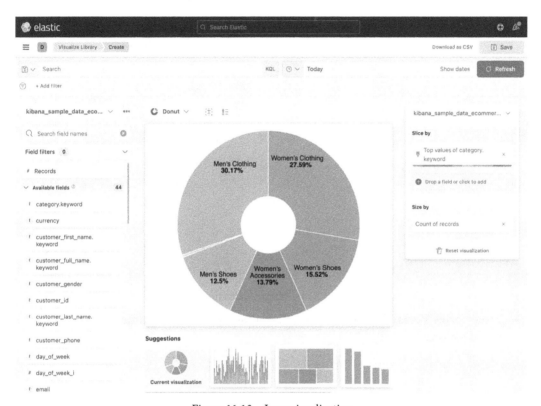

Figure 11.13 – Lens visualization

- **Maps**: This is designed for geodata visualizations:

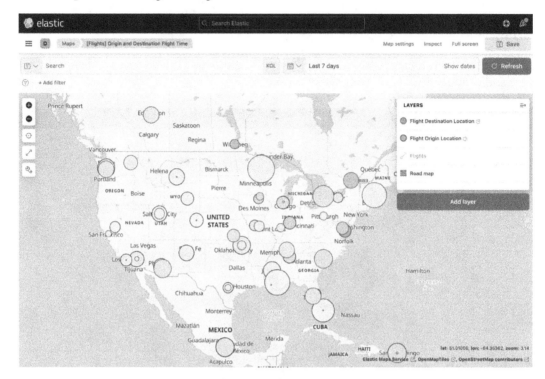

Figure 11.14 – Map visualization

- **Time Series Visual Builder (TSVB)**: This allows you to create time series chart visualizations:

Figure 11.15 – Time series visualization

- **Custom**: This uses Vega (`https://vega.github.io/vega/`), a customizable JavaScript canvas library, to render visualizations.

- **Aggregation based**: This uses the standard visualization based on aggregations. For this kind of family, the built-in visualizations are as follows:

  - An **Area** chart: This is useful for representing stacked timelines.

  - **Data table**: This allows you to create a data table using the aggregation results.

  - **Gauge**: This is useful for showing range values.

  - **Goal**: This is useful for showing the number count.

  - **Heat Map**: This shows data in heat maps.

  - A **Horizontal/Vertical bar** chart: This is the general-purpose bar representation for histograms.

  - A **Line** chart: This is useful for representing time-based hits and comparing them.

  - **Metric**: This represents a numeric metric value.

- **Pie**: This is useful for representing low cardinality values.

- **Tag cloud**: This is useful for representing term values, such as tags and labels.

- **Timelion**: This allows to create a visualization using the Timelion time-series language.

• Other special visualizations, such as the following:

- The **Text/Markdown** widget: This is useful for displaying explanations or instructions for dashboards.

- **Controls**: These are useful for extending filtering.

After selecting a visualization, a custom form is presented on the left, which allows you to populate all the required values. On the right, there is the widget representation that is updated in near-real-time with the result of the queries and aggregations.

After the configuration of the visualization is complete, it must be saved in order to be used as a widget in the dashboard, which can be created via the **Analytics | Dashboard** section:

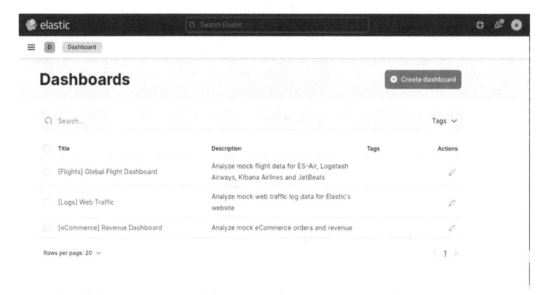

Figure 11.16 – Dashboard page

After selecting **Create dashboard**, you can start editing it by adding your saved visualizations. You can do this by clicking on the **Add from library** button, as shown in the following screenshot:

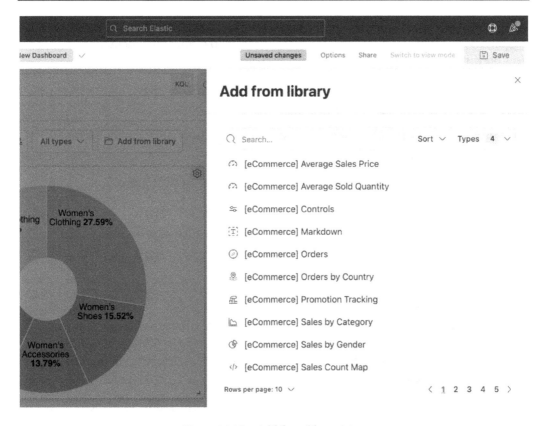

Figure 11.17 – Add from library view

The dashboard top menu depends on the context, but it allows you to do the following:

- Set the dashboard options.
- Save the current dashboard.
- Edit the current dashboard.
- Go full screen (exit with the *Esc* key).
- Share a dashboard or a dashboard snapshot (with the date/time value fixed) using a link.
- If you are using auto refresh dashboards, then you can pause auto refresh using the pause icon. By clicking on the refresh interval, you can change it to suit.
- Change or define the time interval range by clicking on the time range value.

Internally, the Kibana dashboards are stored in an Elasticsearch. `kibana*` special indices; for any kind of asynchronous task, the data is read from this index.

We have only scratched the surface of the powerful Kibana dashboard in this recipe. I suggest that you buy a book related to Kibana, or refer to online documentation or videos on Kibana, as it is a tool that has very rich capabilities.

Using Kibana Dev Tools

Kibana provides a very useful section for developers: Dev Tools. This section contains four tools:

- **Console**: The place where the developer tests and executes commands.
- **Search Profiler**: A tool that is used to profile queries.
- **Grok Debugger**: This is useful for debugging Grok regular expressions.
- **Painless Lab** (beta): This enables you to test and debug Painless scripts.

Getting ready

You will need an up-and-running Elasticsearch installation, similar to the one that we described in the *Downloading and installing Elasticsearch* recipe in *Chapter 1, Getting Started*. Additionally, a working Kibana instance is required, as described in the *Installing Kibana* recipe of this chapter.

If you have used Docker Compose, which is available in the ch01 directory, then everything should be correctly installed.

How to do it...

To use Console, we will perform the following steps:

1. Access the **Dev Tools** section of Kibana, as shown in the following screenshot:

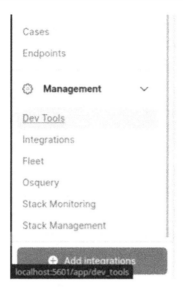

Figure 11.18 – Dev Tools section

2. Now, we can use Console to create, execute, and test queries and other Elasticsearch HTTP APIs that are using it, as shown in the following screenshot:

Figure 11.19 – Example of query executed via Console

Console is one of the most-used components of Kibana because it's the only place where you can put arbitrary REST code to send to Elasticsearch to be able to test all kinds of APIs.

How it works...

Kibana's Console is very similar to the Cerebro interface that we mentioned previously. It allows you to execute every kind of REST API call through the `http/https` interface to Elasticsearch. It can be used for several purposes, including the following:

- Creating complex queries and aggregations: The console interface helps the user by providing code completion and syntax checking during editing.

- Analyzing the returned results: This is very useful for checking particular aggregation responses or the structure of the API answers.

- Testing or debugging queries before embedding them in your application code.

- Executing REST services that are now wrapped in Elasticsearch interfaces, such as repository, snapshot, and restore services.

Console's autocompletion of any query and aggregation helps users to build complex queries quickly.

There's more...

The Kibana Dev Tools also provide support to drill down into the times needed to execute a particular query using the **Search Profiler** section, as shown in the following screenshot:

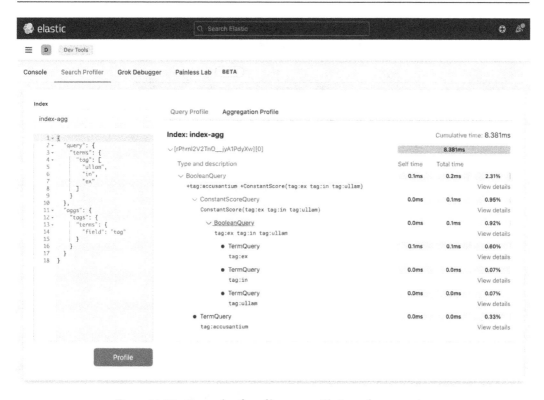

Figure 11.20 – Example of profile query with times for every step

As the execution of a query with some aggregation can be very complex and it can take a lot of time to profile the query, this is the most advanced interface available in Elasticsearch for profiling query executions.

Another section of **Dev Tools** is **Grok Debugger** that allows you to debug complex Grok expressions to save time in using the Grok ingester processor.

Figure 11.21 – Example of debugging a Grok expression

The last section of **Dev Tools** is **Painless Lab**, which is in beta, and allows you to test and debug Painless scripts.

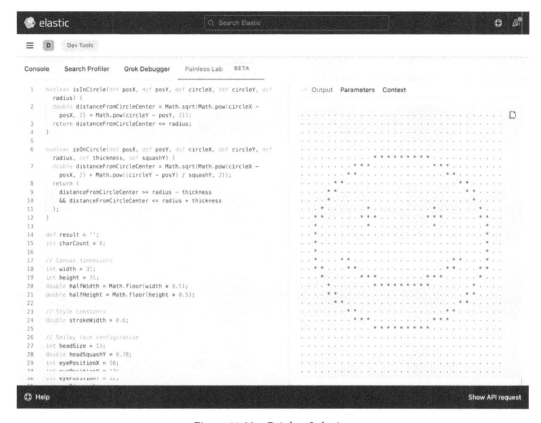

Figure 11.22 – Painless Lab view

Due to the embedded nature of Painless, it is only testable using **Painless Lab**.

See also

You can refer to the following links for more information:

- The official documentation about the Dev Tools section in Kibana is found at
  `https://www.elastic.co/guide/en/kibana/master/devtools-kibana.html`.

- Console official documentation is found at `https://www.elastic.co/guide/en/kibana/master/console-kibana.html`.

- Profiling query and aggregation section official documentation is found at `https://www.elastic.co/guide/en/kibana/master/xpack-profiler.html`.

- Debugging Grok official documentation is found at `https://www.elastic.co/guide/en/kibana/master/xpack-grokdebugger.html`.

- Painless Lab official documentation is found at `https://www.elastic.co/guide/en/kibana/master/painlesslab.html`.

12
Using the Ingest Module

Elasticsearch 8.x introduces a set of powerful functionalities that target the problems that arise during the ingestion of documents via the ingest node.

In *Chapter 1*, *Getting Started*, we discussed that the Elasticsearch node can have different roles and the main important ones are `master`, `data`, and `ingest`; the idea of splitting the `ingest` component from the others is to create a more stable cluster due to problems that can arise when preprocessing documents (mainly due to the custom plugin used in the ingest part, which could require restarting the ingest nodes to be updated).

To create a more stable cluster, the ingest nodes should be isolated by the `master` nodes (and possibly also from the `data` ones) in case some problems occur, such as a crash due to plugins such as the attachment plugin and high loads due to complex type manipulation.

The ingestion node can replace a Logstash installation in simple scenarios.

In this chapter, we will cover the following recipes:

- Pipeline definition
- Inserting an ingest pipeline
- Getting an ingest pipeline
- Deleting an ingest pipeline
- Simulating an ingest pipeline
- Built-in processors
- The grok processor
- Using the ingest attachment plugin
- Using the ingest GeoIP plugin
- Using the enrichment processor

Pipeline definition

The job of `ingest` nodes is to preprocess documents before sending them to the `data` nodes. This process is called a pipeline definition and every single step of this pipeline is a processor definition.

Getting ready

You need an up-and-running Elasticsearch installation, as we described in the *Downloading and installing Elasticsearch* recipe in *Chapter 1, Getting Started*.

To execute these commands, any HTTP client can be used, such as **curl** (`https://curl.haxx.se/`), **Postman** (`https://www.getpostman.com/`), or similar. We will use the Kibana console, as it provides code completion and better character escaping for Elasticsearch.

How to do it...

To define an ingestion pipeline, you need to provide a description and some processors, as follows.

Define a pipeline that adds a field called `user` with the value `john`:

```
{ "description": "Add user john field",
  "processors": [
```

```
        { "set": { "field": "user",   "value": "john" } } ],
    "version": 1
}
```

How it works...

The generic template representation for a pipeline is as follows:

```
{ "description" : "...",
  "processors" : [ ... ],
  "version": 1,
  "on_failure" : [ ... ] }
```

The description field contains a definition of the activities done by this pipeline. It's very useful if you store a lot of pipelines in your cluster.

The processors field contains a list of processor actions. They will be executed in order.

In the preceding example, we have used a simple processor action called set that allows us to set a field with a value. The version field is optional, but it is very useful in keeping track of your pipeline versions. The optional on_failure field allows us to define a list of processors to be applied if there are failures during normal pipeline execution.

There's more...

To prevent failure in the case of missing fields or similar constraints, some processors provide the ignore_failure property.

For example, a pipeline with a rename field that handles the missing field should be defined in this way:

```
{ "description": "my pipeline with handled exceptions",
  "processors": [
    { "rename": {
        "field": "foo", "target_field": "bar",
        "ignore_failure": true } } ] }
```

In case of failure, you can configure an on_failure entry to manage the error.

For example, the error can be saved in a field in the following way:

```
{ "description": "my pipeline with handled exceptions",
  "processors": [
    { "rename": {
        "field": "foo", "target_field": "bar",
        "on_failure": [
          { "set": { "field": "error",
              "value": "{{ _ingest.on_failure_message }}"}
        } ] } } ] }
```

Many of the processors allow us to define an `if` statement using a Painless script or regular expression. This property is very handy to build more complex pipelines.

See also

The official documentation about the conditionals in the pipeline can be found at

- `https://www.elastic.co/guide/en/elasticsearch/reference/master/ingest-conditional-complex.html`.

- `https://www.elastic.co/guide/en/elasticsearch/reference/master/conditionals-with-multiple-pipelines.html`.

- `https://www.elastic.co/guide/en/elasticsearch/reference/master/conditionals-with-regex.html`.

Inserting an ingest pipeline

The power of the pipeline definition is the ability for it to be updated and created without a node restart (compared to Logstash). The definition is stored in a cluster state via the `put` pipeline API.

Now that we've defined a pipeline, we need to provide it to the Elasticsearch cluster.

Getting ready

You need an up-and-running Elasticsearch installation, as we described in the *Downloading and installing Elasticsearch* recipe in *Chapter 1, Getting Started*.

To execute the commands, any HTTP client can be used, such as curl (`https://curl.haxx.se/`), Postman (`https://www.getpostman.com/`), or similar. Use the Kibana console, as it provides code completion and better character escaping for Elasticsearch.

How to do it...

To store or update an ingestion pipeline in Elasticsearch, we will do the following.

Store the `ingest` pipeline using a PUT call:

```
PUT /_ingest/pipeline/add-user-john
{ "description": "Add user john field",
  "processors": [
    { "set": { "field": "user", "value": "john" } } ],
  "version": 1 }
```

The result that's returned by Elasticsearch, if everything is okay, should be as follows:

```
{ "acknowledged" : true }
```

How it works...

The PUT pipeline method works both for creating a pipeline as well as updating an existing one.

The pipelines are stored in a cluster state, and they are immediately propagated to all `ingest` nodes. When the `ingest` nodes receive the new pipeline, they will update their node in an in-memory pipeline representation, and the pipeline changes take effect immediately.

When you store a pipeline in the cluster, pay attention and provide a meaningful name (in the example, `add-user-john`) so that you can easily understand what the pipeline does. The name of the pipeline used in the PUT call will be the ID of the pipeline in other pipeline flows.

After storing your pipeline in Elasticsearch, you can index a document by providing the pipeline name as a query argument, as in the following example:

```
PUT /my_index/my_type/my_id?pipeline=add-user-john
{}
```

The document will be enriched by the pipeline before being indexed:

```
{ "_index" : "my_index", "_id" : "my_id", "_version" : 1,
  "_seq_no" : 0, "_primary_term" : 1, "found" : true,
  "_source" : { "user" : "john" } }
```

Getting an ingest pipeline

After having stored your pipeline, it is common to retrieve its content, so that you can check its definition. This action can be done via the `get` pipeline API.

Getting ready

You need an up-and-running Elasticsearch installation, as we described in the *Downloading and installing Elasticsearch* recipe in *Chapter 1, Getting Started*.

To execute the commands, any HTTP client can be used, such as curl (`https://curl.haxx.se/`), Postman (`https://www.getpostman.com/`), or similar. Use the Kibana console, as it provides code completion and better character escaping for Elasticsearch.

How to do it...

To retrieve an ingestion pipeline in Elasticsearch, we will perform the following steps:

We can retrieve the `ingest` pipeline using a `GET` call:

```
GET /_ingest/pipeline/add-user-john
```

The result that's returned by Elasticsearch, if everything is okay, should be as follows:

```
{ "add-user-john" : {
    "description" : "Add user john field",
    "processors" : [
      { "set" : { "field" : "user", "value" : "john"}}
    ], "version" : 1 } }
```

How it works...

To retrieve an ingestion pipeline, you need its name or ID. For each returned pipeline, all the data is returned: the source and the version, if they are defined.

The `GET` pipeline allows us to use a wildcard in names, so you can do the following:

- Retrieve all pipelines using `*`:

```
GET /_ingest/pipeline/*
```

- Retrieve a partial pipeline:

```
GET /_ingest/pipeline/add-*
```

If you have a lot of pipelines, using a good name convention helps a lot in their management.

There's more...

If you need only a part of the pipeline, such as the version, you can use `filter_path` to filter the pipeline only for the parts that are needed. Take a look at the following code example:

```
GET /_ingest/pipeline/add-user-john?filter_path=*.version
```

It will return only the `version` part of the pipeline:

```
{ "add-user-john" : { "version" : 1 } }
```

Deleting an ingest pipeline

To clean up our Elasticsearch cluster of obsolete or unwanted pipelines, we need to call the `delete` pipeline API with the ID of the pipeline.

Getting ready

You need an up-and-running Elasticsearch installation, as we described in the *Downloading and installing Elasticsearch* recipe in *Chapter 1*, *Getting Started*.

To execute the commands, any HTTP client can be used, such as curl (https://curl.haxx.se/), Postman (https://www.getpostman.com/), or similar. Use the Kibana console, as it provides code completion and better character escaping for Elasticsearch.

How to do it...

To delete an ingestion pipeline in Elasticsearch, we will perform the following step.

We can delete the ingest pipeline using a `DELETE` call:

```
DELETE /_ingest/pipeline/add-user-john
```

The result that's returned by Elasticsearch, if everything is okay, should be as follows:

```
{ "acknowledged" : true }
```

How it works...

The `delete` pipeline API removes the named pipeline from Elasticsearch.

Since the pipelines are kept in memory in every node due to their cluster-level storage and the pipelines are always up and running in the ingest node, it's good practice to keep only the necessary pipelines in the cluster.

The `delete` pipeline API does not allow you to use wildcards in pipeline names or IDs.

Simulating an ingest pipeline

The `ingest` part of every architecture is very sensitive, so the Elasticsearch team has created the possibility of simulating your pipelines without the need to store them in Elasticsearch.

The `simulate` pipeline API allows a user to test, improve, and check functionalities of your pipeline without deployment in the Elasticsearch cluster.

Getting ready

You need an up-and-running Elasticsearch installation, as we described in the *Downloading and installing Elasticsearch* recipe in *Chapter 1, Getting Started*.

To execute the commands, any HTTP client can be used, such as curl (`https://curl.haxx.se/`), postman (`https://www.getpostman.com/`), or similar. Use the Kibana console, as it provides code completion and better character escaping for Elasticsearch.

How to do it...

To simulate an ingestion pipeline in Elasticsearch, we will perform the following step.

Execute a call for passing both the pipeline and a sample subset of a document that you can test the pipeline against:

```
POST /_ingest/pipeline/_simulate
{ "pipeline": {
    "description": "Add user john field",
```

```
    "processors": [
        { "set": {"field": "user", "value": "john"} },
        { "set": {"field": "job", "value": 10} } ],
    "version": 1 },
  "docs": [
    { "_index": "index", "_type": "type", "_id": "1",
      "_source": {"name": "docs1"} },
    { "_index": "index", "_type": "type", "_id": "2",
      "_source": {"name": "docs2"} } ] }
```

The result returned by Elasticsearch, if everything is okay, should be a list of documents with the pipeline processed:

```
{ "docs" : [
    { "doc" : {
        "_index" : "index", "_id" : "1",
        "_source" : { "name" : "docs1",
            "job" : 10, "user" : "john" },
        "_ingest" : { "timestamp" :
"2022-02-20T09:18:31.742033Z" } } },
    { "doc" : {
        "_index" : "index", "_id" : "2",
        "_source" : { "name" : "docs2",
            "job" : 10, "user" : "john" },
        "_ingest" : { "timestamp" :
"2022-02-20T09:18:31.742033Z" } } } ] }
```

How it works...

In a single call, the `simulated` pipeline API is able to test a pipeline on a subset of documents. It internally executes the following steps:

1. It parses the provided pipeline definition, creating an in-memory representation of the pipeline.

2. It reads the provided documents by applying the pipeline.

3. It returns the processed results.

The only required sections are `pipeline one` and `docs` containing a list of documents. The documents (provided in `docs`) must be formatted with metadata fields and the source field, similar to a query result.

There are processors that are able to modify the metadata fields; for example, they are able to change `_index` or `_type` based on its contents. The metadata fields are `_index`, `_id`, `_routing`, and `_parent`.

For *debugging purposes*, it is possible to add the `verbose` URL query argument to return all the intermediate steps of the pipeline. For example, let's say we change the call of the previous simulation in the code:

```
POST /_ingest/pipeline/_simulate?verbose
{ … same of previous example … }
```

The result will be expanded for every pipeline step (inline the comments):

```
{ "docs" : [
    { "processor_results" : [
        {"processor_type" : "set", // first pipeline step
          "status" : "success", // step result
          "doc" : {
            "_index" : "index", "_id" : "1",
            "_source": {"name": "docs1", "user" : "john" },
            "_ingest" : {
              "pipeline" : "_simulate_pipeline", // pipeline
name
              "timestamp" : "2022-02-20T09:21:43.741639Z"
            } } },
        { "processor_type" : "set", // second pipeline step
          "status" : "success", // step result
          "doc" : {
            "_index" : "index", "_id" : "1",
            "_source" : { "name" : "docs1", "job" : 10,
            "user" : "john" },
            "_ingest" : {
              "pipeline" : "_simulate_pipeline", // pipeline
name
              "timestamp" : "2022-02-20T09:21:43.741639Z"
```

```
} } } ] },
... truncated ...
```

There's more...

The `simulate` pipeline API is very handy when a user needs to check a complex pipeline that uses special fields access, such as the following:

- **Ingest metadata fields**: These are special metadata fields, such as `_ingest.timestamp`, and are available during ingestion. This kind of field allows values to be added to the document, as in the following example example:

```
{ "set": {
    "field": "received",
    "value": "{{_ingest.timestamp}}" } }
```

- **Field replace templating**: Using the templating with { { } }, it's possible to inject other fields or join their values:

```
{"set": {
    "field": "full_name",
    "value": "{{name}} {{surname}}" } }
```

The ingest metadata fields (accessible using `_ingest`) are as follows:

- `timestamp`: This contains the current pipeline timestamp.
- `on_failure_message`: This is available only in the `on_failure` block in case of failure. It contains the failure message.
- `on_failure_processor_type`: This is available only in the `on_failure` block in case of failure. It contains the failure processor type that has generated the failure.
- `on_failure_processor_tag`: This is available only in the `on_failure` block in case of failure. It contains the failure tag that has generated the failure.

Built-in processors

Elasticsearch provides a large set of `ingest` processors by default. Their number and functionalities can also change from minor versions to extended versions for new scenarios.

In this recipe, we will look at the most commonly used ones.

Getting ready

You need an up-and-running Elasticsearch installation, as we described in the *Downloading and installing Elasticsearch* recipe in *Chapter 1, Getting Started*.

To execute the commands, any HTTP client can be used, such as curl (https://curl.haxx.se/), Postman (https://www.getpostman.com/), or similar. Use the Kibana console, as it provides code completion and better character escaping for Elasticsearch.

How to do it...

To use several processors in an ingestion pipeline in Elasticsearch, we will perform the following step.

Execute a `simulate` pipeline API call using several processors with a sample subset of a document that you can test the pipeline against:

```
POST /_ingest/pipeline/_simulate
{ "pipeline": {
    "description": "Testing some build-processors",
    "processors": [
        { "dot_expander": {"field": "extfield.innerfield"}},
        { "remove": {"field": "unwanted"}},
        { "trim": { "field": "message" }},
        { "set": {
            "field": "tokens", "value": "{{message}}"}},
        {"split":{"field":"tokens","separator":"\\s+"}},
        { "sort": { "field":"tokens", "order":"desc"}},
        { "convert": { "field": "mynumbertext",
            "target_field": "mynumber",
            "type": "integer" } } ] },
    "docs": [
        { "_index": "index", "_type": "type", "_id": "1",
          "_source": {
            "extfield.innerfield": "booo",
            "unwanted": 32243,
```

```
        "message": "155.2.124.3 GET /index.html 15442 0.038",
        "mynumbertext": "3123" } } ] }
```

The result will be as follows:

```
{ "docs" : [
    { "doc" : { "_index" : "index",   "_id" : "1",
        "_source" : {
            "mynumbertext" : "3123",
            "extfield" : { "innerfield" : "booo" },
            "tokens" : [
                "GET", "155.2.124.3", "15442", "0.038", "/index.
html"
            ],
            "message" : "155.2.124.3 GET /index.html 15442
0.038",
            "mynumber" : 3123 },
        "_ingest" : {
            "timestamp" : "2022-02-20T09:27:28.17516Z"
        } } } ] }
```

How it works...

The preceding example shows how to build a complex pipeline in order to preprocess a document. There are a lot of built-in processors to cover the most common scenarios in log and text processing. More complex ones can be done using scripting.

At the time of writing this book, Elasticsearch provides built-in pipelines. The following are the most used processors:

Name	Description
Append	Appends values to a field. If required, it converts them into an array
Convert	Converts a field value into a different type
Date	Parses a date and uses it as a timestamp for the document
Date index name	Allows us to set the _index name based on the date field
Drop	Drops the following document without raising an error
Fail	Raises a failure
Foreach	Processes the element of an array with the provided processor

Name	Description
Grok	Applies grok pattern extraction
Gsub	Executes a regular expression replace on a field
Join	Joins an array of values using a separator
JSON	Converts a JSON string into a JSON object
Lowercase	Lowercases a field
Remove	Removes a field
Pipeline	Allows us to execute other pipelines
Rename	Renames a field
Script	Allows us to execute a script
Set	Sets the value of a field
Split	Splits a field in an array using a regular expression
Sort	Sorts the values of an array field
Trim	Trims whitespaces from a field
Uppercase	Uppercases a field
Dot expander	Expands a field with a dot in the objects

Figure 12.1 – Commonly used processors in Elasticsearch

See also

In *Chapter 16*, *Plugin Development*, we will cover how to write a custom processor in Java to extend the capabilities of Elasticsearch.

The grok processor

Elasticsearch provides a large number of built-in processors that increases with every release. In the preceding examples, we have seen the `set` and `replace` ones. In this recipe, we will cover one that's mostly used for log analysis: the `grok` processor, which is well known to Logstash users.

Getting ready

You need an up-and-running Elasticsearch installation, as we described in the *Downloading and installing Elasticsearch* recipe in *Chapter 1*, *Getting Started*.

To execute the commands, any HTTP client can be used, such as curl (`https://curl.haxx.se/`), Postman (`https://www.getpostman.com/`), or similar. Use the Kibana console, as it provides code completion and better character escaping for Elasticsearch.

How to do it...

To test a `grok` pattern against some log lines, we will perform the following step.

Execute a call by passing both the pipeline with our `grok` processor and a sample subset of a document to test the pipeline against:

```
POST /_ingest/pipeline/_simulate
{ "pipeline": {
    "description": "Testing grok pattern",
    "processors": [
      { "grok": { "field": "message",
          "patterns": [
            "%{IP:client} %{WORD:method}
%{URIPATHPARAM:request} %{NUMBER:bytes} %{NUMBER:duration}" ] }
} ] },
  "docs": [
    { "_index": "index", "_id": "1",
      "_source": {
        "message": "155.2.124.3 GET /index.html 15442 0.038" }
} ] }
```

The result returned by Elasticsearch, if everything is okay, should be a list of documents with the pipeline processed:

```
{ "docs" : [
    {"doc" : { "_index" : "index", "_id" : "1",
        "_source" : {
          "duration" : "0.038",
          "request" : "/index.html",
          "method" : "GET", "bytes" : "15442",
          "client" : "155.2.124.3",
          "message" : "155.2.124.3 GET /index.html 15442 0.038"
},
        "_ingest" : {
          "timestamp" : "2022-02-20T09:28:23.617626Z"
        } } } ] }
```

How it works...

The `grok` processor allows you to extract structure fields out of a single text field in a document. A grok pattern is like a regular expression that supports aliased expressions that can be reused. It was used mainly in another piece of Elastic software, Logstash, for its powerful syntax for log data extraction.

Elasticsearch has about 120 built-in grok expressions (you can analyze them at `https://github.com/elastic/elasticsearch/tree/master/modules/ingest-common/src/main/resources/patterns`).

Defining a grok expression is quite simple, since the syntax is human-readable. If we want to extract colors from an expression (`pattern`) and check whether their value is in a subset of RED, YELLOW, and BLUE using `pattern_definitions`, we can define a similar processor:

```
POST /_ingest/pipeline/_simulate
{ "pipeline": {
    "description": "custom grok pattern",
    "processors": [
      { "grok": {
          "field": "message",
        "patterns":["my favorite color is %{COLOR:color}"],
          "pattern_definitions": {
            "COLOR": "RED|GREEN|BLUE" } } } ] },
  "docs": [
    { "_source": {"message": "my favorite color is RED"}},
    { "_source": {"message": "happy fail!!" } } ] }
```

The result will be as follows:

```
{ "docs" : [
    { "doc" : { "_index" : "_index", "_id" : "_id",
      "_source" : {
        "message" : "my favorite color is RED",
        "color" : "RED" },
      "_ingest" : {
        "timestamp" : "2022-02-20T09:29:32.23106Z"
      } } },
    { "error" : {
```

```
        "root_cause" : [
            { "type" : "illegal_argument_exception",
                "reason" : "Provided Grok expressions do not match
field value: [happy fail!!]" } ],
            "type" : "illegal_argument_exception",
            "reason" : "Provided Grok expressions do not match
field value: [happy fail!!]" } } ] }
```

In real applications, the failing grok processor exceptions will prevent your document from being indexed. For this reason, when you design your grok pattern, be sure to test it on a large subset.

See also

There are online sites where you can test your grok expressions, such as http://grokdebug.herokuapp.com and http://grokconstructor.appspot.com.

Using the ingest attachment plugin

It's easy to make a cluster non-responsive in Elasticsearch prior to 5.x, by using the attachment mapper. The metadata extraction from a document requires a very high CPU operation and if you are ingesting a lot of documents, your cluster is under-loaded.

To prevent this scenario, Elasticsearch introduced the ingest node. An ingest node can be held under very high pressure without causing problems to the rest of the Elasticsearch cluster.

The attachment processor allows us to use the document extraction capabilities of Tika in an ingest node.

Getting ready

You need an up-and-running Elasticsearch installation, as we described in the *Downloading and installing Elasticsearch* recipe in *Chapter 1, Getting Started*.

To execute the commands, any HTTP client can be used, such as curl (https://curl.haxx.se/), Postman (https://www.getpostman.com/), or similar. Use the Kibana console, as it provides code completion and better character escaping for Elasticsearch.

How to do it...

To be able to use the ingest attachment processor, perform the following steps:

1. You can create a pipeline ingested with the `attachment` processor:

```
PUT /_ingest/pipeline/attachment
{ "description": "Extract data from an attachment via
Tika",
  "processors": [
    { "attachment": { "field": "data"  } } ],
  "version": 1 }
```

If everything is okay, you should receive the following message:

```
{ "acknowledged" : true }
```

2. Now, we can index a document using a pipeline:

```
PUT /my_index/my_type/my_id?pipeline=attachment
{ "data": "e1xydGYxXGFuc2kNCkxvcmVtIGlwc3VtIGRvbG9yI
HNpdCBhbWV0DQpccGFyIH0=" }
```

3. Now, we can recall it:

```
GET /my_index/_doc/my_id
```

The result will be as follows:

```
{ "_index" : "my_index", "_id" : "my_id",
  "_version" : 1, "found" : true,
  "_source" : {
    "data" : "e1xydGYxXGFuc2kNCkxvcmVtIGlwc3VtIGRvbG9yI
HNpdCBhbWV0DQpccGFyIH0=",
    "attachment" : {
      "content_type" : "application/rtf",
      "language" : "ro",
      "content" : "Lorem ipsum dolor sit amet",
      "content_length" : 28 } } }
```

How it works...

The attachment ingest processor is provided by a separate plugin that is provided by default in version 8.x installations. It works like every other processor and the properties that control it are as follows:

- `field`: This is the field that will contain the Base64 representation of the binary data.

- `target_field`: This will hold the attachment information (default `attachment`).

- `indexed_char`: This is the number of characters to be extracted to prevent very huge fields. If it is set to `-1`, all the characters are extracted (default `100000`).

- `properties`: Other metadata fields of the document that need to be extracted. They can be `content`, `title`, `name`, `author`, `keywords`, `date`, `content_type`, `content_length`, and `language` (default `all`).

Using the ingest GeoIP processor

Another interesting processor is the GeoIP plugin, which allows us to map an IP address to a geopoint and other location data. It's provided in every Elasticsearch installation by default from version 7.x.

Getting ready

You need an up-and-running Elasticsearch installation, as we described in the *Downloading and installing Elasticsearch* recipe in *Chapter 1, Getting Started*.

To execute the commands, any HTTP client can be used, such as curl (`https://curl.haxx.se/`), Postman (`https://www.getpostman.com/`), or similar. Use the Kibana console, as it provides code completion and better character escaping for Elasticsearch.

How to do it...

To be able to use the ingest GeoIP processor, perform the following steps:

1. We can create a `pipeline` ingest with the attachment processor, using the following command:

```
PUT /_ingest/pipeline/geoip
{ "description": "Extract geopoint from an IP",
  "processors": [
    { "geoip": { "field": "ip" } } ], "version": 1 }
```

    If everything is okay, you should receive the following message:

```
{ "acknowledged" : true }
```

2. Now, we can index a document using a pipeline:

```
PUT /my_index/_doc/my_id?pipeline=geoip
{ "ip": "8.8.8.8" }
```

3. Then, we can recall it:

```
GET /my_index/_doc/my_id
```

    The result will be as follows:

```
{ "_index" : "my_index", "_id" : "my_id",
  "_version" : 3, "found" : true,
  "_source" : {
    "geoip" : {
      "continent_name" : "North America",
      "country_iso_code" : "US",
      "location" : { "lon" : -97.822, "lat" : 37.751 }
    },
    "ip" : "8.8.8.8" } }
```

How it works...

The GeoIP ingest processor uses data from the `MaxMind` databases to extract information about the geographical location of IP addresses. This processor adds this information by default under the `geoip` field. The GeoIP processor can resolve both IPv4 and IPv6 addresses.

The properties that control it are as follows:

- `field`: This is the field that will contain the IP from which the geo data is extracted.
- `target_field`: This will hold the `geoip` information (the default is `geoip`).
- `database_file`: This is the database file that contains maps from IP to geolocations. The default one is installed during the plugin's installation (`GeoLite2-City.mmdb`).
- `properties`: The `properties` values depend on the database. You should refer to the database description for details on the extracted fields (the default is `all`).
- `first_only` (default `true`): If it is set to `true`, it returns only the `geoip` data of the first entry of `field` (in the case of arrays).

See also

The official documentation about the GeoIP processor and how to use it with other GeoIP2 databases can be found at `https://www.elastic.co/guide/en/elasticsearch/reference/master/geoip-processor.html`.

Using the enrichment processor

It's quite common to enrich your indexed fields with lookups from other sources. Typical examples are as follows:

- Resolving the ID of values referenced as external foreign keys in a database
- Enriching names with the data object to be able to have all the data aggregated in a single document

To be able to solve these use cases, X-Pack provides a special processor called the enrichment processor.

Getting ready

You need an up-and-running Elasticsearch installation, as we described in the *Downloading and installing Elasticsearch* recipe in *Chapter 1, Getting Started*.

To execute the commands, any HTTP client can be used, such as curl (`https://curl.haxx.se/`), Postman (`https://www.getpostman.com/`), or similar. Use the Kibana console, as it provides code completion and better character escaping for Elasticsearch.

How to do it...

To be able to use the ingest enrich processor, perform the following steps:

1. We need to do some data preparation, creating an index and storing data that will be used to enrich our document in the pipeline. For this reason, we will index a document in our `item` index in this way:

```
PUT item/_doc/1
{ "id": "1", "name": "ES Cookbook 8.x", "year": 2022 }
```

2. Now we can create an enrich policy, `item-policy`, that will be used to create our `.enrich-*` index to be used for enriching:

```
PUT /_enrich/policy/item-policy
{ "match": {
    "indices": "item", "match_field": "id",
    "enrich_fields": ["name", "year"] } }
```

The result will be something similar to this:

```
{ "acknowledged" : true }
```

3. Now we need to create our `.enrich-` index executing the previously created policy:

```
POST /_enrich/policy/item-policy/_execute
```

The result will be something similar to this:

```
{ "status" : { "phase" : "COMPLETE" } }
```

4. Finally, we can simulate a pipeline with an enrich step, in this way:

```
POST /_ingest/pipeline/_simulate
{ "pipeline": {
    "processors": [
      { "enrich": {
          "description": "We enrich item_id with the item
data",
          "policy_name": "item-policy",
          "field": "item_id",
          "target_field": "item", "max_matches": "1" }
      } ], "version": 1 },
  "docs": [
```

```
    { "_index": "order", "_id": "1",
        "_source": {"item_id": "1", "quantity": 4 } } ]}
```

If everything is okay, you should receive the following response:

```
{ "docs" : [
    { "doc" : { "_index" : "order", "_id" : "1",
        "_source" : {
            "item" : {
                "name" : "ES Cookbook 8.x",
                "year" : 2022, "id" : "1" },
            "quantity" : 4, "item_id" : "1" },
        "_ingest" : {
            "timestamp" : "2022-02-20T09:36:08.191683Z"
        } } } ] }
```

How it works...

The enrichment workflow is composed of two steps: the data preparation and the pipeline execution (ingestion).

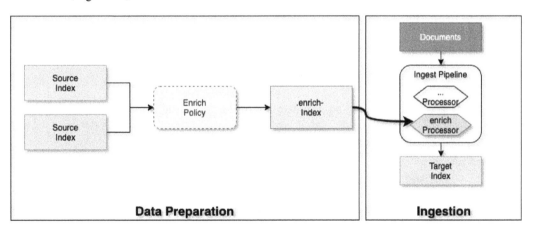

Figure 12.1 – Enrichment workflow

In the data preparation step, we will prepare the data to be used in the ingestion part. This step is composed of the following actions:

1. Ingest the data in indices that we need to use to enrich.

2. Define an enrich policy.

3. Execute the enrich policy.

The definition of an `enrich` policy uses the following REST API format:

```
PUT /_enrich/policy/<policy-name>
```

The `policy-name` parameter will be the name of the policy that will be referred to in the enrich processor. It must be unique.

In the body of the API call, we need to provide the following parameters:

- `policy-type`, which selects the policy behavior. Possible values are as follows:

  - `geo_match` to enrich based on a geo-shape match
  - `match` to enrich based on a `term` query

- `indices`: A single value or a list of indices that contain `match_field`.
- `match_field`: The field to be used for matching the values.
- `enrich_fields`: An array of fields that will be propagated to the matching document in the pipeline. These fields must be present in the source indices.
- `query`: An optional query to be used to filter out documents from source indices.

> **Note**
> A policy cannot be updated. If you need to modify a policy, you need to create a new one with a new name.
>
> You cannot delete a policy if it's referred to in a pipeline.

When a policy is defined, it must be populated with the following REST API before being used:

```
PUT /_enrich/policy/<policy-name>/_execute
POST /_enrich/policy/<policy-name>/_execute
```

After executing the preceding commands, an index will be created with the name `.enrich-<policy-name>`. It's an optimized index with force-merge and has PUT in read-only.

If you update the source indices, you should re-execute the previous command for an update of the enrich index.

After having created your enrich policy and relative index, you can use the enrich processor.

The main parameters of this processor are the following:

- `policy_name`: The name of the policy to use.
- `field`: The field to be used to match the policy.
- `target_field`: The target field to put the result of the match in.
- `max_matches` (default 1): How many matches are to be put in the `target_field`. If the value is 1, a `json` object is stored in the target field, otherwise, a `json` array is used. The upper limit of this value is 128 to prevent large documents.
- `shape_relation` (default `INTERSECTS`): Allows you to customize the shape relation in the case of the `geo_match` and geoshape queries. Refer to the official documentation on spatial relation at `https://www.elastic.co/guide/en/elasticsearch/reference/master/query-dsl-geo-shape-query.html#_spatial_relations`.

This processor also accepts common processor properties.

See also

You can view the following links for further reference, which are related to this recipe:

- The official documentation about the enrich processor can be found at `https://www.elastic.co/guide/en/elasticsearch/reference/master/enrich-processor.html`.
- The enrich data section in the official Elastic documentation at `https://www.elastic.co/guide/en/elasticsearch/reference/master/ingest-enriching-data.html`.
- The blog post about the official announcement of this feature at `https://www.elastic.co/blog/introducing-the-enrich-processor-for-elasticsearch-ingest-nodes`.
- A Medium post about the usage of this processor at `https://quoeamaster.medium.com/elasticsearch-ingest-pipeline-tips-and-tricks-2-enrich-processor-25942e601065`.

13
Java Integration

Elasticsearch functionalities can be easily integrated into any Java application in a couple of ways, via a REST API or via native APIs. In Java, it's easy to call a REST HTTP interface with one of the many libraries available, such as the **Apache HttpComponents** client (see http://hc.apache.org/ for more information). In this field, there's no such thing as the most used library; typically, developers choose the library that best suits their preferences or one that they know very well. From Elasticsearch 6.x onward, Elastic has provided a battle low-/high-level HTTP for clients to use. In version 8.x, Elastic released a modern/functional/strongly typed client and in this chapter, we will mainly use this version for all the examples that are provided.

Each **Java Virtual Machine** (**JVM**) language can also use the native protocol to integrate Elasticsearch with their applications; however, we will not cover this because it has fallen out of use from Elasticsearch 7.x onward. New applications should rely on HTTP. In this chapter, we will learn how to initialize different clients and how to execute the commands that we have seen in the previous chapters. We will not cover every HTTP call in depth, as we have already described them for the REST APIs. The Elasticsearch community recommends using the REST APIs when integrating them, as they are more stable between releases and are well documented.

All the code presented in these recipes is available in this book's code repository and can be built with Maven.

In this chapter, we will cover the following recipes:

- Creating a standard Java HTTP client
- Creating a low-level Elasticsearch client
- Using the Elasticsearch official Java client
- Managing indices
- Managing mappings
- Managing documents
- Managing bulk actions
- Building a query
- Executing a standard search
- Executing a search with aggregations
- Executing a scroll search
- Integrating with DeepLearning4j

Creating a standard Java HTTP client

An HTTP client is one of the easiest clients to create. It's very handy because it allows for the calling, not only of the internal methods as the native protocol does, but also of third-party calls implemented in plugins that can only be called via HTTP.

Getting ready

You need an up-and-running Elasticsearch installation, as we described in the *Downloading and installing Elasticsearch* recipe in *Chapter 1, Getting Started*.

To correctly execute the following commands, you will need an index populated with the ch04/populate_kibana.txt commands that are available in the online code.

A Maven tool or an **Integrated Development Environment** (**IDE**) that natively supports it for Java programming, such as Visual Studio Code, Eclipse, or IntelliJ IDEA, must be installed. Elasticsearch code is targeting Java 17, so it's best practice to have installed JDK 17 or above.

The code for this recipe is in the `chapter_13/http_java_client` directory.

Because Elasticsearch 8.x or above is secured by default, to run all the examples of this chapter, put the credentials in `ES_USER` and `ES_PASSWORD` environment variables when executing the code.

How to do it...

To create an HTTP client, we will perform the following steps:

1. For these examples, we have chosen the Apache `HttpComponents` library, which is one of the most widely used libraries for executing HTTP calls. This library is available in the main Maven repository called `search.maven.org`.

    To enable the compilation in your Maven `pom.xml` project, just add the following code:

    ```xml
    <dependency>
        <groupId>org.apache.httpcomponents</groupId>
        <artifactId>httpclient</artifactId>
        <version>4.5.13</version>
    </dependency>
    ```

2. If we want to instantiate a client and fetch a document with a `get` method, the code (see `AppWithSecurity.java`) will look like the following:

    ```java
    package com.packtpub;
    import org.apache.http.HttpEntity;
    import org.apache.http.HttpHost;
    import org.apache.http.HttpStatus;
    import org.apache.http.auth.AuthScope;
    import org.apache.http.auth.UsernamePasswordCredentials;
    import org.apache.http.client.AuthCache;
    import org.apache.http.client.CredentialsProvider;
    import org.apache.http.client.methods.
    CloseableHttpResponse;
    import org.apache.http.client.methods.HttpGet;
    import org.apache.http.client.protocol.HttpClientContext;
    import org.apache.http.conn.ssl.NoopHostnameVerifier;
    import org.apache.http.ssl.SSLContextBuilder;
    import org.apache.http.conn.ssl.TrustAllStrategy;
    ```

```java
import org.apache.http.impl.auth.BasicScheme;
import org.apache.http.impl.client.BasicAuthCache;
import org.apache.http.impl.client.
BasicCredentialsProvider;
import org.apache.http.impl.client.CloseableHttpClient;
import org.apache.http.impl.client.HttpClients;
import org.apache.http.util.EntityUtils;
import java.io.IOException;
import java.security.KeyManagementException;
import java.security.KeyStoreException;
import java.security.NoSuchAlgorithmException;

public class AppWithSecurity {
    private static String wsUrl =
"https://127.0.0.1:9200";
    public static void main(String[] args) throws
KeyManagementException, NoSuchAlgorithmException,
KeyStoreException {
        CloseableHttpClient client = HttpClients.custom()
                .setSSLContext(new SSLContextBuilder().
loadTrustMaterial(null, TrustAllStrategy.INSTANCE).
build())
                .setSSLHostnameVerifier(NoopHostnameVerifier.
INSTANCE).setRetryHandler(new MyRequestRetryHandler())
                .build();
        HttpGet method = new HttpGet(wsUrl + "/mybooks/_
doc/1");
        // Execute the method.

        HttpHost targetHost = new HttpHost("localhost",
9200, "https");
        CredentialsProvider credsProvider = new
BasicCredentialsProvider();
        credsProvider.setCredentials(AuthScope.ANY,
                new UsernamePasswordCredentials(System.
getenv("ES_USER"), System.getenv("ES_PASSWORD")));
        // Create AuthCache instance
        AuthCache authCache = new BasicAuthCache();
```

```java
        // Generate BASIC scheme object and add it to
local auth cache
        BasicScheme basicAuth = new BasicScheme();
        authCache.put(targetHost, basicAuth);
        // Add AuthCache to the execution context
        HttpClientContext context = HttpClientContext.
create();
        context.setCredentialsProvider(credsProvider);
        method.addHeader("Accept-Encoding", "gzip");
        try {
            CloseableHttpResponse response = client.
execute(method, context);

            if (response.getStatusLine().getStatusCode()
!= HttpStatus.SC_OK) {
                System.err.println("Method failed: " +
response.getStatusLine());
            } else {
                HttpEntity entity = response.getEntity();
                String responseBody = EntityUtils.
toString(entity);
                System.out.println(responseBody);
            }
        } catch (IOException e) {
            System.err.println("Fatal transport error: "
+ e.getMessage());
            e.printStackTrace();
        } finally {
            // Release the connection.
            method.releaseConnection();
        } } }
```

If you run the code, the result output will be as follows:

```
{"_index":"mybooks","_type":"_doc","_id":"1","_
version":1,"_seq_no":0,"_primary_term":1,"found":true,"_
source":{"uuid":"11111","position":1,"title":"Joe
Tester","description":"Joe Testere nice guy","date":"
2021-10-22","price":4.3,"quantity":50}}
```

How it works...

We performed the preceding steps to create and use an HTTP client. Let's look at them in a little more detail.

The first step is to initialize the HTTP client object. In the preceding code, this is done via the following code fragment:

```
CloseableHttpClient client = CloseableHttpClient client =
HttpClients.custom()
                .setSSLContext(new SSLContextBuilder().
loadTrustMaterial(null, TrustAllStrategy.INSTANCE).build())
.setSSLHostnameVerifier(NoopHostnameVerifier.INSTANCE)
                .setRetryHandler(new MyRequestRetryHandler())
                .build();
```

Before using the client, it is good practice to customize it. By default, Elasticsearch 8.0.0 is secured and it uses a self-generated SSL certificate that is not *officially signed*, so it's required to use .setContext and .setSSLHostnameVerifier to accept the self-signed certificate.

In general, the client can also be modified to provide extra functionalities, such as **retry support**. Retry support is very important for designing robust applications; the IP network protocol is never 100% reliable, so it automatically retries an action if something goes bad (HTTP connection closed or server overhead, for example).

In the preceding code, we defined HttpRequestRetryHandler, which monitors the execution and repeats it three times before raising an error.

To be able to perform API calls, we need to authenticate; the authentication is as simple as basicAuth, but works very well for non-complex deployments, as you can see in the following code:

```
HttpHost targetHost = new HttpHost("localhost", 9200, "https");
CredentialsProvider credsProvider = new
BasicCredentialsProvider();
credsProvider.setCredentials(AuthScope.ANY,
                new UsernamePasswordCredentials(System.
getenv("ES_USER"), System.getenv("ES_PASSWORD")));
AuthCache authCache = new BasicAuthCache();
// Generate BASIC scheme object and add it to local auth cache
BasicScheme basicAuth = new BasicScheme();
authCache.put(targetHost, basicAuth);
```

Best security practice is to read the credentials from
Environment variables and not put them in the code.

The create `context` parameter must be used in executing the call, as shown in the
following code:

```
// Add AuthCache to the execution context
HttpClientContext context = HttpClientContext.create();
context.setCredentialsProvider(credsProvider);
CloseableHttpResponse response = client.execute(method,
context);
```

Once we have set up the client, we execute the GET REST call. The used method will be
for `HttpGet` and the URL will be the item named `index/type/id` (similar to the curl
example in the *Getting a document* recipe in *Chapter 3*, *Basic Operations*). To initialize the
method, use the following code:

```
HttpGet method = new HttpGet(wsUrl + "/mybooks/_doc/1");
```

Now we can set up **custom headers** that allow us to pass extra information to the server to
execute a call. Some examples could be API keys or hints about supported formats.

A typical example is using `gzip` data compression over HTTP to reduce bandwidth
usage. To do that, we can add a custom header to the call informing the server that
our client accepts encoding. An example custom header can be made from the phrases
`Accept-Encoding` and `gzip`, as shown in the following code:

```
method.addHeader("Accept-Encoding", "gzip");
```

After configuring the call with all the parameters, we can fire up the request as follows:

```
CloseableHttpResponse response = client.execute(method,
context);
```

Every response object must be validated on its return status: if the call is OK, the return
status should be 200. In the following code, the check is done in the `if` statement, as
follows:

```
if (response.getStatusLine().getStatusCode() != HttpStatus.
SC_OK)
```

If the call is OK and the status code of the response is 200, we can read the answer, as follows:

```
HttpEntity entity = response.getEntity();
String responseBody = EntityUtils.toString(entity);
```

The response is wrapped in HttpEntity, which is a stream.

The HTTP client library provides a helper method called EntityUtils.toString that reads all the content of HttpEntity as a string; otherwise, we'd need to create some code to read from the string and build the string.

Obviously, all the read parts of the call are wrapped in a try-catch block to collect all possible errors created by networking errors.

See also

You can refer to the following URLs for further reference, as they are related to this recipe:

- The Apache HttpComponents library is at http://hc.apache.org/, for a complete reference and more examples about this library.

- How to use SSL certificates is at https://kb.novaordis.com/index. php/Configure_a_Java_HTTP_Client_to_Accept_Self-Signed_ Certificates.

- Elasticsearch's official security plugin is at https://www.elastic.co/guide/ en/elasticsearch/reference/current/security-settings.html.

- The *Getting a document* recipe in *Chapter 3*, *Basic Operations*, which covers the API call that was used in these examples.

Creating a low-level Elasticsearch client

There are two official Elasticsearch clients: the low-level one and the new typed one available from Elasticsearch 8.x (https://github.com/elastic/ elasticsearch-java). The low-level one is used to communicate with Elasticsearch, and its main features are as follows:

- Minimal dependencies

- Load balancing across all available nodes

- Failover in the case of node failures and upon specific response codes

- Failed connection penalization (whether a failed node is retried depends on how many consecutive times it failed; the more failed attempts, the longer the client will wait before trying that same node again)

- Persistent connections

- Trace logging of requests and responses

- Optional automatic discovery of cluster nodes

Getting ready

You need an up-and-running Elasticsearch installation, which can be obtained as described in the *Downloading and installing Elasticsearch* recipe in *Chapter 1, Getting Started*.

To correctly execute the following commands, you will need an index populated with the `ch04/populate_kibana.txt` commands that are available in the online code.

A Maven tool or an IDE that natively supports it for Java programming, such as Visual Studio Code, Eclipse, or IntelliJ IDEA, must be installed.

The code for this recipe is in the `ch13/low_level_client` directory.

How to do it...

To create `RestClient`, we will perform the following steps:

1. We need to add the Elasticsearch HTTP client library that's used to execute HTTP calls. This library is available in the main Maven repository at `search.maven.org`. To enable compilation in your `Maven pom.xml` project, just add the following code:

    ```
    <dependency>
        <groupId>org.elasticsearch.client</groupId>
        <artifactId>elasticsearch-rest-client</artifactId>
        <version>8.0.0</version>
    </dependency>
    ```

2. If we want to instantiate a client and fetch a document with a `get` method, the code will look like the following:

```
package com.packtpub;
import org.apache.http.HttpEntity;
import org.apache.http.HttpHost;
import org.apache.http.HttpStatus;
import org.apache.http.auth.AuthScope;
import org.apache.http.auth.UsernamePasswordCredentials;
import org.apache.http.client.CredentialsProvider;
import org.apache.http.conn.ssl.NoopHostnameVerifier;
import org.apache.http.conn.ssl.TrustAllStrategy;
import org.apache.http.impl.client.
BasicCredentialsProvider;
import org.apache.http.impl.nio.client.
HttpAsyncClientBuilder;
import org.apache.http.ssl.SSLContextBuilder;
import org.apache.http.util.EntityUtils;
import org.elasticsearch.client.*;
import java.io.IOException;
import java.security.*;
import javax.net.ssl.SSLContext;
public class App {
    public static void main(String[] args) throws
KeyManagementException, NoSuchAlgorithmException,
KeyStoreException {
        RestClientBuilder clientBuilder = RestClient.
builder(new HttpHost("localhost", 9200, "https"))
                .setCompressionEnabled(true);
        final CredentialsProvider credentialsProvider =
new BasicCredentialsProvider();
credentialsProvider.setCredentials(AuthScope.ANY,
                new UsernamePasswordCredentials(System.
getenv("ES_USER"), System.getenv("ES_PASSWORD")));
        final SSLContext sslContext = new
SSLContextBuilder().loadTrustMaterial(null,
TrustAllStrategy.INSTANCE).build();
        clientBuilder.setHttpClientConfigCallback(new
RestClientBuilder.HttpClientConfigCallback() {
```

```
        public HttpAsyncClientBuilder
customizeHttpClient(HttpAsyncClientBuilder
httpClientBuilder) {
            return httpClientBuilder
                    .setSSLContext(sslContext)

.setSSLHostnameVerifier(NoopHostnameVerifier.INSTANCE)

.setDefaultCredentialsProvider(credentialsProvider);
        }
    });
    RestClient client = clientBuilder.build();
    try {
        Request request = new Request("GET", "/
mybooks/_doc/1");
        Response response = client.
performRequest(request);
        if (response.getStatusLine().getStatusCode()
!= HttpStatus.SC_OK) {
            System.err.println("Method failed: " +
response.getStatusLine());
        } else {
            HttpEntity entity = response.getEntity();
            String responseBody = EntityUtils.
toString(entity);
            System.out.println(responseBody);
        }
    } catch (IOException e) {
        System.err.println("Fatal transport error: "
+ e.getMessage());
        e.printStackTrace();
    } finally {
        // Release the connection.
        try {
            client.close();
        } catch (IOException e) {
            e.printStackTrace();
        } } } }
```

After compiling and executing the code, the result will be as follows:

```
{"_index":"mybooks","_type":"_doc","_id":"1","_
version":1,"_seq_no":0,"_primary_term":1,"found":true,"_
source":{"uuid":"11111","position":1,"title":"Joe
Tester","description":"Joe Testere nice guy","date":"
2021-10-22", "price":4.3, "quantity":50 }}
```

How it works...

Internally, the Elasticsearch `RestClient` client uses the Apache `HttpComponents` library and wraps it with more convenient methods.

We performed the preceding steps to create and use `RestClient`. Let's look at them in a little more detail:

- The first step is to initialize the `RestClient` object. This is done via the following code fragments:

```
RestClientBuilder clientBuilder = RestClient.builder(new
HttpHost("localhost", 9200, "https"))

// some stuff to setup SSL and auth as seen in previous
example

...

RestClient client = clientBuilder.build();
```

  The `builder` method accepts a multivalue `HttpHost` host (in this way, you can pass a list of HTTP addresses) and returns `RestClientBuilder` under the hood.

- The `RestClientBuilder` object allows client communication to be customized by several methods, such as the following:

  - `setDefaultHeaders(Header[] defaultHeaders)`: This allows the custom headers that must be sent for every request to be provided.

  - `setMaxRetryTimeoutMillis(int maxRetryTimeoutMillis)`: This allows the maximum retry timeout to be defined if there are multiple attempts for the same request.

  - `setPathPrefix(String pathPrefix)`: This allows a custom path prefix to be defined for every request.

  - `setFailureListener(FailureListener failureListener)`: This allows a custom failure listener to be provided, which is called in an instance of node failure. This can be used to provide user-defined behavior in the case of node failure.

- `setHttpClientConfigCallback(RestClientBuilder.HttpClientConfigCallback httpClientConfigCallback)`: This allows the modification of the HTTP client communication, such as adding compression or an encryption layer.

- `setRequestConfigCallback(RestClientBuilder.RequestConfigCallback requestConfigCallback)`: This allows the configuration of request authentications, timeouts, and other properties that can be set at a request level.

- After creating `RestClient`, we can execute some requests against it via several kinds of `performRequest` methods for synchronous calls and `performRequestAsync` methods for asynchronous ones.

  These methods allow you to set parameters, such as the following:

  - `String method`: This is the HTTP method or verb to be used in the call (required).

  - `String endpoint`: This is the API endpoint (required). In the previous example, it is `/test-index/_doc/1`.

  - `Map<String, String> params`: This is a map of values to be passed as query parameters.

  - `HttpEntity entity`: This is the body of the request. It is `org/apache/http/HttpEntity` (see `http://hc.apache.org/httpcomponents-core-ga/httpcore/apidocs/org/apache/http/HttpEntity.html?is-external=true` for more details).

  - `HttpAsyncResponseConsumer<HttpResponse> responseConsumer`: This is used to manage responses in an asynchronous request (see `http://hc.apache.org/httpcomponents-core-ga/httpcore-nio/apidocs/org/apache/http/nio/protocol/HttpAsyncResponseConsumer.html` for more details). By default, it is used to keep all the responses in heap memory (the top memory limit is 100 MB).

  - `ResponseListener responseListener`: This is used to register callbacks during asynchronous calls.

  - `Header... headers`: These are additional headers passed during the call.

  In the preceding example, we executed the GET REST call with the following code:

  ```
  Response response = client.performRequest(request);
  ```

- The `response` object is `org.elasticsearch.client.Response`, which wraps the Apache `HttpComponents` response; for this reason, the code to manage the response is the same as it was in the previous recipe.

> **Note**
>
> `RestClient` is a low-level client; it has no helpers on build queries or actions. For now, using it consists of building the JSON string of the request and then parsing the JSON response string.

See also

You can refer to the following URLs for further reference, which are related to this recipe:

- The official documentation about the REST client at `https://www.elastic.co/guide/en/elasticsearch/client/java-rest/current/index.html` for more usage examples, and more about the Sniffer extension at `https://www.elastic.co/guide/en/elasticsearch/client/java-rest/current/sniffer.html` to support node discovery
- The Apache `HttpComponents` library at `http://hc.apache.org/` for a complete reference and more examples about this library
- The *Getting a document* recipe in *Chapter 3*, *Basic Operations*, which covers the API call used in these examples

Using the Elasticsearch official Java client

The official Java client is built on top of a low-level one and provides strong typed communication with Elasticsearch.

Initially released with Elasticsearch in the latest versions of 7.15 or above, this client is the official Java client for Elasticsearch 8.x. It provides many extra functionalities, such as the following:

- Integration from application classes to JSON instances via an object mapper such as Jackson
- Request/response marshaling/unmarshaling that provides stronger typed programming
- Support for both synchronous and asynchronous calls
- Use of fluent builders and functional patterns to allow writing concise yet readable code when creating complex nested structures
- Built on top of previous low-level client

Getting ready

You need an up-and-running Elasticsearch installation, as we described in the *Downloading and installing Elasticsearch* recipe in *Chapter 1, Getting Started.*

To correctly execute the following commands, you will need an index populated with the ch04/populate_kibana.txt commands available in the online code.

A Maven tool or an IDE that natively supports it for Java programming, such as Visual Studio Code, Eclipse, or IntelliJ IDEA, must be installed.

The code for this recipe is in the ch13/elasticsearch-java-client directory.

Because Elasticsearch 8.x or above is secured by default, to run all the examples of this chapter, put the credentials in ES_USER and ES_PASSWORD environment variables when executing the code.

The following screenshot shows how to put them (**Environment variables**) in IntelliJ IDEA:

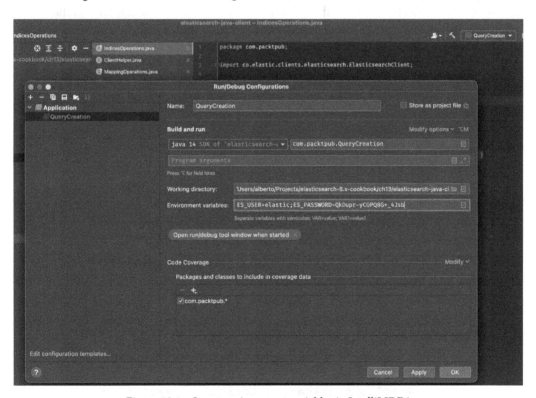

Figure 13.1 – Input environment variables in IntelliJ IDEA

How to do it...

To create a native client, we will perform the following steps:

1. Before starting, we must be sure that Maven loads the Elasticsearch JAR file by adding the following lines to the pom.xml file:

```
<dependency>
    <groupId>co.elastic.clients</groupId>
    <artifactId>elasticsearch-java</artifactId>
    <version>8.0.0</version>
</dependency>
```

I always suggest using the latest available release of Elasticsearch or, in the case of a connection to a specific cluster, using the same version of Elasticsearch that the cluster is using.

2. Now, to create a client, we get the client from the Transport protocol; this is the simplest way to get an Elasticsearch client. We can do this using the following code:

```
package com.packtpub;
import co.elastic.clients.elasticsearch.
ElasticsearchClient;
import co.elastic.clients.json.jackson.
JacksonJsonpMapper;
import co.elastic.clients.transport.
ElasticsearchTransport;
import co.elastic.clients.transport.rest_client.
RestClientTransport;
import org.apache.http.HttpHost;
import org.elasticsearch.client.RestClient;
import org.elasticsearch.client.RestClientBuilder;
import java.io.IOException;
public class InitClientExample {
    public static void main(String[] args) throws
IOException {
        HttpHost httpHost = new HttpHost("localhost",
9200, "http");
        RestClientBuilder restClient = RestClient.
builder(httpHost);
        ElasticsearchTransport transport = new
RestClientTransport(
```

```
        restClient.build(),
        new JacksonJsonpMapper()
    );
  ElasticsearchClient client = new
ElasticsearchClient(transport);
    } }
```

How it works...

Let's look at the steps to create `ElasticsearchClient` in a little more detail.

In your Maven `pom.xml` file, the transport plugin must be defined as follows:

```
<dependency>
    <groupId>co.elastic.clients</groupId>
    <artifactId>elasticsearch-java</artifactId>
    <version>8.0.0</version>
</dependency>
```

We can create one or more instances of `HttpHost` that contain the addresses and ports of the nodes of our cluster, as follows:

```
HttpHost httpHost = new HttpHost("localhost", 9200, "http");
```

`RestClientBuilder` must be provided with all the required `HttpHost` elements, as follows:

```
RestClientBuilder restClient = RestClient.builder(httpHost);
```

We need to create a Transport protocol that is able to manage JSON conversion via Jackson. We will instantiate it via the `RestClientTransport`, as follows:

```
ElasticsearchTransport transport = new RestClientTransport(
    restClient.build(),        // the rest client to be used
    new JacksonJsonpMapper()   // Used to code/decode
application classes
);
```

Now, `ElasticsearchClient` can be initialized with `ElasticsearchTransport`, as shown in the following code:

```
ElasticsearchClient client = new
ElasticsearchClient(transport);
```

If you want to use the `async` client (`ElasticsearchAsyncClient`), you need to initialize it in the same way as the `sync` one:

```
ElasticsearchAsyncClient client = new
ElasticsearchAsyncClient(transport);
```

See also

The official Elasticsearch documentation about `ElasticsearchClient` is at https://www.elastic.co/guide/en/elasticsearch/client/java-api-client/8.0/introduction.html.

Managing indices

In the previous recipe, we learned how to initialize a client to send calls to an Elasticsearch cluster. In this recipe, we will learn how to manage indices via client calls.

Getting ready

You need an up-and-running Elasticsearch installation, which we described how to get in the *Downloading and installing Elasticsearch* recipe in *Chapter 1, Getting Started*.

A Maven tool or an IDE that natively supports it for Java programming, such as Visual Studio Code, Eclipse, or IntelliJ IDEA, must be installed.

The code for this recipe is in the `ch13/elasticsearch-java-client` directory and the referred class is `IndicesOperations`.

How to do it...

An Elasticsearch client maps all index operations under the `indices` object of the client, such as `create`, `delete`, `exists`, `open`, `close`, and `optimize`. The following steps retrieve a client and execute the main operations on the indices:

1. First, we import the required classes, as shown in the following code:

   ```
   import co.elastic.clients.elasticsearch.
   ElasticsearchClient;
   import java.io.IOException;
   import java.security.KeyManagementException;
   import java.security.KeyStoreException;
   import java.security.NoSuchAlgorithmException;
   ```

2. Then, we define an `IndicesOperations` class that manages the index operations, as shown in the following code:

   ```
   public class IndicesOperations {
       private final ElasticsearchClient client;
       public IndicesOperations(ElasticsearchClient client)
   { this.client = client; }
   ```

3. Next, we define a function that is used to check whether the index is there, as shown in the following code:

   ```
   public boolean checkIndexExists(String name) throws
   IOException {
       return client.indices().exists(c -> c.index(name)).
   value(); }
   ```

4. Then, we define a function that can be used to create an index, as shown in the following code:

   ```
   public void createIndex(String name) throws IOException {
       client.indices().create(c -> c.index(name));
   }
   ```

5. We then define a function that can be used to delete an index, as follows:

   ```
   public void deleteIndex(String name) throws IOException {
       client.indices().delete(c -> c.index(name)); }
   ```

6. Then, we define a function that can be used to close an index, as follows:

```
public void closeIndex(String name) throws IOException {
    client.indices().close(c -> c.index(name)); }
```

7. Next, we define a function that can be used to open an index, as follows:

```
public void openIndex(String name) throws IOException {
    client.indices().open(c -> c.index(name)); }
```

8. Then, we test all the previously defined functions, as follows:

```
public static void main(String[] args)
throws InterruptedException, IOException,
NoSuchAlgorithmException, KeyStoreException,
KeyManagementException {
    ClientHelper nativeClient = new ClientHelper();
    ElasticsearchClient client = nativeClient.
getClient();
    IndicesOperations io = new IndicesOperations(client);
    String myIndex = "test";
    if (io.checkIndexExists(myIndex))
        io.deleteIndex(myIndex);
    io.createIndex(myIndex);
    Thread.sleep(1000);
    io.closeIndex(myIndex);
    io.openIndex(myIndex);
    io.deleteIndex(myIndex); }
```

How it works...

Before executing every index operation, a client must be available (we saw how to create one in the previous recipe).

The client has a lot of methods grouped by functionalities, as shown in the following list:

- In the client.* roots, we have record-related operations, such as index, deletion of records, search, and update.

- Under indices().*, we have index-related methods, such as create index and delete index.

- Under cluster().*, we have cluster-related methods, such as state and health.

In this example, we have several index calls, as shown in the following list:

- The `method` call to check that the existence exists. It takes a `GetIndexRequest` Lambda and returns a Boolean value, which contains information about whether the index exists, as shown in the following code:

```
client.indices().exists(c -> c.index(name)).value();
```

- You can create an index with the `create` call, as follows:

```
client.indices().create(c -> c.index(name));
```

- You can close an index with the `close` call, as follows:

```
client.indices().close(c -> c.index(name));
```

- You can open an index with the `open` call, as follows:

```
client.indices().open(c -> c.index(name));
```

- You can delete an index with the `delete` call, as follows:

```
client.indices().delete(c -> c.index(name));
```

We have put a delay of 1 second (`Thread.wait(1000)`) in the code to prevent fast actions on indices because their shard allocations are asynchronous, and they require a few milliseconds to be ready. The best practice is to not use a similar hack, but to poll an index's state before performing further operations, and to only perform those operations when it goes green.

See also

You can refer to the following recipes for further reference, which are related to this recipe:

- The *Creating an index* recipe in *Chapter 3*, *Basic Operations*, for details on index creation

- The *Deleting an index* recipe in *Chapter 3*, *Basic Operations*, for details on index deletion

- The *Opening/closing an index* recipe in *Chapter 3*, *Basic Operations*, for the description of opening/closing index APIs

Managing mappings

After creating an index, the next step is to add some mappings to it. We have already seen how to add a mapping via the REST API in *Chapter 3*, *Basic Operations*. In this recipe, we will look at how to manage mappings via a native client.

Getting ready

You need an up-and-running Elasticsearch installation, which we described how to get in the *Downloading and installing Elasticsearch* recipe in *Chapter 1*, *Getting Started*.

A Maven tool or an IDE that natively supports it for Java programming, such as Visual Studio Code, Eclipse, or IntelliJ IDEA, must be installed.

The code for this recipe is in the ch13/elasticsearch-java-client directory and the referred class is MappingOperations.

How to do it...

In the following steps, we add a mapping to a myindex index via the native client:

1. Import the required classes using the following code:

```java
import co.elastic.clients.elasticsearch.
ElasticsearchClient;
import java.io.IOException;
import java.security.KeyManagementException;
import java.security.KeyStoreException;
import java.security.NoSuchAlgorithmException;
```

2. Define a class to contain our code and to initialize client and index, as follows:

```java
public static void main(String[] args) throws
NoSuchAlgorithmException, KeyStoreException,
KeyManagementException {
    String index = "mytest";
    ElasticsearchClient client = ClientHelper.
createClient();
    IndicesOperations io = new IndicesOperations(client);
    try {
        if (io.checkIndexExists(index))
            io.deleteIndex(index);
        io.createIndex(index);
```

3. Prepare the mapping request to put in the index, as follows:

```
var response = client.indices()
            .putMapping(p -> p.index(index).
   properties("nested1", m -> m.nested(f -> f)));
```

4. We can now check the response, as follows:

```
if (!response.acknowledged()) {
    System.out.println("Something strange happens"); }
```

5. We'll remove index to clean up, as follows:

```
io.deleteIndex(index);
```

How it works...

Before executing a mapping operation, a client must be available, and the index must be created.

In the preceding example, if the index exists, it's deleted and a new one is created, so we are sure that we are starting from scratch. This can be seen in the following code:

```
ElasticsearchClient client = ClientHelper.createClient();
IndicesOperations io = new IndicesOperations(client);
try {
    if (io.checkIndexExists(index))
        io.deleteIndex(index);
    io.createIndex(index);
```

Now, we have a fresh index in which to put the mapping that we need in order to create it. The putMapping mapping client allows us to define via a functional pattern, a mapping request:

```
var response = client.indices()
    .putMapping(p -> p
        .index(index)
        .properties("nested1", m -> m.nested(f -> f)) );
```

The advantage of using a functional pattern is that the request is strongly typed and if in the future there are some changes in the method signature, the compilation will fail; this allows you to build a more maintainable and reliable application than using bare JSON.

If everything is okay, you can check the status in the `response.acknowledged()` method that must be `true` (Boolean value); otherwise, an error is raised.

If you need to update a mapping, then you must execute the same call, but you should only put the fields that you need to add in the mapping.

There's more...

There is another important call that is used to manage the mapping—the `Get` mapping API. The call is like `getMapping`, and returns a `GetMappingResponse` response:

```
GetMappingResponse response = client.indices().getMapping(p ->
p.index(index))
```

`response` contains the mapping information. The data that's returned is structured as it would be in an index map; it contains mapping and the name as `IndexMappingRecord`.

`IndexMappingRecord` is an object that contains all the mapping information and all the sections that we discussed in *Chapter 3*, *Basic Operations*.

See also

You can refer to the following recipes for further reference, which are related to this recipe:

- The *Putting a mapping in an index* recipe in *Chapter 3*, *Basic Operations*, for more details about the `Put` mapping API

- The *Getting a mapping* recipe in *Chapter 3*, *Basic Operations*, for more details about the `Get` mapping API

Managing documents

The native APIs for managing documents (`index`, `delete`, and `update`) are the most important after the search APIs. In this recipe, we will learn how to use them. In the next recipe, we will proceed to bulk actions to improve performance.

Getting ready

You need an up-and-running Elasticsearch installation, which we described how to get in the *Downloading and installing Elasticsearch* recipe in *Chapter 1*, *Getting Started*.

A Maven tool, or an IDE that natively supports it for Java programming such as Visual Studio Code, Eclipse, or IntelliJ IDEA, must be installed.

The code for this recipe is in the `ch13/elasticsearch-java-client` directory and the referred class is `DocumentOperations`.

How to do it...

For managing documents, we will perform the following steps:

1. We'll need to import the required classes to execute all the document CRUD operations via the high-level client, as follows:

```
import co.elastic.clients.elasticsearch.
ElasticsearchClient;
import co.elastic.clients.elasticsearch._types.Script;
import co.elastic.clients.elasticsearch.core.GetResponse;
import co.elastic.clients.elasticsearch.core.
IndexResponse;
import co.elastic.clients.elasticsearch.core.
UpdateResponse;
import java.io.IOException;
import java.security.KeyManagementException;
import java.security.KeyStoreException;
import java.security.NoSuchAlgorithmException;
```

2. In this example, we will map our indexed data in the following data class:

```
public static class Record {
    private String text;
    public String getText() { return text; }
    public void setText(String text) { this.text = text;
}
}
```

3. The following code will create the client and remove the index that contains our data, if it exists:

```
public static void main(String[] args) throws
NoSuchAlgorithmException, KeyStoreException,
KeyManagementException {
    String index = "mytest";
    ElasticsearchClient client = ClientHelper.
createClient();
    IndicesOperations io = new IndicesOperations(client);
    try {
        if (io.checkIndexExists(index))
            io.deleteIndex(index);
```

4. We will call the create index by providing the required mapping, as follows:

```
try {
    client.indices()
            .create(c -> c.index(index).mappings(m
-> m.properties("text", t -> t.text(fn ->
fn.store(true)))));
} catch (IOException e) {
    System.out.println("Unable to create mapping");
}
```

5. Now, we can create a typed document record and store a document in Elasticsearch via the index call, as follows:

```
final Record document = new Record();
document.setText("unicorn");
IndexResponse ir = client.index(c -> c.index(index).
id("2").document(document));
System.out.println("Version: " + ir.version());
```

6. Let's retrieve the stored document via the get call, as follows:

```
GetResponse<DocumentOperations.Record> gr = client.get(c
-> c.index(index).id("2"), Record.class);
System.out.println("Version: " + gr.version());
```

7. We can update the stored document via the `update` call using a `painless` script, as follows:

```
UpdateResponse<DocumentOperations.Record> ur = client.
update(u ->
            u.index(index).id("2")
                        .scriptedUpsert(true)
                        .script(Script.of(s -> s.inline(code ->
        code.source("ctx._source.text = 'v2'")))),
            Record.class
);
System.out.println("Version: " + ur.version());
```

8. We can delete the stored document via the `delete` call, as follows:

```
client.delete(d -> d.index(index).id("2"));
```

9. We can now free up the resources that were used, as follows:

```
io.deleteIndex(index);
} catch (IOException e) {
    e.printStackTrace();}
```

10. The console output result will be as follows:

```
Version: 1
Version: 1
Version: 2
```

The document version, after an update action and if the document is reindexed with new changes, is always incremented by 1.

How it works...

Before executing a document action, the client and the index must be available and document mapping should be created (the mapping is optional because it can be inferred from the indexed document).

To index a document via the native client, the `index` method is created. It requires the index as a mandatory argument. If an ID is provided, it will be used; otherwise, a new one will be created.

The new typed API of the Elasticsearch Java client allows storing classes directly without manually building the JSON. In the preceding example, we have used the `Record` class to design the data format to be stored in Elasticsearch.

Obviously, the `index` method supports all the parameters that we looked at in the *Indexing a document* recipe in *Chapter 3*, *Basic Operations*, such as `parent` and `routing`, for example. In the previous example, the call was as follows:

```
IndexResponse ir = client.index(c -> c
        .index(index)
        .id("2")
        .pipeline("my-pipeline")
        .routing("my-routing")
        .refresh(Refresh.True)
        .document(document));
```

The `IndexResponse` return value can be used in the following ways:

- To check whether the index was successful
- To get the ID of the indexed document if it was not provided during the index action
- To retrieve the document version

To retrieve a document, you need to know the index/ID. The client method is `get`. It requires the usual tuple (`index` and `id`) and the class to be used for unmarshaling (that is, `Record` in the example), but a lot of other methods are available to control the routing (such as `routing` and `parent`) or the fields that we saw in the *Getting a document* recipe in *Chapter 3*, *Basic Operations*. In the previous example, the call is as follows:

```
GetResponse<DocumentOperations.Record> gr = client.get(c ->
c.index(index).id("2"), Record.class);
```

The `GetResponse` return type contains all the requests (if the document exists) and document information (`source`, `version`, `index`, and `id`).

To update a document, you need to know the index/ID and provide a script or a document to be used for the update. The client method is `update`.

In the previous example, there is the following:

```
UpdateResponse<DocumentOperations.Record> ur = client.update(u
->
        u.index(index).id("2")
                .scriptedUpsert(true)
                .script(Script.of(s -> s.inline(code -> code.
source("ctx._source.text = 'v2'")))),
        Record.class);
```

The script code must be a string. If the script language is not defined, the default `painless` method is used.

The returned response contains information about the execution and the new version value to manage concurrency.

To delete a document (without needing to execute a query), we need to know the index/ID tuple, and we can use the `delete` client method to create a delete request. In the previous code, we used the following:

```
client.delete(d -> d.index(index).id("2"));
```

The `delete` request allows all the parameters to be passed to it that we saw in the *Deleting a document* recipe in *Chapter 3*, *Basic Operations*, to control the routing and version.

See also

In our recipes, we have used all the CRUD operations on a document. For more details about these actions, refer to the following:

- The *Indexing a document* recipe in *Chapter 3*, *Basic Operations,* for information on how to index a document

- The *Getting a document* recipe in *Chapter 3*, *Basic Operations,* for information on how to retrieve a stored document

- The *Deleting a document* recipe in *Chapter 3*, *Basic Operations,* for information on how to delete a document

- The *Updating a document* recipe in *Chapter 3*, *Basic Operations,* for information on how to update a document

Managing bulk actions

Executing automatic operations on items via a single call will often be the cause of a bottleneck if you need to index or delete thousands/millions of records. The best practice, in this case, is to execute a bulk action.

We have discussed bulk actions via the REST API in the *Speeding up atomic operations (bulk)* recipe in *Chapter 3, Basic Operations*.

Getting ready

You need an up-and-running Elasticsearch installation, which you can get using the *Downloading and installing Elasticsearch* recipe in *Chapter 1, Getting Started*.

A Maven tool or an IDE that natively supports it for Java programming, such as Visual Studio Code, Eclipse, or IntelliJ IDEA, must be installed.

The code of this recipe is in the `ch13/elasticsearch-java-client` directory and the referred class is `BulkOperations`.

How to do it...

To manage a bulk action, we will perform these steps:

1. We'll need to import the required classes to execute bulk actions via the high-level client, as follows:

    ```java
    import co.elastic.clients.elasticsearch.
    ElasticsearchClient;
    import co.elastic.clients.elasticsearch.core.bulk.
    BulkOperation;
    import java.io.IOException;
    import java.security.KeyManagementException;
    import java.security.KeyStoreException;
    import java.security.NoSuchAlgorithmException;
    import java.util.ArrayList;
    import java.util.List;
    ```

2. Next, we'll create the client, remove the old index if it exists, and create a new one, as follows:

```
public static void main(String[] args) throws
NoSuchAlgorithmException, KeyStoreException,
KeyManagementException {
    String index = "mytest";
    ElasticsearchClient client = ClientHelper.
createClient();
    IndicesOperations io = new IndicesOperations(client);
    try {
        if (io.checkIndexExists(index))
            io.deleteIndex(index);
        try {
            client.indices()
                    .create(c -> c.index(index).
mappings(m -> m

.properties("position", p -> p.integer(i ->
i.store(true)))
                    )
                );
        } catch (IOException e) {
            System.out.println("Unable to create
mapping");
        }
```

3. To execute a bulk operation, we need to collect all the documents in BulkOperation to prepare the list to execute against the client.bulk method, as follows:

```
List<BulkOperation> bulkOperations = new ArrayList<>();
for (int i = 1; i <= 1000; i++) {
    final AppData obj = new AppData();
    obj.setPosition(i);
    bulkOperations.add(BulkOperation.of(o -> o.index(idx
-> idx.index(index).id(Integer.toString(obj.position)).
document(obj))));
}
```

```
System.out.println("Number of actions for index: " +
bulkOperations.size());
```

```
client.bulk(c -> c.operations(bulkOperations));
```

4. We can bulk update the previously-created 1,000 documents via a script, adding the bulk update action to `bulker`, as follows:

```
bulkOperations.clear();
for (int i = 1; i <= 1000; i++) {
    final int id = i;
    bulkOperations.add(BulkOperation.of(o -> o.update(u
-> u.index(index).id(Integer.toString(id)).action(a ->
a.script(s -> s.inline(code -> code.source("ctx._source.
position += 2")))))));
}
System.out.println("Number of actions for update: " +
bulkOperations.size());
```

```
client.bulk(c -> c.operations(bulkOperations));
```

5. We can bulk delete 1,000 documents, adding the bulk delete actions to `bulker`, as follows:

```
bulkOperations.clear();
for (int i = 1; i <= 1000; i++) {
    final int id = i;
    bulkOperations.add(BulkOperation.of(o -> o.delete(u
-> u.index(index).id(Integer.toString(id))))); }
System.out.println("Number of actions for delete: " +
bulkOperations.size());
```

```
client.bulk(c -> c.operations(bulkOperations));
```

6. We can now free up the resources that were used, as follows:

```
    io.deleteIndex(index);
} catch (IOException e) {
    e.printStackTrace();
}
```

The result will be as follows:

```
Number of actions for index: 1000
Number of actions for update: 1000
Number of actions for delete: 1000
```

How it works...

Before executing these bulk actions, a client must be available, and an index must be created. If you wish, you can also create document mapping.

We can consider `BulkRequest` as a collection of `BulkOperation` that can be `index`, `create`, `update`, and `delete` actions.

Generally, when used in code, we can consider it as a list in which we add the actions of the supported types.

To initialize the collection that stores our bulk actions, we use the following code:

```
List<BulkOperation> bulkOperations = new ArrayList<>();
```

The type of `BulkOperation` controls how Elasticsearch will manage the bulk item. Depending on their type, we need to provide some parameters:

- An `index` operation (requires `index` and `document`, but `id` is optional):

```
BulkOperation.of(o -> o.index(idx ->
    idx.index(index)
        .id(Integer.toString(obj.position))
        .document(obj))));
```

- A `create` operation (requires `index`, `id`, and `document`):

```
BulkOperation.of(o -> o.create(idx ->
    idx.index(index)
        .id(Integer.toString(obj.position))
        .document(obj))));
```

- An update operation (requires index, id, and action that can be a document or a script):

```
BulkOperation.of(o -> o.update(idx ->
    idx.index(index)
        .id(Integer.toString(id))
        .action(a -> a.script(s -> s.inline(code -> code.
source("ctx._source.position += 2")))
```

- A delete operation (requires index and id):

```
BulkOperation.of(o -> o.delete(idx ->
    idx.index(index)
        .id(Integer.toString(obj.position))
        .document(obj))));
```

7. After adding all the update actions, we can execute them in bulk using client.bulk, as follows:

```
client.bulk(c -> c.operations(bulkOperations));
```

In this example, to simplify it, I created bulk actions with the same type of actions, but, as I described previously, you can put any supported type of action into the same bulk operation.

Building a query

Before a search, a query must be built. Elasticsearch provides several ways to build these queries. In this recipe, we will learn how to create a query object via QueryBuilder and simple strings.

Getting ready

You need an up-and-running Elasticsearch installation, which you can get as described in the *Downloading and installing Elasticsearch* recipe in *Chapter 1, Getting Started*.

A Maven tool or an IDE that natively supports it for Java programming, such as Visual Studio Code, Eclipse, or IntelliJ IDEA, must be installed.

The code for this recipe is in the ch13/elasticsearch-java-client directory and the referred class is QueryCreation.

How to do it...

To create a query, we will perform the following steps:

1. We need to import `SearchRequest` using the following code:

    ```
    import co.elastic.clients.elasticsearch.core.
    SearchRequest;
    ```

2. Next, we'll create a query using `SearchRequest`, as follows:

    ```
    SearchRequest searchRequest = new SearchRequest.Builder()
            .index(index)
            .query(q ->
                    q.bool(b -> b
                            .must(must ->
                                    must.range(r ->
    r.field("number1").gte(JsonData.of(500)))
                                    ).
                            filter(f -> f.term(t ->
    t.field("number2").value(FieldValue.of(1))))
                    )
            ).build();
    ```

3. Now, we can execute a search, as follows (searching via a native API will be discussed in the following recipes):

    ```
    SearchResponse<Record> response = client.
    search(searchRequest, Record.class);
    assert response.hits().total() != null;
    System.out.println("Matched records of elements: " +
    response.hits().total().value());
    ```

    I've removed the redundant parts that are similar to the example of the previous recipe. The result will be as follows:

    ```
    Matched records of elements: 250 hits
    Matched records of elements: 999 hits
    ```

How it works...

There are several ways to define a query in Elasticsearch.

Generally, a query can be defined as the following:

- `SearchRequest.Builder()`: A helper to build a search request with a query as part of it
- `co.elastic.clients.elasticsearch._types.query_dsl.Query`: A typed object that represents an Elasticsearch query

In the previous example, we created a query via `SearchRequest.QueryBuilder`. The first step is to import `SearchRequest` from the namespace, as follows:

```
import co.elastic.clients.elasticsearch.core.SearchRequest;
```

The query of the example is a `boolean` query with `term` as a filter. The goal of the example is to show how to mix several query types to create a complex query.

The `boolean` query contains a `must` clause with a `range` query. We can start to create the `range` query as follows:

```
b.must(must ->must.range(r -> r.field("number1").gte(JsonData.
of(500))))
```

This `range` query matches all the values that are greater than or equal to (`gte`) 500 in the `number1` field.

We need to define a filter, as shown in the following code. In this case, we have used a `term` query, which is one of the most used kinds of query:

```
filter(f -> f.term(t -> t.field("number2").value(FieldValue.
of(1))))
```

`term` accepts a field and a value, which must be a valid Elasticsearch type.

The previous code is similar to the JSON REST `{"term": {"number2":1}`.

After creating the `range` query, we can add it to a Boolean query in the `must` block and the `filter` query in the `filter` block, as follows:

```
.query(q ->
        q.bool(b -> b
                .must(must ->…).
            filter(f -> …)
    ))
```

In real-world complex queries, you can have a lot of nested queries in a `boolean` query or filter.

Before executing a query, the index must be refreshed so that you don't miss any results.

In our example, this is done using the following code:

```
client.indices().refresh(r -> r.index(index));
```

There's more...

The possible native queries/filters are the same as REST ones and have the same parameters; the only difference is that they are accessible via `builder` methods.

The most common query builders are as follows:

- `matchAll`: This allows all the documents to be matched.
- `match` and `matchPhrase`: These are used to match against text strings.
- `term` and `terms`: These are used to match a term value(s) against a specific field.
- `bool`: This is used to aggregate other queries with Boolean logic.
- `ids`: This is used to match a list of IDs.
- `wildcard`: This is used to match terms with wildcards (`*?`).
- `regexp`: This is used to match terms via a regular expression.
- The span query family (`spanTerms`, `spanTerm`, `spanOr`, `spanNot`, and `spanFirst`, for example): These are a few examples of the span query family, which are used in building span queries.
- `hasChild`, `hasParent`, and `nested`: These are used to manage related documents.

The previous list is not exhaustive, because it will constantly evolve throughout the life of Elasticsearch. New query types will be added to cover new search cases, or they are occasionally renamed, such as a `text` query changing to a `match` query.

Executing a standard search

In the previous recipe, we learned how to build queries. In this recipe, we will execute a query to retrieve some documents.

Getting ready

You need an up-and-running Elasticsearch installation, as we described in the *Downloading and installing Elasticsearch* recipe in *Chapter 1, Getting Started*.

A Maven tool or an IDE that natively supports it for Java programming, such as Visual Studio Code, Eclipse, or IntelliJ IDEA, must be installed.

The code for this recipe is in the `ch13/elasticsearch-java-client` directory and the referred class is the `QueryExample`.

How to do it...

To execute a standard query, we will perform the following steps:

1. We need to import `SearchRequest.QueryBuilder` to create the query, as follows:

    ```
    import co.elastic.clients.elasticsearch.core.
    SearchRequest;
    ```

2. We can create an index and populate it with some data, as follows:

    ```
    String index = "mytest";
    QueryHelper qh = new QueryHelper();
    qh.populateData(index);
    ElasticsearchClient client = qh.getClient();
    ```

3. Now, we will build a query with the `number1` field greater than or equal to `500`, we filter it for `number2` equal to `1`, and we highlight the `name` field as follows:

    ```
    SearchRequest searchRequest = new SearchRequest.
    Builder().index(index)
            .query(q ->
                    q.bool(b -> b
                            .must(must -> must.range(r ->
    r.field("number1").gte(JsonData.of(500)))).
                            filter(f -> f.term(t ->
    t.field("number2").value(FieldValue.of(1)))))))
            .highlight(h -> h.fields("name", h1 ->
    h1.field("name")))).build();
    ```

4. After creating a query, it is enough to execute it using the following code:

```
SearchResponse<QueryHelper.AppData> response = client.
search(searchRequest, QueryHelper.AppData.class);
System.out.println("Matched number of documents: " +
response.hits().total().value());
System.out.println("Maximum score: " + response.hits().
maxScore());;
```

5. When we have `SearchResponse`, we need to check its status and iterate it on hit(s), as follows:

```
for (Hit<QueryHelper.AppData> hit : response.hits().
hits()) {
    System.out.println("hit: " + hit.index() + ":" + hit.
id()); }
```

The result should be similar to the following:

```
Matched number of documents: 251 hits
Maximum score: 1.0
hit: mytest:_doc:499
hit: mytest:_doc:501
hit: mytest:_doc:503
hit: mytest:_doc:505
hit: mytest:_doc:507
hit: mytest:_doc:509
hit: mytest:_doc:511
hit: mytest:_doc:513
hit: mytest:_doc:515
hit: mytest:_doc:517
```

How it works...

The call to execute a search is `client.search`, which returns a `SearchResponse` response, as follows:

```
SearchResponse<QueryHelper.AppData> response = client.
search(searchRequest, QueryHelper.AppData.class);
```

An important parameter of the `search` method is the class that will fit our documents in the response. This approach guarantees a strong typing between Elasticsearch and our business logic.

`SearchRequest.Builder` has a lot of methods that can set all of the parameters that we have already seen in the *Executing a search* recipe in *Chapter 4, Exploring Search Capabilities*. The most used ones are as follows:

- `index`: This allows the indices to be defined.

- `query`: This allows the query that is to be executed to be set.

- `storedFields`: These allow setting fields to be returned (used to reduce the bandwidth by returning only the needed fields).

- `aggregations`: This allows us to compute any added aggregations.

- `highlight`: This allows us to return any added highlighting.

- `scriptFields`: This allows a scripted field to be returned. A scripted field is a field that is computed by server-side scripting using one of the available scripting languages. For example, it can be as follows:

```
.scriptFields("sNum1",
        sf -> sf.script(
                func -> func.inline(
                        inline -> inline
                                .source("_doc.num1.value
 * factor")
                                .params("factor",
  JsonData.of(2.0)) ) ) )
```

After executing a search, a response object is returned.

The response object contains a lot of sections that we analyzed in the *Executing a search* recipe in *Chapter 4, Exploring Search Capabilities*. The most important one is the `hits` section that contains our results. The main accessor methods of this section are as follows:

- `hits.total`: This allows the total number of results to be obtained, as shown in the following code:

```
System.out.println("Matched number of documents: " +
response.getHits().getTotalHits());
```

- maxScore: This gives the maximum score for the documents. It is the same score value as the first SearchHit, as shown in the following code:

```
System.out.println("Maximum score: " + response.
getHits().getMaxScore());
```

- hits: This is an array of SearchHit that contains the results, if available.

The SearchHit is the result object. It has a lot of methods, of which the most important ones are as follows:

- getIndex(): This is the index that contains the document.

- getId(): This is the ID of the document.

- getScore(): This is the query score of the document, if available.

- getVersion(): This is the version of the document, if available.

- getSource(), getSourceAsString(), getSourceAsMap(), and similar: These return the source of the document in different forms, if available.

- getExplanation(): If available (required in the search), this contains the query explanation.

- getFields and getField(String name): These return the fields that were requested if they were passed fields to search for an object.

- getSortValues(): This is the value/values that are used to sort this record. It's only available if sort is specified during the search phase.

- getShard(): This is the shard of the search hit. This value is very important for custom routing.

In the preceding example, we have printed only the index, type, and ID of each hit, as shown in the following code:

```
for (SearchHit hit : response.getHits().getHits()) {
    System.out.println("hit: " + hit.getIndex() + ":" + hit.
getId()); }
```

The number of returned hits, if not defined, is limited to 10. To retrieve more hits, you need to define a larger value in the size method or paginate using the from method.

See also

The *Executing a search* recipe in *Chapter 4, Exploring Search Capabilities*

Executing a search with aggregations

The previous recipe can be extended to support aggregations in order to retrieve analytics on indexed data.

Getting ready

You need an up-and-running Elasticsearch installation, which you can get as described in the *Downloading and installing Elasticsearch* recipe in *Chapter 1, Getting Started*.

A Maven tool or an IDE that natively supports it for Java programming, such as Visual Studio Code, Eclipse, or IntelliJ IDEA, must be installed.

The code for this recipe is in the ch13/elasticsearch-java-client directory, and the referred class is AggregationExample.

How to do it...

To execute a search with aggregations, we will perform the following steps:

1. We need to import the necessary classes for the aggregations using the following code:

    ```java
    import co.elastic.clients.elasticsearch.
    ElasticsearchClient;
    import co.elastic.clients.elasticsearch._types.
    aggregations.StringTermsAggregate;
    import co.elastic.clients.elasticsearch._types.
    aggregations.StringTermsBucket;
    import co.elastic.clients.elasticsearch.core.
    SearchRequest;
    import co.elastic.clients.elasticsearch.core.
    SearchResponse;
    ```

2. We can create an index and populate it with some data that we will use for the aggregations, as follows:

    ```java
    String index = "mytest";
    QueryHelper qh = new QueryHelper();
    qh.populateData(index);
    ElasticsearchClient client = qh.getClient();
    ```

3. We then calculate two different aggregations (terms and extended statistics); we use
    `size(0)` because we don't need the hits, as shown in the following code:

```
SearchRequest searchRequest = new SearchRequest.
Builder().index(index)
        .aggregations("tag", t -> t.terms(terms -> terms.
field("tag")))
        .aggregations("number1", t -> t.extendedStats(agg
-> agg.field("number1")))
        .size(0)
        .build();
```

4. Now, we can execute a search using the following code:

```
SearchResponse<QueryHelper.AppData> response = client.
search(searchRequest, QueryHelper.AppData.class);
```

5. We need to check the response validity and wrap the aggregation results, as shown
    in the following code:

```
System.out.println("Matched number of documents: " +
response.hits().total().value());
StringTermsAggregate termsAggs = response.aggregations().
get("tag").sterms();
System.out.println("Aggregation name: " + termsAggs._
aggregateKind().name());
System.out.println("Aggregation total: " + termsAggs.
buckets().array().size());
for (StringTermsBucket entry : termsAggs.buckets().
array()) {
    System.out.println(" - " + entry.key() + " " + entry.
docCount());
}
var extStats = response.aggregations().get("number1").
extendedStats();
System.out.println("Aggregation name: " + extStats._
aggregateKind().name());
System.out.println("Count: " + extStats.count());
System.out.println("Min: " + extStats.min());
System.out.println("Max: " + extStats.max());
System.out.println("Standard Deviation: " + extStats.
```

```
stdDeviation());
System.out.println("Sum of Squares: " + extStats.
sumOfSquares());
```

The result should be as follows:

```
Number of actions for index: 999

Matched number of documents: 999 hits

Aggregation name: tag

Aggregation total: 4

  - bad 257

  - amazing 255

  - nice 245

  - cool 242

Aggregation name: number1

Count: 999

Min: 2.0

Max: 1000.0

Standard Deviation: 288.38631497813253

Sum of Squares: 3.33833499E8
```

How it works...

The search part is similar to the previous example. In this case, we have used a `matchAll` query, which matches all the documents.

There are several types of aggregation, as we have already seen in *Chapter 5, Text and Numeric Queries*.

The first one, which we created with the `aggregations` method, is a `Terms` aggregation, which collects and counts all `terms` occurrences in buckets, as shown in the following code:

```
.aggregations("tag", t -> t.terms(terms -> terms.field("tag")))
```

The required value for every aggregation is the name, which is passed in the builder constructor. In the case of a `terms` aggregation, the field is required to be able to process the request. There are a lot of other parameters; see the *Executing terms aggregations* recipe in *Chapter 7, Aggregations*, for full details.

The second `aggregations` method call that we created is an extended statistical aggregation based on the `number1` numeric field, as follows:

```
.aggregations("number1", t -> t.extendedStats(agg -> agg.
field("number1")))
```

We can add them on a `SearchRequest.Builder` object via the `aggregations` method, as follows:

```
SearchRequest searchRequest = new SearchRequest.Builder().
index(index)
        .aggregations("tag", t -> t.terms(terms -> terms.
field("tag")))
        .aggregations("number1", t -> t.extendedStats(agg ->
agg.field("number1")))
        .size(0)
        .build();
SearchResponse<QueryHelper.AppData> response = client.
search(searchRequest, QueryHelper.AppData.class);
```

Now, the response holds information about our aggregations. To access them, we need to use the `aggregations` method of the response.

The aggregation's results are contained in a hash-like structure, and you can retrieve them with the names that you have previously defined in the request.

To retrieve the first aggregation results, we need to get them, as follows:

```
StringTermsAggregate termsAggs = response.aggregations().
get("tag").sterms);
```

Every response has a typed class that maps its data. In the case of a `terms` aggregation response composed of strings, the method to retrieve it is `sterms` and the returned type is `StringTermsAggregate`.

Now that we have an aggregation result of type `Terms` (see the *Executing terms aggregations* recipe in *Chapter 7, Aggregations*), we can get the aggregation properties and iterate in buckets, as follows:

```
for (StringTermsBucket entry : termsAggs.buckets().array())
{ System.out.println(" - " + entry.key() + " " + entry.
docCount()); }
```

To retrieve the second aggregation result, because the result is of type
`ExtendedStatsAggregate`, you need to cast to it, as shown in the following code:

```
ExtendedStatsAggregate extStats = response.aggregations().
get("number1").extendedStats();
```

Now, you can access the result properties of this kind of aggregation, as follows:

```
System.out.println("Aggregation name: " + extStats._
aggregateKind().name());
System.out.println("Count: " + extStats.count());
System.out.println("Min: " + extStats.min());
System.out.println("Max: " + extStats.max());
System.out.println("Standard Deviation: " + extStats.
stdDeviation());
System.out.println("Sum of Squares: " + extStats.
sumOfSquares());
System.out.println("Variance: " + extStats.variance());
```

Using aggregations with a native client is quite easy, and you need only pay attention to
the returned aggregation type to execute the correct aggregation return-type cast to access
your results.

See also

You can refer to the following URLs for further reference, which are related to this recipe:

- The *Executing terms aggregations* recipe in *Chapter 7*, *Aggregations*, describes the
  `terms` aggregation in depth

- The *Executing statistical aggregations* recipe in *Chapter 7*, *Aggregations*, for more
  details about statistical aggregations

Executing a scroll search

Pagination with a standard query works very well if you are matching documents with
documents that do not change too often; otherwise, performing pagination with live data
returns unpredictable results. To bypass this problem, Elasticsearch provides an extra
parameter in the query: `scroll`.

Getting ready

You need an up-and-running Elasticsearch installation, which you can get as described in the *Downloading and installing Elasticsearch* recipe in *Chapter 1*, *Getting Started*.

A Maven tool, or an IDE that natively supports it for Java programming, such as Visual Studio Code, Eclipse, or IntelliJ IDEA, must be installed.

The code for this recipe is in the `ch13/elasticsearch-java-client` directory and the referred class is `ScrollQueryExample`.

How to do it...

The search is done as it was shown in the *Execute a standard search* recipe in *Chapter 4*, *Exploring Search Capabilities*. The main difference is the use of a `scroll` timeout, which allows the resulting IDs to be stored in memory for a query for a defined period of time. The steps are like those that are used for a standard search, as you can see from the following steps:

1. We import the `Time` object to define time in a more human way, as follows:

    ```
    import co.elastic.clients.elasticsearch._types.Time;
    ```

2. We execute the search by setting the `scroll` value. We can change the code of the *Execute a standard search* recipe to use `scroll` in the following way:

    ```
    SearchRequest searchRequest = new SearchRequest.
    Builder().index(index)
            .query(q ->
                    q.bool(b -> b
                            .must(must -> must.range(r ->
    r.field("number1").gte(JsonData.of(500)))).
                            filter(f -> f.term(t ->
    t.field("number2").value(FieldValue.of(1))))))
            .size(30)
            .scroll(Time.of(t -> t.time("2m")))
            .build();
    SearchResponse<QueryHelper.AppData> response = client.
    search(searchRequest, QueryHelper.AppData.class);
    ```

3. To manage the scrolling, we need to create a loop until the results are returned, as follows:

```
do {
    for (Hit<AppData> hit : response.hits().hits()) {
        System.out.println("hit: " + hit.index() + ":" +
hit.id());
    }
    final SearchResponse<QueryHelper.AppData> old_
response = response;
    response = client.scroll(s -> s.scrollId(old_
response.scrollId()), QueryHelper.AppData.class);
} while (response.hits().hits().size() != 0); // Zero
hits mark the end of the scroll and the while loop.
```

The loop will iterate on all the results until records are available. The output will be similar to the following:

```
hit: mytest:499
hit: mytest:531
hit: mytest:533
hit: mytest:535
hit: mytest:555
hit: mytest:559
hit: mytest:571
hit: mytest:575
...truncated...
```

How it works...

To use the scrolling result, it's enough to add a `scroll` method with a timeout to the `SearchRequest` object.

When using scrolling, the following behaviors must be kept in mind:

*   The timeout defines the period of time that an Elasticsearch server keeps the results. If you ask for a scroll after the timeout, the server returns an error. The user must be careful with short timeouts.

*   The scroll consumes memory until it ends, or a timeout is raised. Setting too large a timeout without consuming the data results in a big memory overhead. Using a large number of open scrollers consumes a lot of memory proportional to the number of IDs and their related data (score and order, for example) in the results.

- With scrolling, it's not possible to paginate the documents as there is no start. Scrolling is designed to fetch consecutive results.

A standard `SearchRequest` is changed to a scroll in the following way:

```
SearchRequest searchRequest = new SearchRequest.Builder().
index(index)
            .query(…)
            .scroll(Time.of(t -> t.time("2m")))
            .build();
```

The response contains the same results as the standard search, plus a scroll ID, which is required to fetch the next set of results.

To execute the scroll, you need to call the `scroll` client method with a scroll ID and a new timeout. In this example, we are processing all the result documents, as shown in the following code:

```
do { // process the hits
    final SearchResponse<QueryHelper.AppData> old_response =
response;
    response = client.scroll(s -> s.scrollId(old_response.
scrollId()), QueryHelper.AppData.class);
} while (response.hits().hits().size() != 0); // Zero hits mark
the end of the scroll and the while loop.
```

To understand that we are at the end of the scroll, we can check that no results are returned.

There are a lot of scenarios in which `scroll` is very important, but when working on big data solutions when the number of results is very large, it's easy to hit the timeout. In these scenarios, it is important to have a good architecture in which you fetch the results as fast as possible and don't process the results iteratively in the loop, but defer the manipulation result in a distributed way.

In this case, the best solution is to use the `search_after` functionality of Elasticsearch, sorting by `_uid`, as described in the *Using the search_after functionality* recipe in *Chapter 4, Exploring Search Capabilities*.

See also

You can refer to the following recipes for further reference, which are related to this recipe:

- The *Executing a scroll query* recipe in *Chapter 4, Exploring Search Capabilities*
- The *Using the search_after functionality* recipe in *Chapter 4, Exploring Search Capabilities*

Integrating with DeepLearning4j

DeepLearning4J (DL4J) is one of the most used open source libraries in machine learning. It can be found at `https://deeplearning4j.org/`.

The best description for this library is available on its website, which says—*Deeplearning4j is the first commercial-grade, open-source, distributed deep learning library written for Java and Scala. Integrated with Hadoop and Apache Spark, DL4J brings AI to business environments for use on distributed GPUs and CPUs.*

In this recipe, we will see how it's possible to use Elasticsearch as a source for data to be trained in a machine learning algorithm.

Getting ready

You need an up-and-running Elasticsearch installation, as we described in the *Downloading and installing Elasticsearch* recipe in *Chapter 1, Getting Started*.

A Maven tool or an IDE that natively supports Java programming, such as Visual Studio Code, Eclipse, or IntelliJ IDEA, must be installed.

The code for this recipe is in the `ch13/deeplearning4j` directory.

Because Elasticsearch 8.x or above is secured by default, to run all the examples of this chapter, put the credentials in `ES_USER` and `ES_PASSWORD` environment variables when executing the code.

How to do it...

We will use the famous `iris` dataset (`https://en.wikipedia.org/wiki/Iris_flower_data_set`), well known to every data scientist for creating an index with the data to be used in training the deep learning model.

To prepare your index dataset, we need to populate it by executing the `PopulatingIndex` class available in the source code. The `PopulatingIndex` class reads the `iris.txt` file and stores the rows of the dataset with the following object format:

Field Name	Type	Description
f1	float	Feature 1 of the flower
f2	float	Feature 2 of the flower
f3	float	Feature 3 of the flower
f4	float	Feature 4 of the flower
label	int	Label of the flower (valid values 0,1, and 2)

Figure 13.2 – Object formats

To use DL4J as a source data input for your models, you will perform the following steps:

1. Add the various DL4J dependencies to `pom.xml` of the Maven project:

```xml
<dependency>
    <groupId>org.nd4j</groupId>
    <artifactId>nd4j-native-platform</artifactId>
    <version>${nd4j.version}</version>
</dependency>
<!-- ND4J backend. You need one in every DL4J project.
Normally define artifactId as either "nd4j-native-
platform" or "nd4j-cuda-9.2-platform" -->
<dependency>
    <groupId>org.nd4j</groupId>
    <artifactId>${nd4j.backend}</artifactId>
    <version>${nd4j.version}</version>
</dependency>
<!-- Core DL4J functionality -->
<dependency>
    <groupId>org.deeplearning4j</groupId>
    <artifactId>deeplearning4j-core</artifactId>
    <version>${dl4j.version}</version>
</dependency>
```

2. Now, we can write our `ElasticSearchD4J` class to train and test our model. As the first step, we need to initialize the Elasticsearch client:

```
ElasticsearchClient client = ClientHelper.createClient();
String indexName="iris";
```

3. After having the client, we can read our dataset. We will execute a query and collect the `Hit` results using a simple search:

```
SearchRequest searchRequest = new SearchRequest.
Builder().index(indexName).size(1000).build();
SearchResponse<Iris> response = client.
search(searchRequest, Iris.class);
List<Hit<Iris>> hits = response.hits().hits();
```

4. We need to convert the hits in a DL4J dataset. We will do this by creating intermediate arrays and populating them:

```
//Convert the iris data into 150x4 matrix
int row = 150;
int col = 4;
double[][] irisMatrix = new double[row][col];
//Now do the same for the label data
int colLabel = 3;
double[][] labelMatrix = new double[row][colLabel];
for (int r = 0; r < row; r++) {
    // we populate features
    Hit<Iris> source = hits.get(r);
    Iris iris=source.source();
    irisMatrix[r][0] = (double) iris.getF1();
    irisMatrix[r][1] = (double) iris.getF2();
    irisMatrix[r][2] = (double) iris.getF3();
    irisMatrix[r][3] = (double) iris.getF4();
    // we populate labels
    int label = (Integer) iris.getLabel();
    labelMatrix[r][0] = 0.0;
    labelMatrix[r][1] = 0.0;
    labelMatrix[r][2] = 0.0;
    if (label == 0) labelMatrix[r][0] = 1.0;
```

```
        if (label == 1) labelMatrix[r][1] = 1.0;
        if (label == 2) labelMatrix[r][2] = 1.0;
    }
    //Convert the data matrices into training INDArrays
    INDArray training = Nd4j.create(irisMatrix);
    INDArray labels = Nd4j.create(labelMatrix);
    DataSet allData = new DataSet(training, labels);
```

5. Then, split the datasets into two—one for training and one for tests. After having them, we need to normalize the values. These actions can be done with the following code:

```
    allData.shuffle();
    SplitTestAndTrain testAndTrain = allData.
    splitTestAndTrain(0.65);   //Use 65% of data for training
    DataSet trainingData = testAndTrain.getTrain();
    DataSet testData = testAndTrain.getTest();
    //We need to normalize our data. We'll use
    NormalizeStandardize (which gives us mean 0, unit
    variance):
    DataNormalization normalizer = new
    NormalizerStandardize();
    normalizer.fit(trainingData);            //Collect the
    statistics (mean/stdev) from the training data. This does
    not modify the input data
    normalizer.transform(trainingData);      //Apply
    normalization to the training data
    normalizer.transform(testData);          //Apply
    normalization to the test data. This is using statistics
    calculated from the *training* set
```

6. Now, we can design the model to be used for the training:

```
    final int numInputs = 4;
    int outputNum = 3;
    long seed = 6;
    log.info("Build model....");
    MultiLayerConfiguration conf = new
    NeuralNetConfiguration.Builder()
        .seed(seed)
```

```
    .activation(Activation.TANH)
    .weightInit(WeightInit.XAVIER)
    .updater(new Sgd(0.1))
    .l2(1e-4)
    .list()
    .layer(0, new DenseLayer.Builder().nIn(numInputs).
nOut(3)
            .build())
    .layer(1, new DenseLayer.Builder().nIn(3).nOut(3)
            .build())
    .layer(2, new OutputLayer.Builder(LossFunctions.
LossFunction.NEGATIVELOGLIKELIHOOD)
            .activation(Activation.SOFTMAX)
            .nIn(3).nOut(outputNum).build())
    .backprop(true).pretrain(false)
    .build();
```

7. After having defined the model, we can finally train it with our dataset; we use `1000` iterations for the training. The training code is as follows:

```
MultiLayerNetwork model = new MultiLayerNetwork(conf);
model.init();
model.setListeners(new ScoreIterationListener(100));
for(int i=0; i<1000; i++ ) {
    model.fit(trainingData);
}
```

8. Now that we have a trained model, we need to evaluate its accuracy, and we can do that using the test dataset:

```
//evaluate the model on the test set
Evaluation eval = new Evaluation(3);
INDArray output = model.output(testData.getFeatures());
eval.eval(testData.getLabels(), output);
log.info(eval.stats());
```

If you execute the model training, the output will be something similar to the following:

```
18:45:21.831 org.nd4j.linalg.factory.Nd4jBackend - Loaded
[CpuBackend] backend

18:45:25.937 org.nd4j.nativeblas.NativeOpsHolder - Number
of threads used for NativeOps: 4

18:45:27.742 org.nd4j.nativeblas.Nd4jBlas - Number of
threads used for BLAS: 4

18:45:27.748 o.n.l.a.o.e.DefaultOpExecutioner - Backend
used: [CPU]; OS: [Mac OS X]

18:45:27.748 o.n.l.a.o.e.DefaultOpExecutioner - Cores:
[8]; Memory: [14.2GB];

18:45:27.748 o.n.l.a.o.e.DefaultOpExecutioner - Blas
vendor: [MKL]

18:45:27.842 com.packtpub.ElasticSearchD4J - Build
model....

18:45:27.966 o.d.nn.multilayer.MultiLayerNetwork -
Starting MultiLayerNetwork with WorkspaceModes set to
[training: ENABLED; inference: ENABLED], cacheMode set to
[NONE]

18:45:28.040 o.d.o.l.ScoreIterationListener - Score at
iteration 0 is 1.2126128636817546

18:45:28.262 o.d.o.l.ScoreIterationListener - Score at
iteration 100 is 0.3524093414839366

18:45:28.420 o.d.o.l.ScoreIterationListener - Score at
iteration 200 is 0.15881737296024345

18:45:28.584 o.d.o.l.ScoreIterationListener - Score at
iteration 300 is 0.10844747462043258

18:45:28.747 o.d.o.l.ScoreIterationListener - Score at
iteration 400 is 0.09064608464904057

18:45:28.911 o.d.o.l.ScoreIterationListener - Score at
iteration 500 is 0.08178475263018518

18:45:29.079 o.d.o.l.ScoreIterationListener - Score at
iteration 600 is 0.07658190599489766

18:45:29.197 o.d.o.l.ScoreIterationListener - Score at
iteration 700 is 0.07320950281805434

18:45:29.332 o.d.o.l.ScoreIterationListener - Score at
iteration 800 is 0.07086715868248236

18:45:29.444 o.d.o.l.ScoreIterationListener - Score at
iteration 900 is 0.0691510897674625
```

```
18:45:29.714 com.packtpub.ElasticSearchD4J -
==========================Evaluation
Metrics===========================
 # of classes:       3
 Accuracy:           1.0000
 Precision:          1.0000
 Recall:             1.0000
 F1 Score:           1.0000
Precision, recall & F1: macro-averaged (equally weighted
avg. of 3 classes)
==========================Confusion
Matrix===========================
   0   1   2
 ----------
  18   0   0 | 0 = 0
   0  17   0 | 1 = 1
   0   0  18 | 2 = 2
Confusion matrix format: Actual (rowClass) predicted as
(columnClass) N times
============================================================
=========
```

How it works...

Eclipse DL4J is a deep learning programming library written for Java and the JVM. It includes implementations of restricted Boltzmann machines, deep belief networks, deep autoencoders, stacked denoising autoencoders, and recursive neural tensor networks such as word2vec, doc2vec, and GloVe. These algorithms all include distributed parallel versions that integrate with Apache Hadoop and Spark.

DL4J is able to use both CPU and GPU to process deep learning workloads fast.

In the preceding example, we have stored our dataset in Elasticsearch and fetched it to build the DL4J dataset. Using Elasticsearch as dataset storage is very handy because you can use the power of Elasticsearch to analyze, clean, and filter the data before giving it to a machine learning algorithm.

The dataset was shuffled (allData.shuffle();) to provide less bias on the training and test datasets. In this case, we have chosen a three-layer deep learning model and we have trained the model with the data taken by Elasticsearch, iterating the training 1,000 times. The result was a neural network model with an accuracy of 0.98.

This example is very simple, but it shows how it's easy to use Elasticsearch as a data source for machine learning jobs. DL4J is a wonderful library that can be used outside Elasticsearch or can be embedded in a plugin to provide machine learning capabilities to Elasticsearch.

See also

You can refer to the following URLs for further reference, which are related to this recipe:

- The official site of DL4J (`https://deeplearning4j.org/`) for more examples and references about this powerful library.

- A more detailed description of the Iris dataset can be found at `https://en.wikipedia.org/wiki/Iris_flower_data_set`.

14
Scala Integration

Scala is becoming one of the most widely used languages in big data scenarios. This language provides a lot of facilities for managing data, such as immutability and functional programming.

In Scala, you can simply use the libraries that we saw in the previous chapter for Java. However, they are not scalastic, as they don't provide type safety (because many of these libraries take JSON as a string), and it is easy to use asynchronous programming.

In this chapter, we will look at how we can use `elastic4s`, a mature library, to use Elasticsearch in Scala. Its main features are as follows:

- It has type-safe, concise DSL.
- It integrates with standard Scala futures.
- It uses the Scala collections library over the Java collections.
- It returns an option where the Java methods would return `null`.
- It uses Scala durations instead of strings/longs for time values.
- It uses type class for marshaling and unmarshaling classes to/from Elasticsearch documents and is backed by Jackson, Circe, Json4s, and Play JSON implementations.
- It provides a reactive-streams implementation.
- It provides embedded nodes and testkit sub-projects, which are ideal for your tests.

In this chapter, we will mainly look at examples of standard Elastic4s DSL usage and some helpers, such as the `circe` extension, for the easy marshaling/unmarshaling of documents in classes.

In this chapter, we will cover the following recipes:

- Creating a client in Scala
- Managing indices
- Managing mappings
- Managing documents
- Executing a standard search
- Executing a search with aggregations
- Integrating with DeepLearning.scala

Creating a client in Scala

The first step in working with Elastic4s is to create a connection client to call Elasticsearch. Similar to Java, the connection client is native and can be a node or a transport one.

Similar to Java, the connection client can be both a native one and an HTTP one.

In this recipe, we'll initialize an HTTP client because it can be put behind a proxy/balancer to increase the high availability of your solution. This is a good practice.

Getting ready

You need an up-and-running Elasticsearch installation, as described in the *Downloading and installing Elasticsearch* recipe of *Chapter 1*, *Getting Started*.

Additionally, an IDE that supports Scala programming, such as IntelliJ IDEA, with the Scala plugin should be installed globally.

The code for this recipe can be found in the `chapter_14/elastic4s_sample` directory; the referred class is `ClientSample.scala`.

How to do it...

To create an Elasticsearch client and to create/search a document, we will perform the following steps:

1. The first step is to add the `elastic4s` library to the `build.sbt` configuration via the following code:

```
name """"""elastic4s-sam""""""

version "= "0.".0"

scalaVersion "= "2.1".8"

scalacOptions := S"q("-unchec"ed", "-deprecat"on",
"-encod"ng", "u"f8")

libraryDependencies ++= {

  val elastic4sV"= "7.1".0"

  val Log4jVersion"= "2.1".1"

  Seq(

  "   "com.sksamuel.elasti"4s" "% "elastic4s-json-ci"ce" %
elastic4sV,

      // for the http client

  "   "com.sksamuel.elasti"4s" "% "elastic4s-client-esj"va"
% elastic4sV,

      // if you want to use reactive streams

  "   "com.sksamuel.elasti"4s" "% "elastic4s-http-stre"ms"
% elastic4sV,

  "   "ch.qos.logb"ck""% "logback-clas"ic""% "1.".3",

  "   "org.apache.logging.lo"4j""% "log4j-"pi" %
Log4jVersion,

  "   "org.apache.logging.lo"4j""% "log4j-c"re" %
Log4jVersion,

  "   "org.apache.logging.lo"4j""% "log4j-1.2-"pi" %
Log4jVersion

  )}
```

> **Note**
>
> We are using version 7.17.0 of Elastic4s, which works very well with Elasticsearch 8.0.0. This is because, as of the time of writing, a version of Elastic4s that targets version 8.x of Elasticsearch has not yet been released.

2. To use the library, we need to import the client classes and implicits:

```
import com.sksamuel.elastic4s.ElasticDsl._
import com.sksamuel.elastic4s.http.JavaClient
import com.sksamuel.elastic4s.{ElasticClient,
ElasticProperties}
import org.elasticsearch.client.RestClient
```

3. Now, we can initialize the client by providing an Elasticsearch URI, skipping the SSL verification (due to it being self-signed in the local deploy of the certificate), and providing a user/password (as read by the ES_USER/ES_PASSWORD environment variables). Use the following code:

```
object ClientSample extends App {
  lazy val callback = new HttpClientConfigCallback {
    override def customizeHttpClient(
        httpClientBuilder: HttpAsyncClientBuilder
    ): HttpAsyncClientBuilder = {
      val creds = new BasicCredentialsProvider()
      creds.setCredentials(
        AuthScope.ANY,
        new UsernamePasswordCredentials(
          sys.env.getOrElse("ES_USER", ""),
          sys.env.getOrElse("ES_PASSWORD", "")
        ))
      val sslContext = new SSLContextBuilder()
        .loadTrustMaterial(null, TrustAllStrategy.
INSTANCE).build();
      httpClientBuilder
        .setSSLContext(sslContext)
        .setSSLHostnameVerifier(NoopHostnameVerifier.
INSTANCE)
        .setDefaultCredentialsProvider(creds)
    }}
  lazy val client = ElasticClient(
    JavaClient(
      ElasticProperties(s"https://localhost:9200"),
      requestConfigCallback = NoOpRequestConfigCallback,
      httpClientConfigCallback = callback
    ) )
```

4. To index a document, we execute `indexInto` with the document in the following way:

```
client.execute {
    indexInto("bands" ) fields "name" -> "coldplay"   }
```

5. Now, we can search for the document that we indexed earlier by executing a query and waiting for its response:

```
val resp = client.execute {
    search("bands") query "coldplay"
}.await
```

If the document is available, the response will be as follows:

```
RequestSuccess(200,Some({"took":7,"timed_out":false,
"_shards":{"total":1,"successful":1,"skipped":0,
"failed":0},"hits":{"total":{"value":1,"relation":"eq"},
"max_score":0.2876821,"hits":[{"_index":"bands",
"_type":"_doc","_id":"ByX0vHgBwqGK5YGKtDV1","_
score":0.2876821,"_source":{"name":"coldplay"}}]}}),
Map(content-type -> application/json; charset=UTF-8,
content-length -> 276),SearchResponse(7,false,
false,null,Shards(1,0,1),None,null,SearchHits
(Total(1,eq),0.2876821,[Lcom.sksamuel.elastic4s.requests.
searches.SearchHit;@4ad3d266)))
```

How it works...

Elastic4s hides a lot of the boilerplate code, which is required to initialize an Elasticsearch client.

The simplest way to define a connection to Elasticsearch is via `ElasticProperties`. This allows you to provide the following:

- Multiple server endpoints, separated by commas (that is, `http(s)://host:port,host:port(/prefix)?querystring`).

- The other settings are to be provided to the client via `Map[String,String]` (that is, `?cluster.name=elasticsearch`).

After having defined `ElasticProperties`, you can create `ElasticClient`, which is used for every Elasticsearch call:

1. You can initialize `ElasticClient` via `ElasticProperties`, which accepts a string that is similar to a JDBC connection. This is very handy because you can store it as a simple string inside your application's configuration file:

```
val nodes = ElasticProperties("http://host1:9200,http://
host2:9200,http://host3:9200")
val client = ElasticClient(JavaClient(nodes))
```

Note that version 8.x of Elasticsearch is secured by default. For this reason, we need to extend the standard connection with a setup for SSL alongside basic authorization.

2. To do this setup, we need to create an `HttpClientConfigCallback` class with the override of the `customizeHttpClient` method. This allows us to customize the `HttpAsyncClientBuilder` builder, which sets up all the connections of our client:

```
Lazy val callback = new HttpClientConfigCallback {
    override def customizeHttpClient(
        httpClientBuilder: HttpAsyncClientBuilder
    ): HttpAsyncClientBuilder = {
… truncated…
```

3. The user password authorization can be easily provided in the context of HTTP via `BasicCredentialsProvider`. For better security, the best practice is to read them from the environment variables (that is, `ES_USER`/`ES_PASSWORD`) and not to put them in the code:

> **Tip**
> You can put them in the default empty value to simplify the testing process.

```
val creds = new BasicCredentialsProvider()
creds.setCredentials(
  AuthScope.ANY,
  new UsernamePasswordCredentials(
    sys.env.getOrElse("ES_USER", ""),
    sys.env.getOrElse("ES_PASSWORD", "")
  ) )
```

4. In the default Elasticsearch installation, SSL certificates are self-signed (that is, they are not provided by a certificate authority), so you need to remove the certificate validation, for development needs, to be able to connect via HTTPS to your Elasticsearch instance. The following code provides this kind of behavior:

```
val sslContext = new SSLContextBuilder()
  .loadTrustMaterial(null, TrustAllStrategy.INSTANCE)
  .build();
```

5. Now that we have initialized these configurations, we need to connect them to `HttpClientBuilder` in the following way:

```
httpClientBuilder
  .setSSLContext(sslContext)
.setSSLHostnameVerifier(NoopHostnameVerifier.INSTANCE)
  .setDefaultCredentialsProvider(creds)
```

6. The final step is to pass the created `callback` instance to the Elasticsearch client configuration:

```
lazy val client = ElasticClient(
  JavaClient(
    ElasticProperties(s"https://localhost:9200"),
    requestConfigCallback = NoOpRequestConfigCallback,
    httpClientConfigCallback = callback
  ) )
```

See also

For more information, you can refer to the following URLs, which are related to this recipe:

-   The official Elasticsearch documentation for `RestClient` can be found at `https://www.elastic.co/guide/en/elasticsearch/client/java-api/current/transport-client.html`.

-   The official documentation for `elastic4s` is located at `https://github.com/sksamuel/elastic4s` and provides more examples of client initialization.

Managing indices

Now that we have a client, the first thing we need to do is to create a custom index with an optimized mapping for it. Elastic4s provides a powerful DSL to perform this kind of operation.

In this recipe, we will create a custom mapping using the **Domain Syntax Language** (**DSL**), which was developed by the author of Elastic4s. This syntax is designed on top of the Elasticsearch JSON one, so it is very natural and easy to use.

Getting ready

You need an up-and-running Elasticsearch installation, as described in the *Downloading and installing Elasticsearch* recipe of *Chapter 1*, *Getting Started*.

Additionally, an IDE that supports Scala programming, such as IntelliJ IDEA, with the Scala plugin should be installed globally.

The code for this recipe can be found in the ch14/elastic4s_sample directory; the referred class is IndicesExample.

How to do it...

The Elasticsearch client maps all index operations under the admin.indices object of the client.

Here, you will find all the index operations (create, delete, exists, open, close, forceMerge, and more).

The following code retrieves a client and executes the main operations on the indices:

1. We need to import the required classes:

    ```
    import com.sksamuel.elastic4s.ElasticDsl._
    ```

2. We define an IndicesExample class that manages the index operations:

    ```
    object IndicesExample extends App with
    ElasticSearchClientTrait {
    ```

3. We check whether the index exists. If true, we delete it:

    ```
    val indexName = "test"
    if (client.execute { indexExists(indexName) }.await.
    result.isExists) {
        client.execute { deleteIndex(indexName) }.await
    }
    ```

4. We create an index, including a mapping:

```
client.execute {
   createIndex(indexName) shards 1 replicas 0 mapping (
      properties(
   textField("name").termVector("with_positions_offsets").
   stored(true),
         keywordField("tag")
      ))
   }.await
Thread.sleep(2000)
```

5. We can optimize the index to reduce the number of segments:

```
client.execute(forceMerge(indexName)).await
```

6. We close an index as follows:

```
client.execute(closeIndex(indexName)).await
```

7. We open an index as follows:

```
client.execute(openIndex(indexName)).await
```

8. We delete an index as follows:

```
client.execute(deleteIndex(indexName)).await
```

9. We close the client to clean up the resources, as follows:

```
client.close()
```

How it works...

The Elasticsearch **Domain Script Language** (**DSL**) that uses `elastic4s` is very simple and easy to use. It models the standard Elasticsearch functionalities in a way that is more natural to work with. Additionally, it is strong-typed, so it prevents common errors such as typographic errors or value type changes.

To simplify the code in these examples, we have created a trait that contains the code to initialize the `ElasticSearchClientTrait` client.

All the API calls in Elastic4s are asynchronous, so they return `Future`. To materialize the result, we need to add `.wait` to the end of the call.

Under the hood, Elastic4s uses the Java standard Elasticsearch client but wraps it in the DSL. This is so that the methods and the parameters have the same meaning as the standard Elasticsearch documentation.

In the preceding code, we have put a delay of 2 seconds (`Thread.sleep(2000)`) to prevent fast actions on the indices. This is because their shard allocations are asynchronous, and they require several milliseconds to get ready. The best practice is not to have a similar hack, but to poll an index's state before performing further operations, and then only perform those operations when it goes green.

See also

In *Chapter 3*, *Basic Operations*, please refer to the *Creating an index* recipe for details on index creation, the *Deleting an index* recipe for details on index deletion, and the *Opening/closing an index* recipe for a description of open and closed index APIs.

Managing mappings

After creating an index, the next step is to add some mappings to it. We already learned how to include a mapping via the REST API in *Chapter 3*, *Basic Operations*. In this recipe, we will see how to manage mappings via a native client.

Getting ready

You need an up-and-running Elasticsearch installation, as described in the *Downloading and installing Elasticsearch* recipe of *Chapter 1*, *Getting Started*.

Additionally, an IDE that supports Scala programming, such as IntelliJ IDEA, with the Scala plugin should be installed globally.

The code for this recipe can be found in the `ch14/elastic4s_sample` directory; the referred class is `MappingExample`.

How to do it...

In the following code, we add a `mytype` mapping to a `myindex` index via the native client:

1. We need to import the required classes:

```
import com.sksamuel.elastic4s.ElasticDsl._
```

2. We define a class to contain our code and initialize the client and the index:

```
object MappingExample extends App with
ElasticSearchClientTrait {
   val indexName = "myindex"
   if (client.execute { indexExists(indexName) }.await.
result.isExists) {
      client.execute { deleteIndex(indexName) }.await
   }
```

3. We create the index by providing the mapping, as follows:

```
client.execute {
   createIndex(indexName) shards 1 replicas 0 mapping (
      properties(
         textField("name").termVector("with_positions_
offsets").stored(true) ) )
   }.await
   Thread.sleep(2000)
```

4. We add another field in the mapping via a `putMapping` call:

```
client.execute {
   putMapping(indexName).as(
      keywordField("tag")
   )}.await
```

5. Now we can retrieve our mapping to test it:

```
val myMapping = client
   .execute {
      getMapping(indexName)
   }.await.result
```

6. From the mapping, we can extract the `tag` field:

```
val tagMapping = myMapping.head
```

7. We remove the index using the following command:

```
client.execute(deleteIndex(indexName)).await
```

8. Now we can close the client to free up the resources:

```
client.close()
```

How it works...

Before executing a mapping operation, a client must be available:

*   We can include the mapping during index creation via the `mappings` method in the `createIndex` builder:

```
createIndex(indexName) shards 1 replicas 0 mapping (
    properties(
        textField("name").termVector("with_positions_
offsets").stored(true) ) )
```

    The Elastic4s DSL provides a strong-typed definition for mapping fields.

*   If we forget to put a field in the mapping, or if, during our application's life, we need to add a new field, `putMapping` can be called with the new field or a completely new mapping type:

```
putMapping(indexName).as(keywordField("tag"))
```

*   In this way, if the type exists, it is updated; otherwise, it is created. In the admin console, to check that our index types are stored in mappings, we need to retrieve them from the cluster state. The method that we have already seen is the `getMapping` method:

```
val myMapping = client .execute {
    getMapping(indexName)}.await.result
```

*   The returned mapping object is a list of `IndexMapping` elements:

```
case class IndexMappings(index: String, mappings:
Map[String, Any], meta: Map[String, Any] = Map.
empty[String, Any])
```

*   To access our mapping, we take the first result:

```
val tagMapping = myMapping.head
```

See also

For more information, you can refer to the following recipes, which are related to this recipe:

- For more details about the *Putting a Mapping API* recipe, please refer to the *Putting a mapping in an index* recipe of *Chapter 3, Basic Operations*.

- For more details about the *Getting a Mapping API* recipe, please refer to the *Getting a mapping* recipe in *Chapter 3, Basic Operations*.

Managing documents

The APIs for managing documents (such as `index`, `delete`, and `update`) are the most important after the search ones. In this recipe, we will look at how to use them.

Getting ready

You need an up-and-running Elasticsearch installation, as described in the *Downloading and installing Elasticsearch* recipe of *Chapter 1, Getting Started*.

Additionally, an IDE that supports Scala programming, such as IntelliJ IDEA, with the Scala plugin should be installed globally.

The code for this recipe can be found in the `ch14/elastic4s_sample` file; the referred class is `DocumentExample`.

How to do it...

To manage documents, we will perform the following steps:

1. We'll need to import the required classes to execute all the document's CRUD operations:

```
import com.sksamuel.elastic4s.ElasticDsl._
import com.sksamuel.elastic4s.circe._
```

2. We need to create the client and ensure that the index and mapping exists:

```
object DocumentExample extends App with
ElasticSearchClientTrait {
  val indexName = "myindex"
  ensureIndexMapping(indexName)
```

3. Now we can store a document in Elasticsearch via the `indexInto` call:

```scala
client.execute {
  indexInto(indexName) id "0" fields (
    "name" -> "brown",
    "tag" -> List("nice", "simple")
  )}.await
```

4. We can retrieve the stored document via the `get` call:

```scala
val bwn = client.execute {
  get(indexName, "0") }.await
println(bwn.result.sourceAsString)
```

5. We can update the stored document via the `update` call using a script in Painless:

```scala
client.execute {
  updateById(indexName, "0").script("ctx._source.name =
'red'") }.await
```

6. We can check whether our update has been applied:

```scala
val red = client.execute {
  get(indexName, "0")}.await
println(red.result.sourceAsString)
```

7. The console output result will be as follows:

```
{"name":"brown","tag":["nice","simple"]}
{"name":"red","tag":["nice","simple"]}
```

The document version, following an update action, or if the document has been reindexed with new changes, is always incremented by 1.

How it works...

Before executing a document action, a client and the index must be available, and document mapping should be created (the mapping is optional because it can be inferred from the indexed document).

To index a document, `elastic4s` allows us to provide the document content in several ways, such as the following:

- `fields`
- A sequence of tuples (`String, Any`), as shown in the preceding example
- `Map[String, Any]`
- `Iterable[(String, Any)]`
- `doc/source`
- A string
- A type class that derives `Indexable[T]`

Of course, it's possible to add all the parameters that we saw in the *Indexing a document* recipe of *Chapter 3*, *Basic Operations*, such as `parent`, `routing`, and more.

The return value, `IndexReponse`, is the object that's returned from the Java call.

To retrieve a document, we need to know the `index/id` values; the method is `get`. It requires the `index` value and the `id` value. A lot of other methods are available to control the routing (such as `routing`, `parent`) or fields, as we learned in the *Getting a document* recipe in *Chapter 3*, *Basic Operations*. In the preceding example, the call is as follows:

```
val bwn = client.execute { get(indexName, "0") }
```

The return type, `GetResponse`, contains all the requests (if the document exists) and the document information (such as `source`, `version`, `index`, and `id`).

To update a document, it's necessary to know the `index/id` values and provide a script or a document to be used for the update. The client method is `update`. In the preceding example, we have used a script:

```
client.execute {
  updateById(indexName, "0").script("ctx._source.name = 'red'")
}
```

The script code must be a string. If a scripting language is not defined, the default language, Painless, is used.

The returned response contains information about the execution and the new version value to manage concurrency.

To delete a document (without the need to execute a query), we must know the index/ id values. We can use the client method, `delete`, to create a `delete` request. In the preceding code, we used the following:

```
client.execute { deleteById(indexName, "0") }
```

The `delete` request allows all the parameters as we saw in the *Deleting a document* recipe of *Chapter 3*, *Basic Operations*, to control the routing and versions, and to be passed to it.

There's more...

Scala programmers love type classes, automatic marshaling/unmarshaling from case classes, and the strong type management of data. Here, `elastics4` provides additional support for the common JSON serialization library, such as the following:

- **Circe** (https://circe.github.io/circe/). To use this library, you need to add the following dependency:

  a) **Jackson** (https://github.com/FasterXML/jackson-module-scala). To use this library, you need to add the following dependency:

  ```
  "com.sksamuel.elastic4s" %% "elastic4s-jackson" %
  elastic4sV
  ```

  b) **Json4s** (http://json4s.org/). To use this library, you need to add the following dependency:

  ```
  "com.sksamuel.elastic4s" %% "elastic4s-json4s" %
  elastic4sV
  ```

For example, if you want to use Circe, perform the following steps:

1. You need to import the `circe` implicit:

    ```
    import com.sksamuel.elastic4s.circe._
    import io.circe.generic.auto._
    import com.sksamuel.elastic4s.Indexable
    ```

2. You need to define the `case` class, which needs to be deserialized:

```
case class Place(id: Int, name: String)
case class Cafe(name: String, place: Place)
```

3. You need to force the implicit serializer:

```
implicitly[Indexable[Cafe]]
```

4. Now you can index the case classes directly:

```
client.execute {
  indexInto(indexName).id(cafe.name).source(cafe)
}.await
```

See also

In the preceding recipes, we have used all the CRUD operations in a document. For more details about these actions, please refer to the following recipes:

*   The *Indexing a document* recipe of *Chapter 3*, *Basic Operations*
*   The *Getting a document* recipe of *Chapter 3*, *Basic Operations*, on how to retrieve a stored document
*   The *Deleting a document* recipe of *Chapter 3*, *Basic Operations*
*   The *Updating a document* recipe of *Chapter 3*, *Basic Operations*

Executing a standard search

Of course, the most common action in Elasticsearch is searching. Elastic4s leverages the query DSL, which brings a type-safe definition for queries to Scala. One of the most common advantages of this functionality is that, as Elasticsearch evolves, in Scala code via `elastic4s`, you can have deprecation. Alternatively, your compilation might break, requiring you to update your code.

In this recipe, we will learn how to execute a search, retrieve the results, and convert them into typed domain objects (classes) without the need to write a serializer/deserializer for our data.

Getting ready

You need an up-and-running Elasticsearch installation, as described in the *Downloading and installing Elasticsearch* recipe of *Chapter 1, Getting Started*.

Additionally, an IDE that supports Scala programming, such as IntelliJ IDEA, with the Scala plugin should be installed globally.

The code for this recipe can be found in the `ch14/elastic4s_sample` file; the referred class is `QueryExample`.

How to do it...

To execute a standard query, we will perform the following steps:

1. We need to import the classes and implicits that are required to index and search the data:

    ```
    import com.sksamuel.elastic4s.ElasticDsl._
    import com.sksamuel.elastic4s.Indexable
    import com.sksamuel.elastic4s.circe._
    import io.circe.generic.auto._
    ```

2. We will create an index and populate it with some data. We will use bulk calls for speedup:

    ```
    object QueryExample extends App with
    ElasticSearchClientTrait {
      val indexName = "myindex"
      case class Place(id: Int, name: String)
      case class Cafe(name: String, place: Place)
      implicitly[Indexable[Cafe]]
      ensureIndexMapping(indexName)
      client.execute {
        bulk(
          indexInto(indexName)
            .id("0")
            .source(Cafe("nespresso", Place(20, "Milan"))),
          indexInto(indexName)
            .id("1")
            .source(Cafe("java", Place(60, "Rome"))),
    // truncated
    ```

```
        )
    }.await
    Thread.sleep(2000)
```

3. We can use a `bool` filter to search for documents where `name` is equal to `java` and `place.id` is greater than or equal to `80`:

```
    val resp = client.execute {
        search(indexName).bool(
            must(termQuery("name", "java"), rangeQuery("place.
id").gte(80)))
        }.await
```

4. When we have the `response` parameter, we need to check its count so that we can convert it back into a list of classes:

```
    println(resp.result.size)
    println(resp.result.to[Cafe].toList)
```

The result should be the following:

```
    List(Cafe(java,Place(80,Chicago)),
    Cafe(java,Place(89,London)))
```

How it works...

The `Elastic4s` query DSL wraps the Elasticsearch query in a more human-readable way.

The `search` method allows us to define a complex query via DSL. The result is a wrapper of the original Java result, which provides some helpers to be more productive.

The common methods of the Java result are available at a top level, but they also provide two interesting methods: `to` and `safeTo`.

They are able to convert the results into case classes via the implicit conversions that are available in the scope. In the case of the `to[T]` method, the result is an iterator of `T` (in the preceding example, we have the conversion back into a `List` of `Cafe`). In the case of `safeTo[T]`, the result is `Either[Throwable, T]`; in this way, it's possible to collect the conversion errors/exceptions.

Using the type class in Scala allows you to write cleaner and easy-to-understand code. It also reduces errors due to string management in Elasticsearch.

See also

The *Executing a search* recipe in *Chapter 4*, *Exploring Search Capabilities*, has more detailed information about how to execute a query.

Executing a search with aggregations

The next step after searching in Elasticsearch is to execute the aggregations. The Elastic4s DSL also provides support for aggregation so that it can be built in a safer typed way.

Getting ready

You need an up-and-running Elasticsearch installation, as described in the *Downloading and installing Elasticsearch* recipe of *Chapter 1*, *Getting Started*.

Additionally, an IDE that supports Scala programming, such as IntelliJ IDEA, with the Scala plugin should be installed globally.

The code for this recipe can be found in the `ch14/elastic4s_sample` file; the referred class is `AggregationExample`.

How to do it...

To execute a search with aggregations, we will perform the following steps:

1. We need to import the classes that are needed for the aggregations:

    ```
    import com.sksamuel.elastic4s.ElasticDsl._
    import com.sksamuel.elastic4s.requests.searches.aggs.
    responses.bucket.Terms
    ```

2. We will create an index and populate it with some data that will be used for the aggregations:

    ```
    object AggregationExample extends App with
    ElasticSearchClientTrait {
      val indexName = "myindex"
      ensureIndexMapping(indexName)
      populateSampleData(indexName, 1000)
    ```

3. We already know how to execute a search with aggregation (we don't have the results, so `size=0`) using `termsAgg` in the `tag` field with several sub-aggregations (extended statistics on the `price` and `size` fields, and geobounds on `location`):

```
val resp = client
    .execute {
      search(indexName) size 0 aggregations (termsAgg(
        "tag", "tag") size 100 subAggregations (
        extendedStatsAgg("price", "price"),
        extendedStatsAgg("size", "size"),
        geoBoundsAggregation("centroid") field "location"
    ))}.await.result
```

4. The `resp` variable contains our query result. We can extract the aggregation results from it and show some values:

```
val tagsAgg = resp.aggregations.result[Terms]("tag")
println(s"Result Hits: ${resp.size}")
println(s"number of tags: ${tagsAgg.buckets.size}")
println(
   s"max price of first tag ${tagsAgg.buckets.head.key}:
${tagsAgg.buckets.head.extendedStats("price").max}"
)
println(
   s"min size of first tag ${tagsAgg.buckets.head.key}:
${tagsAgg.buckets.head.extendedStats("size").min}"
)
```

5. Finally, we can clean up the used resources:

```
client.execute(deleteIndex(indexName)).await
```

The result should look similar to the following:

```
number of tags: 5
max price of first tag awesome: 10.799999999999999
min size of first tag awesome: 0.0
```

How it works...

Elastic4s provides a powerful DSL for more type-safe aggregations.

In the preceding example, initially, we used `termsAgg` to aggregate the buckets by their tag settings to collect at least 100 buckets (`termsAgg("tag", "tag") size 100`). Following this, we have two types of sub-aggregations:

- `extendedStatsAggr`: This is used to collect extended statistics regarding the price and size fields.
- `geoBoundsAggregation`: This is used to compute the center of the document's results.

The Elastic4s DSL provides all the official Elasticsearch aggregations.

Also, the aggregation result contains helpers for managing aggregations, such as automatic casing for some types. The most commonly used are as follows:

- `Terms`: This wraps a generic terms aggregation result.
- `MissingAggResult`: This wraps a missing aggregation result.
- `CardinalityAggResult`: This wraps a cardinality aggregation result.
- `ExtendedStatsAggResult`: This wraps an extended statistics result.
- `AvgAggResult`: This wraps an average metric aggregation result.
- `MaxAggResult`: This wraps a maximum metric aggregation result.
- `SumAggResult`: This wraps a sum metric aggregation result.
- `MinAggResult`: This wraps a minimum metric aggregation result.
- `HistogramAggResult`: This wraps a histogram aggregation result.
- `ValueCountAggResult`: This wraps a count aggregation result.

If the aggregation result is not part of these aggregations results, a helper method, `result[T]:T`, allows you to retrieve a casted aggregation result.

See also

For more information, you can refer to the following recipes, which are related to this recipe:

- Please refer to the *Executing term aggregations* recipe of *Chapter 7, Aggregations*, which describes term aggregations.

- For more details about statistical aggregations, please refer to the *Executing statistical aggregations* recipe of *Chapter 7, Aggregations*.

Integrating with DeepLearning.scala

In the previous chapter, we learned how to use DeepLearning4j with Java. This library can be used natively in Scala to provide deep learning capabilities for our Scala applications.

In this recipe, we will learn how to use Elasticsearch as a source of training data in a machine learning algorithm.

Getting ready

You need an up-and-running Elasticsearch installation, as described in the *Downloading and installing Elasticsearch* recipe of *Chapter 1, Getting Started*.

Additionally, Maven, or an IDE that natively supports Java programming, such as Eclipse or IntelliJ IDEA, must be installed.

The code for this recipe can be found in the `ch14/deeplearningscala` directory.

We will use the `iris` dataset (`https://en.wikipedia.org/wiki/Iris_flower_data_set`) that we used in *Chapter 13, Java Integration*. To prepare your `iris` index dataset, we need to populate it by executing the `PopulatingIndex` class, which is available in the source code of *Chapter 13, Java Integration*.

How to do it...

To use DeepLearning4J as a source data input for your models, you will perform the following steps:

1. We need to add the DeepLearning4J dependencies to `build.sbt`:

```
    "org.nd4j" % "nd4j-native-platform" % nd4jVersion,
    // ND4J backend. You need one in every DL4J project.
Normally define artifactId as either "nd4j-native-
platform" or "nd4j-cuda-9.2-platform" -->
    "org.nd4j" % "nd4j-native-platform" % nd4jVersion,
    "org.deeplearning4j" % "deeplearning4j-core" %
dl4jVersion,
    // ParallelWrapper & ParallelInference live here
    "org.deeplearning4j" % "deeplearning4j-parallel-
wrapper"% dl4jVersion
```

2. Now we can write our `DeepLearning4J` class to train and test our model. Initialize the Elasticsearch client:

```scala
object DeepLearning4s extends App with LazyLogging with
ElasticSearchClientTrait{
lazy val indexName = "iris"
```

3. After having the client, we can read our dataset. We will execute a query and collect the hit results using a simple search:

```scala
case class Iris(label: Int, f1: Double, f2: Double, f3:
Double, f4: Double)
implicitly[Indexable[Iris]]
val response = client.execute {
  search(indexName).size(1000)}.await
val hits = response.result.to[Iris].toArray
```

4. We need to convert the hits into a DeepLearning4J dataset. To do this, create intermediate arrays and populate them:

```scala
//Convert the iris data into 150x4 matrix
val irisMatrix: Array[Array[Double]] = hits.map(r =>
Array(r.f1, r.f2, r.f3, r.f4))
//Now do the same for the label data
val labelMatrix: Array[Array[Double]] = hits.map { r =>
  r.label match {
    case 0 => Array(1.0, 0.0, 0.0)
    case 1 => Array(0.0, 1.0, 0.0)
    case 2 => Array(0.0, 0.0, 1.0)
  }}
val training = Nd4j.create(irisMatrix)
val labels = Nd4j.create(labelMatrix)
val allData = new DataSet(training, labels)
```

5. We need to split the datasets into two parts—one for training and one for tests. Then, we need to normalize the values. These actions can be done with the following code:

```
allData.shuffle()

val testAndTrain = allData.splitTestAndTrain(0.65) //Use
65% of data for training

val trainingData = testAndTrain.getTrain

val testData = testAndTrain.getTest

//We need to normalize our data. We'll use
NormalizeStandardize (which gives us mean 0, unit
variance):

val normalizer = new NormalizerStandardize

normalizer.fit(trainingData) //Collect the statistics
(mean/stdev) from the training data. This does not modify
the input data

normalizer.transform(trainingData) //Apply normalization
to the training data

normalizer.transform(testData) //Apply normalization to
the test data. This is using statistics calculated from
the *training* set
```

6. Now we can design the model to be used for training:

```
val numInputs = 4

val outputNum = 3

val seed = 6

logger.info("Build model....")

val conf = new NeuralNetConfiguration.Builder()
    .seed(seed)
    .activation(Activation.TANH)
    .weightInit(WeightInit.XAVIER)
    .updater(new Sgd(0.1))
    .l2(1e-4).list
    .layer(0, new DenseLayer.Builder().nIn(numInputs).
nOut(3).build)
    .layer(1, new DenseLayer.Builder().nIn(3).nOut(3).
build)
    .layer(2, new OutputLayer.Builder(LossFunctions.
LossFunction.NEGATIVELOGLIKELIHOOD)
```

```
        .activation(Activation.SOFTMAX).nIn(3).
    nOut(outputNum).build)
      .backprop(true).pretrain(false).build
```

7. After having defined the model, we can finally train it with our dataset—we use 1,000 iterations for the training. The code is as follows:

```
//run the model
val model = new MultiLayerNetwork(conf)
model.init()
model.setListeners(new ScoreIterationListener(100))
```

8. Now that we have a trained model, we need to evaluate its accuracy. We can do that by using the test dataset (this depends on your hardware; the processing can take anything from a few seconds to a few minutes):

```
//evaluate the model on the test set
val eval = new Evaluation(3)
val output = model.output(testData.getFeatures)
eval.eval(testData.getLabels, output)
logger.info(eval.stats)
```

How it works...

Elasticsearch can be used as a data store—using compact Scala code significantly reduces the amount of code required to fetch the dataset and use it. The best way to manage a dataset is to create a data model:

1. In this case, we created the Iris class for this purpose:

```
case class Iris(label: Int, f1: Double, f2: Double, f3:
Double, f4: Double)
```

2. We use Circe (https://circe.github.io/circe/) to derive the JSON encoder and decoder for Elasticsearch search hits in an array of Iris objects:

```
implicitly[Indexable[Iris]]
val response = client.execute {
   search(indexName).size(1000)}.await
val hits = response.result.to[Iris].toArray
```

Using this approach, the code required to convert our data is reduced (thanks to the help of case classes and the Circe library).

3. The final step to create our dataset is to convert the `Iris` object into an array of values to be given to the algorithm. We have achieved this by using a functional map of our Elasticsearch hits:

```
val irisMatrix: Array[Array[Double]] = hits.map(r =>
  Array(r.f1, r.f2, r.f3, r.f4))
```

4. The same can be done with the labels. However, in this case, we had to generate a three-dimensional array depending on the label value:

```
val labelMatrix: Array[Array[Double]] = hits.map { r =>
  r.label match {
    case 0 => Array(1.0, 0.0, 0.0)
    case 1 => Array(0.0, 1.0, 0.0)
    case 2 => Array(0.0, 0.0, 1.0)
  }}
```

Using an `Iris` model to populate our arrays, the code is simpler and more readable. Another advantage is that, in the future, it allows you to replace hits with streamable structures without requiring massive code refactoring (a typical use case is using Apache Kafka for streaming and evaluating the model).

After having built the dataset, you can design your model, train it, and evaluate the quality of your hits—this approach is independent of the machine learning library that you use. In our case, we have reused the Java code of the previous chapter and used it in Scala.

See also

For further information, you can refer to the following URLs that are related to this recipe:

- Please visit the official site of DeepLearning4J (`https://deeplearning4j.org/`) for more examples and references regarding this powerful library.

- The official documentation for Circe can be viewed at `https://circe.github.io/circe/`.

- A more detailed description of the Iris dataset can be viewed at `https://en.wikipedia.org/wiki/Iris_flower_data_set`.

15
Python Integration

In the previous chapter, we learned how to use a native client to access the Elasticsearch server via Java. This chapter is dedicated to the Python language and how to manage common tasks via its clients.

Apart from Java, the Elasticsearch team supports official clients for Perl, PHP, Python, .NET, and Ruby (see the announcement post on the Elasticsearch blog at `http://www.`
`elasticsearch.org/blog/unleash-the-clients-ruby-python-php-`
`perl/`). These clients have a lot of advantages over other implementations. A few of them are as follows:

- They are strongly tied to the Elasticsearch API. These clients are direct translations of the native Elasticsearch REST interface – the Elasticsearch team.

- They handle dynamic node detection and failovers. They are built with a strong networking base for communicating with the cluster.

- They have full coverage of the REST API. They share the same application approach for every language that they are available in, so switching from one language to another is fast.

- They are easily extensible.

The Python client plays very well with other Python frameworks, such as Django, web2py, and Pyramid. It allows very fast access to documents, indices, and clusters.

In this chapter, I'll try to describe the most important functionalities of Elasticsearch's official Python client; for additional examples, I suggest that you take a look at the online GitHub repository and documentation at `https://github.com/elastic/elasticsearch-py` and the related documentation at `https://elasticsearch-py.readthedocs.io/en/master/`.

In this chapter, we will cover the following recipes:

- Creating a client
- Managing indices
- Managing mappings
- Managing documents
- Executing a standard search
- Executing a search with aggregations
- Integrating with Numpy and scikit-learn
- Using AsyncElasticsearch
- Using Elasticsearch with FastAPI

Creating a client

The official Elasticsearch clients are designed to manage a lot of issues that are typically required to create solid REST clients, such as `retry` if there are network issues, autodiscovery of other nodes in the cluster, and data conversions for communicating on the HTTP layer.

In this recipe, we'll learn how to instantiate a client with varying options.

Getting ready

You will need an up-and-running Elasticsearch installation, which we described how to get in the *Downloading and installing Elasticsearch* recipe in *Chapter 1, Getting Started*.

A Python 3.x distribution should be installed. On Linux and macOS systems, it's already provided in the standard installation. To manage Python, you must also install the necessary `pip` packages (`https://pypi.python.org/pypi/pip/`).

The full code for this recipe can be found in the `ch15/code/client_creation.py` file.

How to do it...

To create a client, perform the following steps:

1. Before you can use the Python client, you need to install it (possibly in a Python virtual environment). The client is officially hosted on PyPi (http://pypi.python.org/) and it's easy to install with the pip command, as shown in the following code:

```
pip install elasticsearch
```

This standard installation only provides HTTP.

2. After installing the package, you can instantiate the client. It resides in the Python elasticsearch package and must be imported to instantiate the client.

3. If you don't pass arguments to the Elasticsearch class, it instantiates a client that connects to the localhost and port 9200 (the default Elasticsearch HTTP one), as shown in the following code:

```
import elasticsearch
```

4. If your cluster is composed of more than one node, you can pass the list of nodes as a round-robin connection between them and distribute the HTTP load, as follows:

```
es = elasticsearch.Elasticsearch(["search1:9200",
"search2:9200"])
```

5. Often, the complete topology of the cluster is unknown; if you know at least one node IP, you can use the sniff_on_start=True option, as shown in the following code. This option activates the client's ability to detect other nodes in the cluster:

```
from elasticsearch.connection import
RequestsHttpConnection
es = elasticsearch.Elasticsearch(sniff_on_start=True,
connection_class=RequestsHttpConnection)
```

How it works...

To communicate with an Elasticsearch cluster, a client is required.

This client manages all the communication layers from your application to an Elasticsearch server using HTTP REST calls.

The Elasticsearch Python client allows you to use one of the following library implementations:

- `requests`: The default implementation, the `requests` library is one of the most used libraries for performing HTTP requests in Python (`https://pypi.python.org/pypi/requests`).

- `urllib3`: This is also a very common Python HTTP library that's provided as an option transport layer (`https://pypi.python.org/pypi/urllib3`).

The Elasticsearch Python client requires a server to connect to. If one is not defined, it tries to use one on the local machine (`localhost:9200`). If you have more than one node, you can pass a list of servers for it to connect to.

The client automatically tries to balance operations on all cluster nodes. This is a very powerful functionality that's provided by the Elasticsearch client.

To improve the list of available nodes, it is possible to set the client to auto-discover new nodes. I recommend using this feature because you will often find yourself with a cluster with a lot of nodes and will need to shut down some of them for maintenance. The options that can be passed to the client to control discovery are as follows:

- `sniff_on_start`: The default value is `False`, which allows you to obtain the list of nodes from the cluster at startup time.

- `sniffer_timeout`: The default value is `None`; it is the number of seconds between the cluster nodes being automatically sniffed.

- `sniff_on_connection_fail`: The default value is `False`, which controls whether a connection failure triggers a `sniff` of cluster nodes.

> **Note**
>
> The `sniff` capability will only work if the Python script you're using is running in the same network as your Elasticsearch nodes. If you're using Elasticsearch via Docker, K8S, or behind reverse proxies, it won't work because the returned IPs of Elasticsearch nodes can't usually be reached.

The default client configuration uses the HTTP protocol via the `requests` library. If you want to use other transport protocols, you need to pass the type of the transport class to the `transport_class` variable. The currently implemented classes are as follows:

- `RequestsHttpConnection`: This is the default value – that is, a wrapper around the `requests` library.

- `Urllib3HttpConnection`: This is an alternative transport layer based on the `urllib3` library.

If there are no special requirements, the best practice is to use the default transport protocol based on `requests`.

See also

The official documentation about the Python Elasticsearch client, available at `https://elasticsearch-py.readthedocs.io/en/master/index.html`, provides a more detailed explanation of the several options that are available to initialize the client.

Managing indices

In the previous recipe, we learned how to initialize a client to send calls to an Elasticsearch cluster. In this recipe, we will learn how to manage indices via client calls.

Getting ready

You will need an up-and-running Elasticsearch installation, as described in the *Downloading and installing Elasticsearch* recipe in *Chapter 1, Getting Started*.

You will also need the Python-installed packages from the *Creating a client* recipe in this chapter.

The full code for this recipe can be found in the `ch15/code/indices_management.py` file.

How to do it...

In Python, managing the life cycle of your indices is very easy. To do this, perform the following steps:

1. First, you must initialize a client reading credential from environment variables (`ES_USER/ES_PASSWORD`) and disable `ssl` key verification due to the self-signed SSL certificate, as shown in the following code:

```
import elasticsearch
import os
import ssl
es = elasticsearch.Elasticsearch(
```

```
    hosts=os.environ.get("ES_HOST", "https://
localhost:9200"),
    basic_auth=(
        os.environ.get("ES_USER", "elastic"),
        os.environ.get("ES_PASSWORD", "password"),
    ),
    ssl_version=ssl.TLSVersion.TLSv1_2,
    verify_certs=False)
index_name = "my_index"
```

2. You need to check whether the index exists, and if it does, delete it. You can set this up using the following code:

```
if es.indices.exists(index=index_name):
    es.indices.delete(index=index_name)
```

3. All the `indices` methods are available in the `client.indices` namespace. You can create and wait for an index to be created using the following code:

```
es.indices.create(index=index_name)
es.cluster.health(wait_for_status="yellow")
```

4. You can close/open the index using the following code:

```
es.indices.close(index=index_name)
es.indices.open(index=index_name)
es.cluster.health(wait_for_status="yellow")
```

5. You can optimize an index by reducing the number of segments via the following code:

```
es.indices.forcemerge(index=index_name)
```

6. You can delete an index using the following code:

```
es.indices.delete(index=index_name)
```

How it works...

The first step of every Elasticsearch operation is to initialize the client, which can be done via the `hosts` parameter or `cluster_id` in case you are using a managed Elasticsearch cluster.

> **Note**
>
> From Elasticsearch Python client version 8.x or above, all the argument parameters that are passed to the Python functions must be provided in **keyword** mode. It's no longer possible to pass positional arguments to Elasticsearch client methods.

The Elasticsearch Python client has two special managers: one for indices (`<client>.indices`) and one for the cluster (`<client>.cluster`).

For every operation that needs to work with indices, the first value is generally the name of the index. If you need to execute an action on several indices in one go, the indices must be concatenated with a comma, , (that is, `index1,index2,indexN`). It's also possible to use glob patterns to define multi-indexes, such as `index*`. Let's take a look:

1. To create an index, the call requires `index_name` and other optional parameters, such as index settings and mappings, as shown in the following code (we'll see this advanced feature in the next recipe):

```
es.indices.create(index=index_name)
```

2. Index creation can take some time (from a few milliseconds to a few seconds); it is an asynchronous operation and it depends on the complexity of the cluster, the speed of the disk, the network congestion, and so on. To ensure that this action is completed, you need to check that the cluster's health has turned `yellow` or `green`, as follows:

```
es.cluster.health(wait_for_status="yellow")
```

It's good practice to wait until the cluster's status is `yellow` (at least) after performing operations that involve creating and opening indexes since these actions are asynchronous.

3. To close an index, you must use the `<client>.indices.close` method, and then specify the name of the index to close, as follows:

```
es.indices.close(index=index_name)
```

4. To open an index, you must use the is `client>.indices.open` method, and then specify the name of the index to open, as shown in the following code:

```
es.indices.open(index=index_name)
es.cluster.health(wait_for_status="yellow")
```

5. Similar to index creation, once an index has been opened, it is good practice to wait until the index is fully open before executing an operation on it; otherwise, errors will occur regarding the commands being executed on the index. You can do this by checking the cluster's health.

6. To improve the performance of an index, Elasticsearch allows you to optimize it by removing deleted documents (documents are marked as deleted but are not purged from the segment's index for performance reasons) and reducing the number of segments. To optimize an index, `<client>.indices.forcemerge` must be called on the index, as shown in the following code:

```
es.indices.forcemerge(index=index_name)
```

- Finally, if you want to delete the index, you can call `<client>.indices.delete` while specifying the name of the index to remove.

Remember that deleting an index removes everything related to it, including all the data, and this action cannot be reversed.

There's more...

The Python client wraps the Elasticsearch API in groups such as the following:

- `<client>.indices`: This wraps all the REST APIs related to index management.
- `<client>.ingest`: This wraps all the REST APIs related to ingest calls.
- `<client>.cluster`: This wraps all the REST APIs related to cluster management.
- `<client>.cat`: This wraps the CAT API, a subset of the API that returns a textual representation of traditional JSON calls.
- `<client>.nodes`: This wraps all the REST APIs related to nodes management.
- `<client>.snapshot`: This allows you to execute a snapshot and restore data from Elasticsearch.
- `<client>.tasks`: This wraps all the REST APIs related to task management.
- `<client>.remote`: This wraps all the REST APIs related to remote information.
- `<client>.xpack`: This wraps all the REST APIs related to `xpack` information and usage.

Standard document operations (CRUD) and search operations are available at the top level of the client.

See also

- The *Creating an index* recipe in *Chapter 3, Basic Operations*

- The *Deleting an index* recipe in *Chapter 3, Basic Operations*

- The *Opening/closing an index* recipe in *Chapter 3, Basic Operations*, for more details about the actions that can be used to save cluster/node memory

Managing mappings

After creating an index, the next step is to add some mappings to it. We learned how to include a mapping via the REST API in *Chapter 3, Basic Operations*.

Getting ready

You will need an up-and-running Elasticsearch installation, which we described how to get in the *Downloading and installing Elasticsearch* recipe in *Chapter 1, Getting Started*.

You will also need the Python packages that we installed in the *Creating a client* recipe in this chapter.

The code for this recipe can be found in the `ch15/code/mapping_management.py` file.

How to do it...

After initializing a client and creating an index, you can manage its indices. To do so, you must do the following:

1. Create a mapping.
2. Retrieve a mapping.

Perform the following steps:

1. First, initialize the client with auth, as in *Managing indices* section of this chapter.
2. Then, create an index, as follows:

```
index_name = "my_index"
if es.indices.exists(index=index_name):
    es.indices.delete(index=index_name)
es.indices.create(index=index_name)
es.cluster.health(wait_for_status="yellow")
```

3. Next, include the mapping, as follows:

```
es.indices.put_mapping(
  index=index_name, body={"properties": {
    "uuid": {"type": "keyword"},
    "title": {"type": "text", "term_vector": "with_
positions_offsets"},
      "parsedtext": {"type": "text", "term_vector": "with_
positions_offsets"},
      "nested": {"type": "nested", "properties": {"num":
{"type": "integer"},
          "name": {"type": "keyword"},
          "value": {"type": "keyword"}}},
      "date": {"type": "date"},
      "position": {"type": "integer"},
      "name": {"type": "text", "term_vector": "with_
positions_offsets"}}})
```

4. Retrieve the mapping, as follows:

```
mappings = es.indices.get_mapping(index_name, type_name)
```

5. Finally, delete the index, as follows:

```
es.indices.delete(index_name)
```

How it works...

You learned how to initialize the client and create an index in the previous recipe.

To create a mapping, you must use the `<client>.indices.create_mapping` method call while specifying an `index_name`, the type name, and the mapping, as shown in the following code. How to create a mapping was fully covered in *Chapter 3, Managing Mapping*. It is easy to convert the standard Python types into JSON and vice versa:

```
es.indices.put_mapping(index=index_name, {...})
```

If an error occurs in the mapping process, an exception is raised. The `put_mapping` API has two behaviors: create and update.

In Elasticsearch, you cannot remove a property from a mapping. Schema manipulation allows us to enter new properties with the `put_mapping` call.

To retrieve a mapping with the get_mapping API, use the <client>.indices. get_mapping method while providing an indexname and type name, as shown in the following code:

```
mappings = es.indices.get_mapping(index=index_name)
```

The returned object is the dictionary describing the mapping.

See also

- The *Putting a mapping in an index* recipe in *Chapter 3*, *Basic Operations*
- The *Getting a mapping* recipe in *Chapter 3*, *Basic Operations*

Managing documents

The APIs for managing a document (index, update, and delete) are the most important after the search APIs. In this recipe, you will learn how to use them in a standard way and use bulk actions to improve performance.

Getting ready

You will need an up-and-running Elasticsearch installation, which we described how to get in the *Downloading and installing Elasticsearch* recipe in *Chapter 1*, *Getting Started*.

You will also need the Python packages that we installed in the *Creating a client* recipe in this chapter.

The full code for this recipe can be found in the ch15/code/document_ management.py file.

How to do it...

The three main operations you can use to manage documents are as follows:

- index: This operation stores a document in Elasticsearch. It is mapped to the index API call.
- update: This allows you to update values in a document. This operation is composed internally (via Lucene) by deleting the previous document and reindexing the document with the new values. It is mapped to the update API call.
- delete: This deletes a document from the index. It is mapped to the delete API call.

With the Elasticsearch Python client, these operations can be performed by following these steps:

1. First, initialize a client and create an index with the mapping as in previous recipe, as follows:

```
from elasticsearch.helpers import bulk
from utils import create_and_add_mapping
import elasticsearch
import os
import ssl
es = elasticsearch.Elasticsearch( … truncated … )
index_name = "my_index"
if es.indices.exists(index=index_name):
    es.indices.delete(index=index_name)
create_and_add_mapping(es, index_name)
```

2. Then, index some documents (you must manage the parent/child), as follows:

```
es.index(index=index_name, id="2.1",
    body={"name": "data2", "value": "value2",
        "join_field": {"name": "metadata", "parent": "2"}},
    routing=2)
```

3. Next, update a document, as follows:

```
es.update(index=index_name,  id=2,
    body={"script": 'ctx._source.position += 1'})
```

4. Then, delete a document, as follows:

```
es.delete(index=index_name,  id=3)
```

5. Next, bulk insert some documents, as follows:

```
bulk(es, [
    {"_index": index_name, "_id": "1",
        "source": {
            "name": "Joe Tester",
            "parsedtext": "Joe Testere nice guy",
            "uuid": "11111", "position": 1,
```

```
                          "date": datetime(2018, 12, 8)}},
                {"_index": index_name, "_id": "1",
                    "source": {"name": "Bill Baloney", "parsedtext":
            "Bill Testere nice guy", "uuid": "22222", "position": 2,
                            "date": datetime(2018, 12, 8)}}
            ])
```

6. Finally, remove the index, as follows:

```
    es.indices.delete(index=index_name)
```

How it works...

To simplify this example, after instantiating the client, a function from the `utils` package is called, which sets up the index and places the mapping, as follows:

```
from utils import create_and_add_mapping
create_and_add_mapping(es, index_name)
```

This function contains the code for creating the mapping from the previous recipe.

The method that's used to index a document is `<client>.index`, and it requires the name of the index, the type of the document, and the body of the document, as shown in the following code (if the ID is not given, it will be autogenerated):

```
es.index(index=index_name, id=1,
    body={"name": "Joe Tester", "parsedtext": "Joe Testere nice
guy", "uuid": "11111",
        "position": 1,
        "date": datetime(2018, 12, 8), "join_field": {"name":
"book"}})
```

It also accepts all the parameters that we saw in the REST index API call in the *Indexing a document* recipe in *Chapter 3*, *Basic Operations*. The most common parameters that are passed to this function are as follows:

- `id`: This provides an ID that is used to index the document.

- `routing`: This provides shard routing to index the document in the specified shard.

- `parent`: This provides a parent ID that is used to put the child document in the correct shard.

To update a document, you can use the `<client>.update` method, which requires the following parameters:

- `index_name`.

- `type_name`.

- `id` of the document.

- `script` or `document` to update the document.

- `* lang`, which is optional, and indicates the language to be used. This is usually `painless`.

If you want to increment a position by 1, you can write something similar to the following:

```
es.update(index=index_name,  d=2, body={"script": 'ctx._source.
position += 1'})
```

The call accepts all the parameters that we discussed in the *Updating a document* recipe in *Chapter 3*, *Basic Operations*.

To delete a document, you can use the `<client>.delete` method, which requires the following parameters:

- `index_name`

- `type_name`

- `id` of the document

If you want to delete a document with `id=3`, you can write something similar to the following:

```
es.delete(index=index_name,  id=3)
```

Remember that all the Elasticsearch actions that work on documents are never seen instantly in the search. If you want to search without having to wait for the automatic refresh to occur (every 1 second), you need to manually call the refresh API on the index.

To execute bulk indexing, the Elasticsearch client provides a `helper` function, which accepts a connection, an iterable list of documents, and the bulk size. The bulk size (the default is 500) defines the number of actions to send via a single bulk call. The parameters that must be passed to correctly control the indexing of the document are put in the document with `_` prefix. The documents that are to be provided to the bulker must be formatted as a standard search result with the body in the `source` field, as follows:

```
from elasticsearch.helpers import bulk
bulk(es, [
    {"_index": index_name,   "_id": "1",
     "source": { "name": "Joe Tester",
         "parsedtext": "Joe Testere nice guy",
         "uuid": "11111", "position": 1,
         "date": datetime(2018, 12, 8)}},
    {"_index": index_name, "_id": "1",
     "source": {"name": "Bill Baloney", "parsedtext": "Bill
Testere nice guy", "uuid": "22222", "position": 2,
               "date": datetime(2018, 12, 8)}}])
```

See also

- The *Indexing a document* recipe in *Chapter 3, Basic Operations*
- The *Getting a document* recipe in *Chapter 3, Basic Operations*
- The *Deleting a document* recipe in *Chapter 3, Basic Operations*
- The *Updating a document* recipe in *Chapter 3, Basic Operations*
- The *Speeding up atomic operations (Bulk operations)* recipe in *Chapter 3, Basic Operations*

Executing a standard search

After document insertion, the most commonly executed action in Elasticsearch is the search action. The official Elasticsearch client APIs for searching are similar to the ones for the REST API.

Getting ready

You will need an up-and-running Elasticsearch installation, which we described how to get in the *Downloading and installing Elasticsearch* recipe in *Chapter 1, Getting Started*.

You will also need the Python packages that were installed in the *Creating a client* recipe in this chapter.

The code for this recipe can be found in the ch15/code/searching.py file.

How to do it...

To execute a standard query, the client search method must be called by passing the query parameters, as we saw in *Chapter 4, Exploring Search Capabilities*. The required parameters are index_name, type_name, and the query's DSL. In this recipe, you will learn how to call a match_all query, a term query, and a filter query. Perform the following steps:

1. First, initialize the client and populate the index as in previous sections, as follows:

```
from utils import create_and_add_mapping, populate
from pprint import pprint
import elasticsearch
import os
import ssl
es = elasticsearch.Elasticsearch( … truncated … )
index_name = "my_index"
if es.indices.exists(index=index_name):
    es.indices.delete(index=index_name)
create_and_add_mapping(es, index_name)
populate(es, index_name)
```

2. Then, execute a search with a match_all query and print the results, as follows:

```
results = es.search(index=index_name,
    body={"query": {"match_all": {}}})
pprint(results)
```

3. Then, execute a search with a term query and print the results, as follows:

```
results = es.search(index=index_name,
    body={
        "query": {
```

```
        "term": {"name": {"boost": 3.0, "value": "joe"}}}
    })
pprint(results)
```

4. Next, execute a search with a `bool` filter query and print the results, as follows:

```
results = es.search(index=index_name,
    body={"query": {
        "bool": {
            "filter": {
                "bool": {
                    "should": [
                        {"term": {"position": 1}},
                        {"term": {"position": 2}}]}}}}})
pprint(results)
```

5. Finally, remove the index, as follows:

```
es.indices.delete(index=index_name)
```

How it works...

The idea behind the Elasticsearch official clients is that they should offer a common API that is more similar to REST calls. In Python, it is very easy to use the query DSL, as it provides an easy mapping from the Python dictionary to JSON objects and vice versa.

In the preceding example, before calling the search, you need to initialize the index and put some data in it; this is done using the two helpers available in the `utils` package, which is available in the `ch_15` directory.

These two methods are as follows:

- `create_and_add_mapping(es, index_name)`: This initializes the index and inserts the correct mapping to perform the search. The code for this function was taken from the *Managing mappings* recipe in this chapter.

- `populate(es, index_name)`: This populates the index with data. The code for this function was taken from the previous recipe.

After initializing some data, you can execute queries against it. To execute a search, you must call the `search` method on the client. This method accepts all the parameters that were described for REST calls in the *Searching* recipe in *Chapter 4*, *Exploring Search Capabilities*.

The actual method signature for the `search` method is as follows:

```
@query_params(
    "_source",
    "_source_excludes",
    "_source_includes",
    ... truncated...
)
def search(self, body=None, index=None, doc_type=None,
params=None, headers=None):
```

The `index` value could be one of the following:

- An index name or an alias name.

- A list of index (or alias) names as a string separated by commas (that is, `index1`, `index2`, `indexN`).

- `_all`, the special keyword that indicates all the indices.

The body is the search DSL, as we saw in *Chapter 4, Exploring Search Capabilities*. In the preceding example, we have the following:

- A `match_all` query (see the *Matching all the documents* recipe in *Chapter 4, Exploring Search Capabilities*) to match all the index-type documents, as follows:

```
results = es.search(index=index_name, body={"query":
{"match_all": {}}})
```

- A `term` query that matches a name term, `joe`, with `boost 3.0`, as shown in the following code:

```
results = es.search(index=index_name,
    body={
        "query": {
            "term": {"name": {"boost": 3.0, "value":
"joe"}}}
    })
```

- A filtered query with a query (`match_all`) and an `or` filter with two `term` filters matching positions 1 and 2, as shown in the following code:

```
results = es.search(index=index_name,
    body={"query": {
```

```
        "bool": {
            "filter": {
                "bool": {
                    "should": [
                            {"term": {"position": 1}},
                            {"term": {"position": 2}}]}
            }}}})
```

The returned result is a JSON dictionary, which we discussed in *Chapter 4, Exploring Search Capabilities*.

If some hits are matched, they are returned in the hits field. The standard number of results that are returned is 10. To return more results, you need to paginate the results with the from and start parameters.

In *Chapter 4, Exploring Search Capabilities*, you can find a list of definitions for all the parameters that are used in the search.

See also

Please refer to the following recipes for more information related to this recipe:

- The *Executing a search* recipe in *Chapter 4, Exploring Search Capabilities*, for a detailed description of some search parameters
- The *Matching all the documents* recipe in *Chapter 4, Exploring Search Capabilities*, for a description of the match_all query

Executing a search with aggregations

Searching for results is the main activity of a search engine, so aggregations are very important because they often help augment the results.

Aggregations can be executed along with the search by performing analytics on the results that are returned.

Getting ready

You will need an up-and-running Elasticsearch installation, which we described how to get in the *Downloading and installing Elasticsearch* recipe in *Chapter 1, Getting Started*.

You will also need the Python packages that we installed in the *Creating a client* recipe of this chapter.

The code for this recipe can be found in the ch15/code/aggregation.py file.

How to do it...

To extend a query with aggregations, you need to define an aggregation section, as we saw in *Chapter 7, Aggregations*. In the case of the official Elasticsearch client, you can add the aggregation DSL to the search dictionary to provide aggregations. To set this up, perform the following steps:

1. First, initialize the client and populate the index as in previous sections.

2. Then, execute a search with the `terms` aggregation, as follows:

```
results = es.search(
    index=index_name,
    body={"size": 0,
        "aggs": {
            "pterms": {"terms": {"field": "name", "size":
10}}
        }
    })
pprint(results)
```

3. Next, execute a search with the `date_histogram` aggregation, as follows:

```
results = es.search(
    index=index_name,
    body={"size": 0,
            "aggs": {
                "date_histo": {"date_histogram": {"field":
"date", "interval": "month"}}}})
pprint(results)
es.indices.delete(index=index_name)
```

How it works...

As described in *Chapter 7, Aggregations*, the aggregations are calculated during the search in a distributed way. When you send a query to Elasticsearch with the aggregations defined, it adds another step to the query's processing, which allows the aggregations to be computed.

In the preceding example, we saw two kinds of aggregations: a terms aggregation and a date histogram aggregation.

The first is used to count terms, and it is often seen in sites that provide facet filtering on term aggregations of results, such as producers, geographic locations, and so on. This is shown in the following code:

```
results = es.search(
    index=index_name,
    body={"size": 0,
        "aggs": {
            "pterms": {"terms": {"field": "name", "size":
10}}
        }
    })
```

The term aggregation requires a field to count on. The default number of buckets for the field that's returned is 10. This value can be changed by defining the size parameter.

The second kind of aggregation that is calculated is the date histogram, which provides hits based on a datetime field. This aggregation requires at least two parameters: the datetime field, to be used as the source, and the interval field, to be used for computation, as follows:

```
results = es.search(
    index=index_name,
    body={"size": 0,
        "aggs": {
            "date_histo": {"date_histogram": {"field":
"date", "interval": "month"}}
        }
    })
```

The search results are standard search responses, as we saw in *Chapter 7, Aggregations*.

See also

Please refer to the following recipes for more information related to this recipe:

- The *Executing the terms aggregation* recipe in *Chapter 7, Aggregations*, on aggregating term values

- The *Executing the date histogram aggregation* recipe in *Chapter 7, Aggregations*, on computing the histogram aggregation on date/time fields

Integrating with NumPy and scikit-learn

Elasticsearch can easily be integrated with many Python machine learning libraries. One of the most used libraries for working with datasets is NumPy. A NumPy array is a building block dataset that's used for many Python machine learning libraries. In this recipe, you will see how it's possible to use Elasticsearch as a dataset for the `scikit-learn` library (https://scikit-learn.org/).

Getting ready

You will need an up and running Elasticsearch installation, as described in the *Downloading and installing Elasticsearch* recipe in *Chapter 1, Getting Started*.

The code for this recipe can be found in the `ch15/code` directory. The file we'll be using in the following section is called `kmeans_example.py`.

We will be using the `iris` dataset (https://en.wikipedia.org/wiki/Iris_flower_data_set), which we used in *Chapter 13, Java Integration*. To prepare the `iris` dataset, you need to populate it by executing the `PopulatingIndex` class, which is available in the source code for *Chapter 13, Java Integration*.

How to do it...

We are going to use Elasticsearch as a data source to build our dataset and execute a clusterization using the `KMeans` algorithm provided by `scikit-learn`. To do so, perform the following steps:

1. First, you need to add the required machine learning libraries to the `requirements.txt` file and install them via `pip install -r requirements.txt`:

    ```
    elasticsearch
    requests
    pandas
    matplotlib
    sklearn
    ```

2. Now, you can initialize the Elasticsearch client and fetch the samples:

    ```
    # importing the libraries
    import numpy as np
    import matplotlib.pyplot as plt
    import pandas as pd
    ```

```
import elasticsearch
import os
import ssl
es = elasticsearch.Elasticsearch(
    hosts=os.environ.get("ES_HOST", "https://
localhost:9200"),
    basic_auth=(
        os.environ.get("ES_USER", "elastic"),
        os.environ.get("ES_PASSWORD", "password"),
    ),
    ssl_version=ssl.TLSVersion.TLSv1_2,
    verify_certs=False,
)
index_name = "iris"
result = es.search(index="iris", size=100)
```

3. Now, you can read the dataset by iterating on Elasticsearch hit results:

```
x = []
for hit in result["hits"]["hits"]:
    source = hit["_source"]
    x.append(np.array([source['f1'], source['f2'],
source['f3'], source['f4']]))
x = np.array(x)
```

4. Once you've loaded the dataset, you can execute a clusterization using the KMeans algorithm:

```
from sklearn.cluster import KMeans
kmeans = KMeans(n_clusters=3, init='k-means++', max_
iter=300, n_init=10, random_state=0)
y_kmeans = kmeans.fit_predict(x)
```

5. Now that the clusters have been computed, you need to show them to verify the result. To do this, use the matplotlib.pyplot module:

```
plt.scatter(x[y_kmeans == 0, 0], x[y_kmeans == 0, 1],
s=100, c='red', label='Iris-setosa')
plt.scatter(x[y_kmeans == 1, 0], x[y_kmeans == 1, 1],
s=100, c='blue', label='Iris-versicolour')
```

```
plt.scatter(x[y_kmeans == 2, 0], x[y_kmeans == 2, 1],
s=100, c='green', label='Iris-virginica')
```

```
# Plotting the centroids of the clusters
```

```
plt.scatter(kmeans.cluster_centers_[:, 0], kmeans.
cluster_centers_[:, 1], s=100, c='yellow',
label='Centroids')
```

```
plt.legend()
```

```
plt.show()
```

The final output will be the following plot:

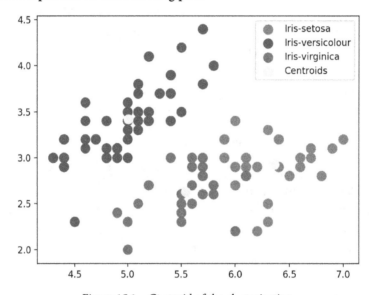

Figure 15.1 – Centroid of the clusterization

How it works...

Elasticsearch is a very powerful data store for machine learning datasets. It allows you to use its query capabilities to filter the data and retrieve what is needed to build a dataset that can be used for machine learning activities.

As shown in the preceding code, to fetch the data from Elasticsearch with Python, you can use just a few lines of code – one line to initialize the client and another to retrieve the results:

```
es = elasticsearch.Elasticsearch(
    hosts=os.environ.get("ES_HOST", "https://localhost:9200"),
```

```
    basic_auth=(
        os.environ.get("ES_USER", "elastic"),
        os.environ.get("ES_PASSWORD", "password"),
    ),
    ssl_version=ssl.TLSVersion.TLSv1_2,
    verify_certs=False,
)
index_name = "iris"
result = es.search(index="iris", size=100)
```

When you have the result hits, it's easy to iterate on them and extract the NumPy array that's required for the machine learning libraries.

In the preceding code, a NumPy array was generated for every sample by iterating on all the hits:

```
x = []
for hit in result["hits"]["hits"]:
    source = hit["_source"]
    x.append(np.array([source['f1'], source['f2'],
source['f3'], source['f4']]))
x = np.array(x)
```

The resulting Numpy array, x, can be used as input for every statistical or machine learning library.

Using Elasticsearch to manage datasets provides more advantages than a typical approach to using CSV or data files, including the following:

- You can automatically distribute your datasets to everyone that can connect to Elasticsearch.

- The samples are already in the correct type format (int, double, or string). You don't need to convert the values due to file reading.

- You can easily filter the samples without reading all the data in memory using Elasticsearch queries.

- You can use Kibana for data exploration before creating your machine models.

See also

Please refer to the following URLs for more information related to this recipe:

- A more detailed description of the Iris dataset can be found at `https://en.wikipedia.org/wiki/Iris_flower_data_set`.

- The official site of scikit-learn (`https://scikit-learn.org/stable/`) for more examples and references to this library.

- The official NumPy site (`https://www.numpy.org/`) for more details about this library.

Using AsyncElasticsearch

Python is not a language that's famous for its performance due to the **Global Interpreter Lock** (`GIL` – `https://realpython.com/python-gil/`). To speed up a program that is using I/O, a new module must be created called `asyncio` (`https://docs.python.org/3/library/asyncio.html`) that allows you to write concurrent code using the `async`/`await` syntax. This is the modern approach to writing Python applications and many frameworks are using it by default, such as Flask (`https://flask-aiohttp.readthedocs.io/en/latest/`), FastAPI (`https://fastapi.tiangolo.com/`), and Starlette (`https://www.starlette.io/`).

The Elasticsearch Python client (version 7.9.x or above) allows you to write concurrent code that uses `asyncio`.

Getting ready

You will need an up and running Elasticsearch installation, as we described in the *Downloading and installing Elasticsearch* recipe in *Chapter 1, Getting Started*.

The code for this recipe can be found in the `ch15/fastapi-es` directory. You need to have installed poetry (`https://python-poetry.org/docs/`), as well as the dependencies via `poetry install` and the file that we'll be using in the following section; that is, `app/async_es_example.py`.

How to do it...

In this recipe, you will be using Elasticsearch's `asyncio` module. To do so, perform the following steps:

1. First, install the necessary libraries for the `pyproject.toml` project by using the following command:

    ```
    poetry install
    ```

    This command will create a virtual environment and will install all the required libraries.

2. Now, initialize the `AsyncElasticsearch` client:

    ```
    import ssl
    import os
    from elasticsearch import AsyncElasticsearch
    es = AsyncElasticsearch(… truncated …)
    ```

3. At the end of the file, you must define the application flow:

    ```
    if __name__ == "__main__":
        loop = asyncio.get_event_loop()
        loop.run_until_complete(main())
        loop.run_until_complete(es.close())
    ```

4. To be able to call the function asynchronously, we need to define an async `main` function that contains our code and initializes the index, as follows:

    ```
    async def main():
        index_name = "my_index"
        if await es.indices.exists(index=index_name):
            await es.indices.delete(index=index_name)
        await create_and_add_mapping(es, index_name)
        await populate(es, index_name)
    ```

5. Now, you can execute a search putting an `await` statement before the client call:

```
results = await es.search(index=index_name,
    body={"query": {"match_all": {}}})
pprint(results)
```

6. Finally, remove the index, as follows:

```
await es.indices.delete(index=index_name)
```

How it works...

Migrating the code from being standard synchronized to asynchronous is very easy. You don't need to change the code, but you need to ensure the following:

-   Every function that uses asynchronous code has the `async` keyword before the `def` keyword:

```
async def main():
```

-   Every executed call within the Elasticsearch client must be `await`:

```
await es.indices.exists(index_name)
```

> **Note**
> It's very important to `await` an async function; otherwise, it will not be executed.

Due to these minimal changes, migrating to asynchronous code is very quick and easy.

The most important part is the execution flow at the end of the file, which does the following:

-   It inits the event loop that processes the `async` methods:

```
loop = asyncio.get_event_loop()
```

-   It executes our async `main` function and waits for it to be terminated:

```
loop.run_until_complete(main())
```

- It executes at the end the `close` method of the client to ensure that everything has been processed (internally, the client creates connection pools and can also manage n-flight async Elasticsearch calls) in the client:

```
loop.run_until_complete(es.close())
```

It's a best practice to close the client when you're finished.

See also

Please refer to the following URLs for more information related to this recipe:

- The Elasticsearch Python client page describes how to integrate `async` (`https://elasticsearch-py.readthedocs.io/en/master/async.html`).

- The official site of the Python documentation about `asyncio` (`https://docs.python.org/3/library/asyncio.html`) provides more details about this library.

Using Elasticsearch with FastAPI

FastAPI (`https://fastapi.tiangolo.com/`) is a new Python web framework based on `asyncio` (`https://docs.python.org/3/library/asyncio.html`) and Starlette (`https://www.starlette.io/`) that's powerful and enjoyable to use. The following features make FastAPI worth trying:

- FastAPI is one of the fastest Python web frameworks (`https://www.techempower.com/benchmarks/#section=data-r20&hw=ph&test=query&l=yykm7z-cn&a=2`).

- It has detailed and easy-to-use developer docs that can be found at `https://fastapi.tiangolo.com/tutorial/`.

- It creates an OpenAPI interface by default.

- You can *type hint* your code and get free data validation and conversion.

- You can create plugins easily using dependency injection.

Due to its async nature, FastAPI is very easy to integrate with AsyncElasicsearch.

Getting ready

You will need an up and running Elasticsearch installation, as we described in the *Downloading and installing Elasticsearch* recipe in *Chapter 1, Getting Started*.

The code for this recipe can be found in the ch15/fastapi-es directory. You need to have installed poetry (https://python-poetry.org/docs/), as well as the dependencies via poetry install and the file we'll be using in the following section; that is, the app/main.py file.

We will be using the index with the iris data from the previous recipes.

How to do it...

In this recipe, you are going to create a simple FastAPI web application to serve our Elasticsearch data using the AsyncElasticsearch client. To do so, perform the following steps:

1. First, install the required libraries for the pyproject.toml project by using the following command:

    ```
    poetry install
    ```

    This command will create a virtual environment and install all the required libraries.

2. Initialize the AsyncElasticsearch client, as you did in the previous recipe:

    ```
    import ssl
    import os
    from elasticsearch import AsyncElasticsearch
    es = AsyncElasticsearch(… truncated …)
    ```

3. Next, initialize the FastAPI application:

    ```
    from fastapi import FastAPI
    app = FastAPI(
        title=title,
        description=title,
        debug=os.getenv("DEBUG", "0"),
        version=os.getenv("VERSION", "0.1")
    )
    ```

4. Now, register a hook to close the Elasticsearch `async` client at the end of the `FastAPI` application:

```
@app.on_event("shutdown")
async def app_shutdown():
    await es.close()
```

5. Next, register the endpoint that will be used to serve our Elasticsearch results:

```
@app.get("/iris/sample/{size}")
async def read_item(size: int = 10):
    result = await es.search(index="iris", size=size)
    return {"items": result["hits"]["hits"]}
```

6. To simplify your life, create a redirect from the site root to the documentation endpoint:

```
@app.get("/", include_in_schema=False)
async def index_page():
    return RedirectResponse(url='/docs')
```

7. At the end of the file, define the application flow:

```
if __name__ == "__main__":
    uvicorn.run("main:app", host="0.0.0.0",
        port=int(os.getenv("API_PORT", 8000)),
        proxy_headers=True, lifespan="on",
        root_path=os.getenv("ROOT_PATH", ""),
        log_level=os.getenv("LOG_LEVEL", "debug") )
```

- You can start the web service using the following commands:

```
poetry shell
poetry app/main.py
```

If you navigate to `http://0.0.0.0:80` via your web browser, you'll see the following web interface:

Figure 15.2 – Example of a FastAPI-generated OpenAPI frontend

How it works...

Once you have populated your Elasticsearch, you need to provide your data or analytics to your final users. These users can use Elasticsearch or Kibana to access the data, but it's very common to provide a microservice that can manage the data and your business logic in Elasticsearch.

FastAPI is one of the most modern web frameworks that allows you to build microservices with few lines of code. It uses `asyncio` code by default, so it's very easy to integrate `AsyncElasticsearch` code with this framework.

In the preceding example, you used the same code from the previous recipes and integrated it into a REST microservice by doing the following:

1. First, you created a `FastAPI` app that will be enriched with the endpoints that you want to expose:

    ```
    from fastapi import FastAPI
    app = FastAPI(
        title=title,
        description=title,
        debug=os.getenv("DEBUG", "0"),
        version=os.getenv("VERSION", "0.1")
    )
    ```

    All the parameters to initialize the `FastAPI` instance are optional, but I suggest that you set `title` and `description` at a minimum to get a better web interface view.

2. The FastAPI app allows you to register functions to answer events using `@app.on_event`, which is the ideal place to put the Elasticsearch `close` client method:

    ```
    @app.on_event("shutdown")
    async def app_shutdown():
        await es.close()
    ```

    This code ensures that, on application shutdown, the Elasticsearch client is closed.

3. Registering web endpoints is very easy: you must decorate the `async` function with `@app.get` and the URL to be used (that is, `"/iris/sample/{size}"`):

    ```
    @app.get("/iris/sample/{size}")
    async def read_item(size: int = 10):
    ```

```
result = await es.search(index="iris", size=size)
return {"items": result["hits"]["hits"]}
```

In the preceding example, you can see how, by using {size} in the URL, the value is automatically extracted and passed to the function, and FastAPI automatically converts the Python dictionary into JSON when it's returned to the function.

Because you are using the AsyncElasticsearch client, you must put await in front of es.search.

4. Finally, you used uvicorn (https://www.uvicorn.org/) to serve the web application. In this case, you provided the binding address (that is, 0.0.0.0) and port (that is, 8000), as well as the entry point of the FastAPI app:

```
uvicorn.run(
        "main:app",
        host="0.0.0.0",
        port=8000,
        lifespan="on"
    )
```

As you can see, FastAPI allows you to build REST microservices with a very small portion of Python code (mainly by adding some decorators to your code) and integrates naturally with the Elasticsearch async client.

See also

Please refer to the following URLs for more information related to this recipe:

*   The official site for FastAPI can be found at https://fastapi.tiangolo.com/.

*   The official site for the uvicorn library can be found at https://www.uvicorn.org/.

*   Some templates so that you can start using FastAPI with support for Docker and other tools can be found at https://github.com/yxtay/python-project-template and https://github.com/arthurhenrique/cookiecutter-fastapi.

16
Plugin Development

Elasticsearch is designed to be extended with plugins to improve its capabilities. In the previous chapters, we installed and used many of them (new queries, REST endpoints, and scripting plugins).

Plugins are application extensions that can add many features to Elasticsearch. They can have several usages, including the following:

- Adding a new scripting language (that is, Python and JavaScript plugins)
- Adding new aggregation types
- Adding a new ingest processor
- Extending Lucene-supported analyzers and tokenizers
- Using native scripting to speed up the computation of scores, filters, and field manipulation
- Extending node capabilities, for example, creating a node plugin that can execute your logic
- Monitoring and administering clusters

In this chapter, the Java language will be used to develop a native plugin, but it is possible to use any **Java virtual machine** (**JVM**) language that generates JAR files.

The standard tools for building and testing Elasticsearch components are built on top of Gradle (`https://gradle.org/`). All our custom plugins will use Gradle to build them.

In this chapter, we will cover the following recipes:

- Creating a plugin
- Creating an analyzer plugin
- Creating a REST plugin
- Creating a cluster action
- Creating an ingest plugin

Creating a plugin

Native plugins allow several aspects of the Elasticsearch server to be extended, but they require good knowledge of Java.

In this recipe, we will see how to set up a working environment to develop native plugins.

Getting ready

You need an up and running Elasticsearch installation, as we described in the *Downloading and installing Elasticsearch* recipe in *Chapter 1, Getting Started*.

Gradle or an **integrated development environment** (**IDE**) that supports Java programming with Gradle (version 7.3.x used in the examples), such as Eclipse, Visual Studio Code, or IntelliJ IDEA, is required. Java JDK 17 or above needs to be installed.

The code for this recipe is available in the `ch16/simple_plugin` directory.

How to do it...

Generally, Elasticsearch plugins are developed in Java using the Gradle build tool (`https://gradle.org/`) and deployed as a ZIP file.

To create a simple JAR plugin, we will perform the following steps:

1. To correctly build and serve a plugin, some files must be defined:

   - `build.gradle` and `settings.gradle` are used to define the build configuration for Gradle.

   - `LICENSE.txt` defines the plugin license.

   - `NOTICE.txt` is a copyright notice.

2. A `build.gradle` file is used to create a plugin that contains the following code:

```
buildscript {
  repositories {
    mavenCentral()
  }
  dependencies {
    classpath "org.elasticsearch.gradle:build-
tools:8.0.0"
  }
}
repositories {
  mavenCentral()
}
group = 'org.elasticsearch.plugin'
version = '8.0.0-SNAPSHOT'
apply plugin: 'java'
apply plugin: 'idea'
apply plugin: 'elasticsearch.esplugin'
apply plugin: 'elasticsearch.yaml-rest-test'
esplugin {
  name 'simple-plugin'
  description 'A simple plugin for ElasticSearch'
  classname 'org.elasticsearch.plugin.simple.
SimplePlugin'
  // license of the plugin, may be different than the
above license
  licenseFile rootProject.file('LICENSE.txt')
  // copyright notices, may be different than the above
notice
```

```
        noticeFile rootProject.file('NOTICE.txt')
}
// In this section you declare the dependencies for your
production and test code
// Note, the two dependencies are not really needed as
the buildscript dependency gets them in already
// they are just here as an example
dependencies {
    implementation 'org.elasticsearch:elasticsearch:8.0.0'
    yamlRestTestImplementation 'org.elasticsearch.
test:framework:8.0.0'
}
// ignore javadoc linting errors for now
tasks.withType(Javadoc) {
    options.addStringOption('Xdoclint:none', '-quiet')
}
```

3. The `settings.gradle` file is used for the project name, as follows:

```
rootProject.name = 'simple-plugin'
```

4. The `src/main/java/org/elasticsearch/plugin/simple/`
   `SimplePlugin.java` class is an example of the basic (the minimum required)
   code that needs to be compiled for executing a plugin, as follows:

```
package org.elasticsearch.plugin.simple;
import org.elasticsearch.plugins.Plugin;
public class SimplePlugin extends Plugin {
}
```

How it works...

Several parts make up the development life cycle of a plugin, such as designing, coding, building, and deploying. To speed up the build and deployment steps, which are common to all plugins, we need to create a `build.gradle` file.

The preceding `build.gradle` file is a standard for developing Elasticsearch plugins. This file is composed of the following:

- A `buildscript` section that depends on the Gradle building tools for Elasticsearch, as follows:

```
buildscript {
  repositories {
//    mavenLocal() // if you want use your local maven
artifact
    mavenCentral()
  }
  dependencies {
    classpath "org.elasticsearch.gradle:build-
tools:8.0.0"
  }
}
```

> **Note**
>
> The version of the plugin (that is, `"org.elasticsearch.gradle:build-tools:8.0.0"`) will be the target version of Elasticsearch.

- The group and the version of the plugin, used for artifact deployment, as follows:

```
group = 'org.elasticsearch.plugin'
version = '8.0.0-SNAPSHOT'
```

- A list of Gradle plugins that must be activated, as follows:

```
apply plugin: 'java' // for java support
apply plugin: 'idea' // for Intellij IDEA support
apply plugin: 'elasticsearch.esplugin' // Elasticsearch
plugin support
apply plugin: 'elasticsearch.yaml-rest-test' //
Elasticsearch Yaml Rest Test
```

- The core information that's needed to populate the plugin description: that is, information that is used to generate the `plugin-descriptor.properties` file that will be available in the final distribution ZIP and used by Elasticsearch to load the plugin. The most important parameter is `classname`, which is the main entry point of the plugin:

```
esplugin {
  name 'simple-plugin'
  description 'A simple plugin for ElasticSearch'
  classname 'org.elasticsearch.plugin.simple.
SimplePlugin'
  // license of the plugin, may be different than the
above license
  licenseFile rootProject.file('LICENSE.txt')
  // copyright notices, may be different than the above
notice
  noticeFile rootProject.file('NOTICE.txt')
}
```

- To compile the code, some dependencies are required, as follows:

```
dependencies {
  implements 'org.elasticsearch:elasticsearch:8.0.0'
}
```

After configuring Gradle, we can start writing the main plugin class.

Every `plugin` class must be derived from a plugin one and it must have a public scope, otherwise, it cannot be loaded dynamically from the JAR, as follows:

```
package org.elasticsearch.plugin.simple;
import org.elasticsearch.plugins.Plugin;
public class SimplePlugin extends Plugin {}
```

After having defined all the files that are required to generate a ZIP release of our plugin, it is enough to invoke the `gradle clean check` command. This command will compile the code and create a `zip` package in the `build/distributions/` directory of your project: the final ZIP file can be deployed as a plugin on your Elasticsearch cluster (see *Chapter 1, Installing Plugins in Elasticsearch*, as reference).

In this recipe, we configured a working environment to build, deploy, and test plugins. In the following recipes, we will reuse this environment to develop several plugin types.

There's more...

Compiling and packaging a plugin are not enough to define a good life cycle for your plugin; a test phase for testing your plugin functionalities needs to be provided.

Testing the plugin functionalities with test cases reduces the number of bugs that can affect the plugin when it's released.

The Elasticsearch test framework is designed to simplify different test scenarios such as **Unit Test** and **Integration Test** with running node instances. To enable these functionalities, make sure to put the following line in your dependencies:

```
dependencies {
  yamlRestTestImplementation
'org.elasticsearch.test:framework:8.0.0'
}
```

The extension of Elasticsearch for Gradle has everything to set up unit tests (https://en.wikipedia.org/wiki/Unit_testing) and integration tests (https://en.wikipedia.org/wiki/Integration_testing).

Creating an analyzer plugin

Elasticsearch provides a large set of analyzers and tokenizers to cover general needs out of the box. Sometimes, we need to extend the capabilities of Elasticsearch by adding new analyzers.

Typically, you can create an analyzer plugin when you need to do the following:

- Add standard Lucene analyzers/tokenizers that are not provided by Elasticsearch.
- Integrate third-party analyzers.
- Add custom analyzers.

In this recipe, we will add a new custom English analyzer, similar to the one provided by Elasticsearch.

Getting ready

You need an up and running Elasticsearch installation, as we described in the *Downloading and installing Elasticsearch* recipe in *Chapter 1, Getting Started*.

Gradle or an **integrated development environment (IDE)** that supports Java programming with Gradle (version 7.3.x used in the examples), such as Eclipse, Visual Studio Code, or IntelliJ IDEA, is required. Java JDK 17 or above needs to be installed.

The code for this recipe is available in the `ch16/analysis_plugin` directory.

How to do it...

An analyzer plugin is generally composed of the following two classes:

- A `Plugin` class, which implements the `org.elasticsearch.plugins.AnalysisPlugin` class
- An `AnalyzerProviders` class, which provides an analyzer

To create an analyzer plugin, we will perform the following steps:

1. The `plugin` class is similar to the ones we've seen in previous recipes, but it includes a method that returns the analyzers, as follows:

```
package org.elasticsearch.plugin.analysis;
import org.apache.lucene.analysis.Analyzer;
import org.elasticsearch.index.analysis.AnalyzerProvider;
import org.elasticsearch.index.analysis.
CustomEnglishAnalyzerProvider;
import org.elasticsearch.indices.analysis.AnalysisModule;
import org.elasticsearch.plugins.Plugin;
import java.util.HashMap;
import java.util.Map;
public class AnalysisPlugin extends Plugin implements
org.elasticsearch.plugins.AnalysisPlugin {
    @Override
    public Map<String, AnalysisModule.
AnalysisProvider<AnalyzerProvider<? extends Analyzer>>>
getAnalyzers() {
        Map<String, AnalysisModule.
AnalysisProvider<AnalyzerProvider<? extends Analyzer>>>
analyzers = new HashMap<>();
        analyzers.put(CustomEnglishAnalyzerProvider.NAME,
CustomEnglishAnalyzerProvider::getCustomEnglishAnalyz
erProvider);
```

```
            return analyzers;
        }
    }
```

2. The `AnalyzerProvider` class provides the initialization of our analyzer, and
    passes the parameters that are provided by the settings, as follows:

```
import org.apache.lucene.analysis.en.EnglishAnalyzer;
import org.apache.lucene.analysis.CharArraySet;
import org.elasticsearch.common.settings.Settings;
import org.elasticsearch.env.Environment;
import org.elasticsearch.index.IndexSettings;
public class CustomEnglishAnalyzerProvider extends
AbstractIndexAnalyzerProvider<EnglishAnalyzer> {
    public static String NAME = "custom_english";
    private final EnglishAnalyzer analyzer;
    public CustomEnglishAnalyzerProvider(IndexSettings
indexSettings, Environment env, String name, Settings
settings,
                                                boolean
ignoreCase) {
        super(indexSettings, name, settings);
        analyzer = new EnglishAnalyzer(
                Analysis.parseStopWords(env, settings,
EnglishAnalyzer.getDefaultStopSet(), ignoreCase),
                Analysis.parseStemExclusion(settings,
CharArraySet.EMPTY_SET));
    }
    public static CustomEnglishAnalyzerProvider
getCustomEnglishAnalyzerProvider(IndexSettings
indexSettings,

Environment env, String name,

Settings settings) {
        return new
CustomEnglishAnalyzerProvider(indexSettings, env, name,
settings, true);
    }
    @Override
```

```
        public EnglishAnalyzer get() { return this.analyzer;
    } }
```

After building the plugin and installing it on an Elasticsearch server, our analyzer is accessible as any native Elasticsearch analyzer.

How it works...

Creating an analyzer plugin is quite simple. The general workflow is as follows:

1. Wrap the analyzer initialization in a provider.

2. Register the analyzer provider in the plugin.

3. In the preceding example, we registered a `CustomEnglishAnalyzerProvider` class, which extends the `EnglishAnalyzer` class:

```
public class CustomEnglishAnalyzerProvider extends
AbstractIndexAnalyzerProvider<EnglishAnalyzer> {
```

4. We need to provide a name to `analyzer`, as follows:

```
public static String NAME = "custom_english";
```

5. We instantiate a private scope Lucene analyzer (in *Chapter 2*, *Managing Mapping*, we discussed custom Lucene analyzer usage) to be provided on request with the `get` method, as follows:

```
    @Override
    public EnglishAnalyzer get() {
        return this.analyzer; }
```

6. The `CustomEnglishAnalyzerProvider` constructor can be injected via Google Guice, with settings that can be used to provide cluster defaults via index settings or `elasticsearch.yml`, as follows:

```
public CustomEnglishAnalyzerProvider(IndexSettings
indexSettings, Environment env, String name, Settings
settings, boolean ignoreCase) {
```

7. To make it work correctly, we need to set up the parent constructor via the `super` call, as follows:

```
super(indexSettings, name, settings);
```

8. Now, we can initialize the internal analyzer, which must be returned by the `get` method, as follows:

```
analyzer = new EnglishAnalyzer(
                Analysis.parseStopWords(env, settings,
    EnglishAnalyzer.getDefaultStopSet(), ignoreCase),
                Analysis.parseStemExclusion(settings,
    CharArraySet.EMPTY_SET));
```

This analyzer accepts the following:

- A list of stopwords that can be loaded via the settings or set by the default ones

- A list of words that must be excluded by the stemming step

9. To easily wrap the analyzer, we need to create a `static` method that can be called to create the analyzer. We'll use it in the plugin definition, as follows:

```
    public static CustomEnglishAnalyzerProvider
getCustomEnglishAnalyzerProvider(IndexSettings
indexSettings,

Environment env, String name,

Settings settings) {
        return new
CustomEnglishAnalyzerProvider(indexSettings, env, name,
settings, true); }
```

10. Finally, we can register our analyzer in the plugin. To do so, our plugin must derive from `AnalysisPlugin` so that we can override the `getAnalyzers` method, as follows:

```
    @Override
    public Map<String, AnalysisModule.
AnalysisProvider<AnalyzerProvider<? extends Analyzer>>>
getAnalyzers() {
        Map<String, AnalysisModule.
AnalysisProvider<AnalyzerProvider<? extends Analyzer>>>
analyzers = new HashMap<>();
        analyzers.put(CustomEnglishAnalyzerProvider.NAME,
CustomEnglishAnalyzerProvider::getCustomEnglishAnalyzer
Provider);
        return analyzers; }
```

The : : operator of Java 8 allows us to provide a function that will be used for the construction of our `AnalyzerProvider`.

There's more...

A plugin extends several Elasticsearch functionalities. To provide them with this requires extending the correct plugin interface. In Elasticsearch 7.x, the following are the main plugin interfaces:

- `ActionPlugin`: This is used for REST and cluster actions.
- `AnalysisPlugin`: This is used for extending all the analysis stuff, such as **analyzers**, **tokenizers**, **tokenFilters**, and **charFilters**.
- `ClusterPlugin`: This is used to provide new deciders.
- `DiscoveryPlugin`: This is used to provide custom node name resolvers.
- `EnginePlugin`: This is used to provide a new custom engine for indices.
- `IndexStorePlugin`: This is used to provide a custom index store.
- `IngestPlugin`: This is used to provide new ingest processors.
- `MapperPlugin`: This is used to provide new mappers and metadata mappers.
- `ReloadablePlugin`: This allows you to create plugins that reload their state.
- `RepositoryPlugin`: This allows the provision of new repositories to be used in backup/restore functionalities.
- `ScriptPlugin`: This allows the provision of new scripting languages, scripting contexts, or native scripts (Java-based ones).
- `SearchPlugin`: This allows an extension of all the search functionalities: Highlighter, aggregations, suggesters, and queries.

If your plugin needs to extend more than a single functionality, it can extend from several plugin interfaces at once.

Creating a REST plugin

In the previous recipe, we read how to build an analyzer plugin that extends the query capabilities of Elasticsearch. In this recipe, we will see how to create one of the most common Elasticsearch plugins. This kind of plugin allows the standard REST calls to be extended with custom ones to easily improve the capabilities of Elasticsearch.

In this recipe, we will see how to define a **REST entry point** and create its action; in the next one, we'll see how to execute this action distributed in shards.

Getting ready

You need an up and running Elasticsearch installation, as we described in the *Downloading and installing Elasticsearch* recipe in *Chapter 1, Getting Started*.

Gradle or an **integrated development environment** (**IDE**) that supports Java programming with Gradle (version 7.3.x used in the examples), such as Eclipse, Visual Studio Code, or IntelliJ IDEA, is required. Java JDK 17 or above needs to be installed.

The code for this recipe is available in the `ch16/rest_plugin` directory.

How to do it...

To create a REST entry point, we need to create the action and then register it in the plugin. We will perform the following steps:

1. We create a REST `simple` action (`RestSimpleAction.java`) as follows:

```java
public class RestSimpleAction extends BaseRestHandler {
    public RestSimpleAction(Settings settings,
RestController controller) {
    }
    @Override
    public List<Route> routes() {
        return unmodifiableList(asList(
                new Route(POST, "/_simple"),
                new Route(POST, "/{index}/_simple"),
                new Route(POST, "/_simple/{field}"),
                new Route(GET, "/_simple"),
                new Route(GET, "/{index}/_simple"),
                new Route(GET, "/_simple/{field}")));
    }
    @Override
    public String getName() {return "simple_rest"; }
    @Override
    protected RestChannelConsumer
prepareRequest(RestRequest request, NodeClient client) {
        final SimpleRequest simpleRequest = new
```

```
SimpleRequest(Strings.splitStringByCommaToArray(request.
param("index")));
        simpleRequest.setField(request.param("field"));
        return channel -> client.
execute(SimpleAction.INSTANCE, simpleRequest, new
RestBuilderListener<SimpleResponse>(channel) {
            @Override
            public RestResponse
buildResponse(SimpleResponse simpleResponse,
XContentBuilder builder) {
                try {
                    builder.startObject();
                    builder.field("ok", true);
                    builder.array("terms",
simpleResponse.getSimple().toArray());
                    builder.endObject();
                } catch (Exception e) {
                    onFailure(e);
                }
                return new BytesRestResponse(OK,
builder); } }); } }
```

2. We need to register it in the plugin with the following lines:

```
public class RestPlugin extends Plugin implements
ActionPlugin {
    @Override
    public List<RestHandler> getRestHandlers(Settings
settings, RestController restController,

ClusterSettings clusterSettings, IndexScopedSettings
indexScopedSettings,

SettingsFilter settingsFilter,

IndexNameExpressionResolver indexNameExpressionResolver,

Supplier<DiscoveryNodes> nodesInCluster) {
        return Arrays.asList(new
RestSimpleAction(settings, restController));
```

```
        }
    @Override
    public List<ActionHandler<? extends ActionRequest, ?
extends ActionResponse>> getActions() {
        return Arrays.asList(new
ActionHandler<>(SimpleAction.INSTANCE,
TransportSimpleAction.class)); } }
```

3. Now, we can build the plugin via `gradle clean check` and manually install the ZIP. If we restart the Elasticsearch server, we should see the plugin loaded.

4. We can test out custom REST via `curl`, as follows:

```
curl -XPUT http://127.0.0.1:9200/mytest
curl -XPUT http://127.0.0.1:9200/mytest2
curl 'http://127.0.0.1:9200/_simple?field=mytest&pretty
```

5. The result will be something similar to the following:

```
{"ok" : true,
  "terms":["mytest_[mytest2][0]","mytest_[mytest][0]"]}
```

How it works...

Adding a REST action is very easy: We need to create a `RestXXXAction` class that handles the calls.

The REST action is derived from the `BaseRestHandler` class and needs to implement the `handleRequest` method.

The constructor is very important. So, let's start by writing the following:

```
public RestSimpleAction(Settings settings, RestController
controller) {}
```

The public constructor takes the following parameters:

- `Settings`: This can be used to load custom settings for your REST action.
- `RestController`: This is used to register the advanced REST action with the controller.

The list of actions that must be registered in an override `routers` methods is as follows:

```
@Override
public List<Route> routes() {
    return unmodifiableList(asList(
            new Route(POST, "/_simple"),
            new Route(POST, "/{index}/_simple"),
            new Route(POST, "/_simple/{field}"),
            new Route(GET, "/_simple"),
            new Route(GET, "/{index}/_simple"),
            new Route(GET, "/_simple/{field}")));}
```

To register an action, a `Route` class must be instantiated with the following parameters:

- The REST method (GET/POST/PUT/DELETE/HEAD/OPTIONS)
- The URL **entry point**

After having defined the constructor, if an action is fired, the `prepareRequest` class method is called as follows:

```
RestChannelConsumer
@Override
protected RestChannelConsumer prepareRequest(RestRequest
request, NodeClient client) {
```

This method is the core of the REST action. It processes the request and sends back the result. The following parameters are passed to the method:

- `RestRequest`: This is the REST request that hits the Elasticsearch server.
- `NodeClient`: This is the client used to communicate in the cluster.

The returned value is `RestChannelConsumer`, which is a `FunctionalInterface` interface type that accepts a `RestChannel` request—it's a simple Java **Lambda**.

A `prepareRequest` method is usually composed of these phases:

- Process the REST request and build an inner Elasticsearch request object.
- Call the client with the Elasticsearch request.
- If it is okay, process the Elasticsearch response and build the resulting JSON.
- If there are errors, send back the JSON error response.

In the following example, we created `SimpleRequest` class that processes the request:

```
final SimpleRequest simpleRequest = new SimpleRequest(Strings.
splitStringByCommaToArray(request.param("index")));
simpleRequest.setField(request.param("field"));
```

As you can see, it accepts a list of indices (we split the classic comma-separated list of indices via the `Strings.splitStringByCommaToArray` helper) and we had the `field` parameter if available.

Now that we have `SimpleRequest`, we can send it to the cluster and get back `SimpleResponse` via the Lambda closure as follows:

```
return channel -> client.execute(SimpleAction.INSTANCE,
simpleRequest, new RestBuilderListener<SimpleResponse>(channel)
{
```

`client.execute` accepts an action, a request, and a `RestBuilderListener` class that maps a future response. We can now process the response via the definition of a `buildResponse` method.

`buildResponse` receives a `Response` object that must be converted to a JSON result via `XContentBuilder` as follows:

```
@Override
public RestResponse buildResponse(SimpleResponse
simpleResponse, XContentBuilder builder) {
```

The builder is the standard JSON `XContentBuilder` class that we have already seen in *Chapter 13, Java Integration*.

After having processed the cluster response and built the JSON, we can send the REST response as follows:

```
builder.startObject()
builder.field("ok", true);
builder.array("terms", simpleResponse.getSimple().toArray());
builder.endObject();
```

Obviously, if something goes wrong during the JSON creation, an exception must be raised, such as the following:

```
try {
    // JSON creation
```

```
    } catch (Exception e) {
        onFailure(e);
    }
```

We will discuss `SimpleRequest` in the next recipe.

See also

Google Guice (`https://github.com/google/guice`) is used for dependency injection. Refer to its official documentation for more insights about the dependency injection system used by Elasticsearch.

Creating a cluster action

In the previous recipe, we saw how to create a REST **entry point**, but to execute the action at the cluster level, we will need to create a cluster action.

An Elasticsearch action is generally executed and distributed in the cluster and, in this recipe, we will see how to implement this kind of action. The cluster action will be very bare; we send a string with a value to every shard and the shards echo a result string, concatenating the string with the shard number.

Getting ready

You need an up and running Elasticsearch installation, as we described in the *Downloading and installing Elasticsearch* recipe in *Chapter 1, Getting Started*.

Gradle or an **integrated development environment** (**IDE**) that supports Java programming with Gradle (version 7.3.x used in the examples), such as Eclipse, Visual Studio Code, or IntelliJ IDEA, is required. Java JDK 17 or above needs to be installed.

The code for this recipe is available in the `ch16/rest_plugin` directory.

How to do it...

In this recipe, we will see that a REST call is converted to an internal cluster action. To execute an internal cluster action, the following classes are required:

- A `Request` and `Response` class to communicate with the cluster.
- A `RequestBuilder` class is used to execute a request to the cluster.
- An `Action` class used to register the action and bound `Request`, `Response`, and `RequestBuilder`.

- A `Transport*Action` to bind the request and response to `ShardResponse`: it manages the *reduce* part of the query.

- A `ShardResponse` to manage the shard results.

We will perform the following steps:

1. We write a `SimpleRequest` class as follows:

```
public class SimpleRequest extends
BroadcastRequest<SimpleRequest> {
    private String field;
    SimpleRequest() {}
    public SimpleRequest(String... indices) {
        super(indices);
    }
    public SimpleRequest(StreamInput in) throws
IOException {
        super(in);
        field = in.readString();
    }
    public void setField(String field) {
        this.field = field;
    }
    public String getField() { return field; }
    @Override
    public void writeTo(StreamOutput out) throws
IOException {
        super.writeTo(out);
        out.writeString(field); } }
```

2. The `SimpleResponse` class is very similar to the `SimpleRequest` class.

3. To bind the request and the response, an action (`SimpleAction`) is required as follows:

```
public class SimpleAction extends
ActionType<SimpleResponse> {
    public static final SimpleAction INSTANCE = new
SimpleAction();
    public static final String NAME = "custom:indices/
simple";
```

```
        private SimpleAction() {
            super(NAME, SimpleResponse::new);
        }
    }
```

4. The `Transport` class is the core of the action. It's quite long, so we'll only present the main important parts as follows:

```
public class TransportSimpleAction
        extends
TransportBroadcastByNodeAction<SimpleRequest,
SimpleResponse, ShardSimpleResponse> {
    private final IndicesService indicesService;
    @Inject
    public TransportSimpleAction(ClusterService
clusterService,
                                    TransportService
transportService, IndicesService indicesService,
                                    ActionFilters
actionFilters, IndexNameExpressionResolver
indexNameExpressionResolver) {
        super(SimpleAction.NAME, clusterService,
transportService, actionFilters,
                indexNameExpressionResolver,
SimpleRequest::new, ThreadPool.Names.SEARCH);
        this.indicesService = indicesService;
    }
    @Override
    protected SimpleResponse newResponse(SimpleRequest
request, int totalShards, int successfulShards, int
failedShards,

List<ShardSimpleResponse> shardSimpleResponses,

List<DefaultShardOperationFailedException> shardFailures,
                                    ClusterState
clusterState) {
        Set<String> simple = new HashSet<String>();
        for (ShardSimpleResponse shardSimpleResponse :
shardSimpleResponses) {
```

```
                        simple.addAll(shardSimpleResponse.
getTermList());
            }
        return new SimpleResponse(totalShards,
successfulShards, failedShards, shardFailures, simple);
    }
    @Override
    protected void shardOperation(SimpleRequest request,
ShardRouting shardRouting, Task task,
            ActionListener<ShardSimpleResponse> listener)
{
        IndexService indexService = indicesService.
indexServiceSafe(shardRouting.shardId().getIndex());
        IndexShard indexShard = indexService.
getShard(shardRouting.shardId().id());
        indexShard.store().directory();
        Set<String> set = new HashSet<String>();
        set.add(request.getField() + "_" + shardRouting.
shardId());
        listener.onResponse(new
ShardSimpleResponse(shardRouting, set));
    }
    @Override
    protected ShardSimpleResponse
readShardResult(StreamInput in) throws IOException {
        return ShardSimpleResponse.readShardResult(in);
    }
    @Override
    protected SimpleRequest readRequestFrom(StreamInput
in) throws IOException {
        return new SimpleRequest(in);
    }
    @Override
    protected ShardsIterator shards(ClusterState
clusterState, SimpleRequest request, String[]
concreteIndices) {
        return clusterState.routingTable().
allShards(concreteIndices);
    }
```

```
    @Override
    protected ClusterBlockException
checkGlobalBlock(ClusterState state, SimpleRequest
request) {
        return state.blocks().
globalBlockedException(ClusterBlockLevel.METADATA_READ);
    }
    @Override
    protected ClusterBlockException
checkRequestBlock(ClusterState state, SimpleRequest
request, String[] concreteIndices) {
        return state.blocks().
indicesBlockedException(ClusterBlockLevel.METADATA_READ,
concreteIndices);
    }
}
```

How it works...

In this example, we used an action that is executed in every cluster node and for every shard that is selected on that node.

As you have seen, to execute a cluster action, the following classes are required:

- A couple of Request/Response class to interact with the cluster
- A task action on the cluster level
- A shard Response class to interact with the shards
- A Transport class to manage the map/reduce shard part that must be invoked by the REST call

These classes must extend one of the supported actions, for example:

- TrasportBroadcastAction: For actions that must be spread across the entire cluster.
- TransportClusterInfoAction: For actions that need to read information at the cluster level.
- TransportMasterNodeAction: For actions that must be executed only by the master node (such as index and mapping configuration). For a simple acknowledgment on the master, there is also the AcknowledgedRequest response.

- `TransportNodeAction`: For actions that must be executed on nodes (that is, all the node statistic actions).

- `TransportBroadcastReplicationAction`, `TransportReplicationAction`, `TransportWriteAction`: For actions that must be executed by a particular replica, first on primary and then on secondary ones.

- `TransportInstanceSingleOperationAction`: For actions that must be executed as a singleton in the cluster.

- `TransportSingleShardAction`: For actions that must be executed only in a shard (that is, GET actions). If it fails on a shard, it automatically tries on the shard replicas.

- `TransportTasksAction`: For actions that need to interact with cluster tasks.

In our example, we have defined an action that will be broadcast to every node and for every node, it collects its shard result and then it aggregates as follows:

```
public class TransportSimpleAction
        extends TransportBroadcastByNodeAction<SimpleRequest,
SimpleResponse, ShardSimpleResponse> {
```

All the request/response classes extend an `ActionResponse` class, so the following two methods for serializing their content must be provided:

- A contructor, which reads from a `StreamInput`, a class that encapsulates common input stream operations. This method allows the deserialization of the data we transmit on the wire. In the preceding example, we read a string with the following code:

```
public SimpleResponse(StreamInput in) throws IOException
{
    super(in);
    int n = in.readInt();
    simple = new HashSet<String>();
    for (int i = 0; i < n; i++) {
        simple.add(in.readString());
    }
}
```

- writeTo, which writes the contents of the class to be sent via the network. StreamOutput provides convenient methods to process the output. In the following example, we serialized the StreamOutput string:

```
@Override
public void writeTo(StreamOutput out) throws IOException
{
    super.writeTo(out);
    out.writeInt(simple.size());
    for (String t : simple) {
        out.writeString(t);
    }
}
```

In both actions, super must be called to allow the correct serialization of parent classes.

Every internal action in Elasticsearch is designed as a request/response pattern:

1. To complete the request/response action, we must define an action that binds the request with the correct response and a builder to construct it. To do so, we need to define an Action class as follows:

```
public class SimpleAction extends
ActionType<SimpleResponse> {
```

2. This Action object is a singleton object. We obtain it by creating a default static instance and private constructors as follows:

```
public static final SimpleAction INSTANCE = new
SimpleAction();
```

3. The NAME static string is used to uniquely identify the action at the cluster level.

```
public static final String NAME = "custom:indices/
simple";
```

4. To complete the Action definition, in the definition of the super constructor, we need to define a lambda to create a Response, which is used to create a new empty response as follows:

```
private SimpleAction() {
    super(NAME, SimpleResponse::new);
}
```

When the action is executed, the request and the response are serialized and sent to the cluster. To execute our custom code at the cluster level, a transport action is required.

The transport actions are usually defined as map and reduce jobs. The map part consists of executing the action on several shards and then reducing parts consisting of collecting all the results from the shards in a response that must be sent back to the requester. To speed up the process in Elasticsearch 5.x or above, all the shard's responses that belong in the same node are reduced in place to optimize the I/O and the network usage.

The transport action is a long class with many methods, but the most important ones are `ShardOperation` (map part) and `newResponse` (reduce part).

The original request is converted to a distributed `ShardRequest` that is processed by the `shardOperation` method as follows:

```
@Override
protected void shardOperation(SimpleRequest request,
ShardRouting shardRouting, Task task,
          ActionListener<ShardSimpleResponse> listener) {
```

To obtain the internal shard, we need to ask `IndexService` to return a shard based on the wanted index.

The shard request contains the index and the ID of the shard that must be used to execute the action as follows:

```
IndexService indexService = indicesService.
indexServiceSafe(shardRouting.shardId().getIndex());
IndexShard indexShard = indexService.getShard(shardRouting.
shardId().id());
```

The `IndexShard` object allows the execution of every possible shard operation (`search`, `get`, `index`, and many others). By means of this method, we can execute every data shard manipulation that we want.

> **Tip**
> Custom shard action can execute the application's business operation in a distributed and fast way.

1. In the following example, we have created a simple set of values:

```
indexShard.store().directory();
Set<String> set = new HashSet<String>();
```

```
set.add(request.getField() + "_" + shardRouting.
shardId());
```

2. The final step of our shard operation is to create a response to send back to the reduce step. In creating `ShardResponse`, we need to return the result plus information about the index and the shard that executed the action via `listener` as follows:

```
listener.onResponse(new ShardSimpleResponse(shardRouting,
set));
```

3. The distributed shard operations are collected in the reduce step (the `newResponse` method). This step aggregates all the shard results and sends back the result to the original `Action` as follows:

```
@Override
protected SimpleResponse newResponse(SimpleRequest
request, int totalShards, int successfulShards, int
failedShards,

List<ShardSimpleResponse> shardSimpleResponses,

List<DefaultShardOperationFailedException> shardFailures,
                                        ClusterState
clusterState) {
```

Other than the shard's result, the methods receive the status of the shard level operation and they are collected in three values: `successfulShards`, `failedShards`, and `shardFailures`.

4. The request result is a set of collected strings, so we create an empty set to collect the term's results as follows:

```
Set<String> simple = new HashSet<String>();
return new SimpleResponse(totalShards, successfulShards,
failedShards, shardFailures, simple);
```

5. Then you collect the results that we need to iterate on the shard responses as follows:

```
for (ShardSimpleResponse shardSimpleResponse :
shardSimpleResponses) {
    simple.addAll(shardSimpleResponse.getTermList());
}
```

6. The final step is to create the response by collecting the previous result and response status as follows:

```
return new SimpleResponse(totalShards, successfulShards,
failedShards, shardFailures, simple);
```

A cluster action needs to be created when there are low-level operations that we want to execute very quickly, such as special aggregations, server-side join, or a complex manipulation that requires several Elasticsearch calls to be executed. Writing custom Elasticsearch actions is an advanced Elasticsearch feature, but it can create new business use scenarios that can level up the capabilities of Elasticsearch.

See also

Refer to the *Creating a REST plugin* recipe in this chapter for how to interface the cluster action with a REST call.

Creating an ingest plugin

Elasticsearch 5.x introduced the ingest node that allows the modification, via a pipeline, to the records before ingesting in Elasticsearch. We have already seen in *Chapter 12, Using the Ingest Module*, that a pipeline is composed of one or more processor actions. In this recipe, we will see how to create a custom processor that stores in a field the initial character of another one.

Getting ready

You need an up and running Elasticsearch installation, as we described in the *Downloading and installing Elasticsearch* recipe in *Chapter 1, Getting Started*.

Gradle or an **integrated development environment** (**IDE**) that supports Java programming with Gradle (version 7.3.x used in the examples), such as Eclipse, Visual Studio Code, or IntelliJ IDEA, is required. Java JDK 17 or above needs to be installed.

The code for this recipe is available in the `ch16/ingest_plugin` directory.

How to do it...

To create an ingest processor plugin, we need to create the processor and then register it in the `plugin` class. We will perform the following steps:

1. We create the processor, and its factory, as follows:

    The class declaration and internal attributes:

    ```
    public final class InitialProcessor extends
    AbstractProcessor {
        public static final String TYPE = "initial";
        private final String field;
        private final String targetField;
        private final String defaultValue;
        private final boolean ignoreMissing;
        public InitialProcessor(String tag, String
    description, String field, String targetField, boolean
    ignoreMissing, String defaultValue) {
            super(tag, description);
            this.field = field;
            this.targetField = targetField;
            this.ignoreMissing = ignoreMissing;
            this.defaultValue = defaultValue;
        }
    ```

    The helper methods to access private attributes:

    ```
        String getField() { return field; }
        String getTargetField() { return targetField; }
        String getDefaultField() { return defaultValue; }
        boolean isIgnoreMissing() {return ignoreMissing;}
        @Override
        public String getType() { return TYPE; }
    ```

    The `execute` function, which is the core of the processors:

    ```
        @Override
        public IngestDocument execute(IngestDocument
    document) {
            if (document.hasField(field, true) == false) {
                if (ignoreMissing) { return document;
                } else {
    ```

```
                        throw new IllegalArgumentException("field
[" + field + "] not present as part of path [" + field +
"]");
                }
        }
        // We fail here if the target field point to an
array slot that is out of range.
        // If we didn't do this then we would fail if we
set the value in the target_field
        // and then on failure processors would not see
that value we tried to rename as we already
        // removed it.
        if (document.hasField(targetField, true)) {
                throw new IllegalArgumentException("field ["
+ targetField + "] already exists");
        }
        Object value = document.getFieldValue(field,
Object.class);
        if( value!=null && value instanceof String ) {
                String myValue=value.toString().trim();
                if(myValue.length()>1){
                        try {
                                document.setFieldValue(targetField,
myValue.substring(0,1).toLowerCase(Locale.getDefault()));
                        } catch (Exception e) {
                                // setting the value back to the
original field shouldn't as we just fetched it from that
field:
                                document.setFieldValue(field, value);
                                throw e;
                        }
                }
        }
        return document;
    }
```

The factory used to initialize the processor:

```
    public static final class Factory implements
Processor.Factory {
        @Override
        public Processor create(Map<String, Processor.
Factory> processorFactories, String tag, String
description, Map<String, Object> config) throws Exception
{
            String field = ConfigurationUtils.
readStringProperty(TYPE, tag, config, "field");
            String targetField = ConfigurationUtils.
readStringProperty(TYPE, tag,
                    config, "target_field");
            String defaultValue = ConfigurationUtils.
readOptionalStringProperty(TYPE, tag,
                    config, "defaultValue");
            boolean ignoreMissing = ConfigurationUtils.
readBooleanProperty(TYPE, tag,
                    config, "ignore_missing", false);
            return new InitialProcessor(tag, description,
field, targetField, ignoreMissing, defaultValue);
        }
    }
}
```

2. We need to register it in the Plugin class with the following lines:

```
public class InitialIngestPlugin extends Plugin
implements IngestPlugin {
    @Override
    public Map<String, Processor.Factory>
getProcessors(Processor.Parameters parameters) {
        return Collections.singletonMap(InitialProcessor.
TYPE, new InitialProcessor.Factory());
    }}
```

3. Now we can build the plugin via `gradlew clean check` and manually install the ZIP. If we restart the Elasticsearch server, we should see the plugin loaded as follows:

```
… truncated …
[2021-05-01T20:55:39,740][INFO ][o.e.p.PluginsService
] [iMacParo] loaded module [x-pack-watcher]
[2021-05-01T20:55:39,741][INFO ][o.e.p.PluginsService
] [iMacParo] loaded plugin [initial-processor]
[2021-05-01T20:55:39,780][INFO ][o.e.e.NodeEnvironment
] [iMacParo] using [1] data paths, mounts [[/System/
Volumes/Data (/dev/disk1s1)]], net usable_space [16.9gb],
net total_space [931.6gb], types [apfs]
... truncated ...
```

4. We can test our custom ingest plugin via the Simulate Ingest API with a `curl` as follows:

```
curl -XPOST -H "Content-Type: application/
json" 'http://127.0.0.1:9200/_ingest/pipeline/_
simulate?verbose&pretty' -d '{
"pipeline": {
    "description": "Test my custom plugin",
    "processors": [
    {
        "initial": {
        "field": "user",
        "target_field": "user_initial"
    } } ], "version": 1 },
"docs": [
    { "_source": { "user": "john" } },
    { "_source": { "user": "Nancy" } } ] }'
```

5. The result will be something similar to the following:

```
{ "docs" : [
    { "processor_results" : [
        { "processor_type" : "initial",
        "status" : "success",
        "doc" : {
```

```
            "_index" : "_index", "_id" : "_id",
            "_source" : {
                "user_initial" : "j", "user" : "john" },
            "_ingest" : {
                "pipeline" : "_simulate_pipeline",
                "timestamp" : "2021-05-01T18:57:58.316032Z"
} } } ] },
        { "processor_results" : [
            { "processor_type" : "initial",
            "status" : "success",
            "doc" : {
                "_index" : "_index", "_id" : "_id",
                "_source" : {
                    "user_initial" : "n", "user" : "Nancy"},
                "_ingest" : {
                    "pipeline" : "_simulate_pipeline",
                    "timestamp" : "2021-05-01T18:57:58.316046Z"
} } } ] } ] }
```

How it works...

First, you need to define the class that will manage your custom processor, which extends `AbstractProcessor`:

```
public final class InitialProcessor extends AbstractProcessor {
```

The `processor` class needs to know the fields on which it operates (`tag` name and `description` are mandatory). They are kept in the internal state of the processor as follows:

```
public InitialProcessor(String tag, String description, String
field, String targetField, boolean ignoreMissing, String
defaultValue) {
    super(tag, description);
    this.field = field;
    this.targetField = targetField;
    this.ignoreMissing = ignoreMissing;
    this.defaultValue = defaultValue;}
```

The core of the processor is the `execute` function, which contains our processor login as follows:

```
@Override
public IngestDocument execute(IngestDocument document) {
```

The `execute` function comprises the following steps:

1. Check whether the `source` field exists as follows:

    ```
    if (!document.hasField(field, true)) {
        if (ignoreMissing) { return document; } else {
            throw new IllegalArgumentException("field [" +
    field + "] not present as part of path [" + field + "]");
        } }
    ```

2. Check whether the `target` field does not exist as follows:

    ```
    if (document.hasField(targetField, true)) {
        throw new IllegalArgumentException("field [" +
    targetField + "] already exists"); }
    ```

3. We extract the value from `document` and check whether it's valid as follows:

    ```
    Object value = document.getFieldValue(field, Object.
    class);
    if( value!=null && value instanceof String ) {
    ```

4. Now, we can process the value and set in the `target` field as follows:

    ```
    String myValue=value.toString().trim();
    if(myValue.length()>1){
        try {
            document.setFieldValue(targetField, myValue.
    substring(0,1).toLowerCase(Locale.getDefault()));
        } catch (Exception e) {
            // setting the value back to the original field
    shouldn't as we just fetched it from that field:
            document.setFieldValue(field, value);
            throw e;
        } }
    return document;
    ```

5. To be able to initialize the processor for its definition, we need to define a `Factory` object as follows:

```
public static final class Factory implements Processor.
Factory {
```

6. The `Factory` object contains the `create` method that receives the registered processors, `processorTag`, and its configuration, which must be read as follows:

```
@Override
public Processor create(Map<String, Processor.Factory>
processorFactories, String tag, String description,
Map<String, Object> config) throws Exception {
    String field = ConfigurationUtils.
readStringProperty(TYPE, tag, config, "field");
    String targetField = ConfigurationUtils.
readStringProperty(TYPE, tag,
        config, "target_field");
    String defaultValue = ConfigurationUtils.
readOptionalStringProperty(TYPE, tag,
        config, "defaultValue");
    boolean ignoreMissing = ConfigurationUtils.
readBooleanProperty(TYPE, tag,
        config, "ignore_missing", false);
```

7. After having recovered, we can initialize the processor parameters as follows:

```
    return new InitialProcessor(tag, description, field,
targetField, ignoreMissing, defaultValue);
}
```

8. To be used as a custom processor, it needs to be registered in the plugin. This is done by extending the plugin as `IngestPlugin` as follows:

```
public class InitialIngestPlugin extends Plugin
implements IngestPlugin {
```

9. Now, we can register the `Factory` plugin in the `getProcessors` method as follows:

```
@Override
public Map<String, Processor.Factory>
getProcessors(Processor.Parameters parameters) {
```

```
    return Collections.singletonMap(InitialProcessor.
TYPE, new InitialProcessor.Factory());
}
```

Implementing an ingestion processor via a plugin is quite simple, and it's an incredibly powerful feature. With this approach, a user can create a custom step in enrichment pipelines.

See also

You can refer to the following URLs for further reference, which are related to this recipe:

- The official Elasticsearch documentation about the Ingest pipeline at `https://www.elastic.co/guide/en/elasticsearch/reference/current/ingest.html`, and *Chapter 12, Using the Ingest Module*

- The official Elasticsearch page of Ingest plugins at `https://www.elastic.co/guide/en/elasticsearch/plugins/current/ingest.html`

17
Big Data Integration

Elasticsearch has become a common component in big data architectures because it provides several of the following features:

- It allows you to search for massive amounts of data quickly.

- For common aggregation operations, it provides real-time analytics on big data.

- It's easier to use an Elasticsearch aggregation than a Spark one.

- If you need to move on to a fast data solution, starting from a subset of documents after a query is faster than doing a full rescan of all your data.

The most common big data software that's used for processing data is now Apache Spark (`http://spark.apache.org/`), which is considered the evolution of the obsolete Hadoop MapReduce for moving the processing from disk to memory.

In this chapter, we will see how to integrate Elasticsearch in Spark, both for write and read data. At the end, we will see how to use Apache Pig to write data in Elasticsearch in a simple way.

In this chapter, we will cover the following recipes:

- Installing Apache Spark
- Indexing data using Apache Spark
- Indexing data with meta using Apache Spark
- Reading data with Apache Spark
- Reading data using Spark SQL
- Indexing data with Apache Pig
- Using Elasticsearch with Alpakka
- Using Elasticsearch with MongoDB

Installing Apache Spark

To use Apache Spark, we need to install it. The process is very easy because its requirements are not the traditional Hadoop ones that require Apache ZooKeeper and **Hadoop Distributed File System (HDFS)**.

Apache Spark can work in a standalone node installation similar to Elasticsearch.

Getting ready

You need a Java virtual machine installed. Generally, version 8.x or above is used. The maximum Java version supported by Apache Spark is 11.x.

How to do it...

To install Apache Spark, we will perform the following steps:

1. Download a binary distribution from `https://spark.apache.org/downloads.html`. For generic usage, I would suggest that you download a standard version using the following request:

```
wget https://dlcdn.apache.org/spark/spark-3.2.1/spark-
3.2.1-bin-hadoop3.2.tgz
```

2. Now, we can extract the Spark distribution using `tar`, as follows:

```
tar xfvz spark-3.2.1-bin-hadoop3.2.tgz
```

3. Now, we can test whether Apache Spark is working by executing a test, as follows:

```
cd spark-3.2.1-bin-hadoop3.2
./bin/run-example SparkPi 10
```

We will have the following output:

```
22/03/12 11:37:16 WARN NativeCodeLoader: Unable to load
native-hadoop library for your platform... using builtin-
java classes where applicable
Using Spark's default log4j profile: org/apache/spark/
log4j-defaults.properties
22/03/12 11:37:16 INFO SparkContext: Running Spark
version 3.2.1
22/03/12 11:37:16 INFO SparkContext: Submitted
application: Spark Pi
22/03/12 11:37:16 INFO SecurityManager: Changing view
acls to: alberto
… truncated …
22/03/12 11:37:18 INFO DAGScheduler: Job 0 finished:
reduce at SparkPi.scala:38, took 0.911597 s
Pi is roughly 3.141939141939142
22/03/12 11:37:18 INFO SparkUI: Stopped Spark web UI at
http://imacparo:4040
… truncated …
22/03/12 11:37:18 INFO ShutdownHookManager:
Deleting directory /private/var/folders/0h/
fkvg8wz54d30g1_9b9k3_7zr0000gn/T/spark-dd06cd19-5a0a-
4da3-88b1-5ca4ddfad132
22/03/12 11:37:18 INFO ShutdownHookManager:
Deleting directory /private/var/folders/0h/
fkvg8wz54d30g1_9b9k3_7zr0000gn/T/spark-19bef937-6653-
4840-81fe-e87ccfbf0a5e
```

How it works...

Apache Spark as a standalone node is very easy to install. Like Elasticsearch, it requires only a Java virtual machine installed on the system. The installation process is very easy – you only need to unpack the archive and there will be a complete working installation.

In the preceding steps, we also tested whether the Spark installation was working. Spark is written in Scala and the default binaries target version 2.11.x. Major Scala versions are not compatible, so you need to pay attention to ensure that both Spark and Elasticsearch Hadoop are using the same version.

When executing a Spark job, the simplified steps are as follows:

1.   The Spark environment is initialized.
2.   Spark `MemoryStore` and `BlockManager` masters are initialized.
3.   A `SparkContext` function for the execution is initialized.
4.   `SparkUI` is activated at `http://0.0.0.0:4040`.
5.   The job is taken.
6.   An execution graph, a **Directed Acyclic Graph** (**DAG**), is created for the job.
7.   Every vertex in the DAG is a stage, and a stage is split into tasks that are executed in parallel.
8.   After executing the stages and tasks, the processing ends.
9.   The result is returned.
10. The `SparkContext` function is stopped.
11. The Spark system is shut down.

There's more...

One of the most powerful tools of Spark is the shell (the Spark shell). It allows you to enter commands and execute them directly on the Spark cluster. To access the Spark shell, you need to invoke it using `./bin/spark-shell`.

When invoked, the output will be something like this:

```
Using Spark's default log4j profile: org/apache/spark/log4j-
defaults.properties
Setting default log level to "WARN".
To adjust logging level use sc.setLogLevel(newLevel). For
SparkR, use setLogLevel(newLevel).
22/03/12 15:31:46 WARN NativeCodeLoader: Unable to load native-
hadoop library for your platform... using builtin-java classes
where applicable
Spark context Web UI available at http://blackparone:4040
Spark context available as 'sc' (master = local[*], app id =
local-1647095506655).
```

```
Spark session available as 'spark'.
Welcome to


      ____              __
     / __/__  ___ _____/ /__
    _\ \/ _ \/ _ `/ __/  '_/
   /___/ .__/\_,_/_/ /_/\_\   version 3.2.1
      /_/

Using Scala version 2.12.15 (OpenJDK 64-Bit Server VM, Java
1.8.0_312)
Type in expressions to have them evaluated.
Type :help for more information.
scala>
```

Now, it's possible to insert the command-line commands that are to be executed in the cluster.

Indexing data using Apache Spark

Now that we have installed Apache Spark, we can configure it to work with Elasticsearch and write some data in it.

Getting ready

You need an up-and-running Elasticsearch installation, as we described in the *Downloading and installing Elasticsearch* recipe in *Chapter 1, Getting Started*.

You also need a working installation of Apache Spark.

To simplify the configuration, we disable the HTTP **Secure Sockets Layer** (**SSL**) self-signed certificate that updates the section of config/elasticsearch.yml to false, as shown in the following code:

```
# Enable encryption for HTTP API client connections, such as
Kibana, Logstash, and Agents
xpack.security.http.ssl:
  enabled: false
  keystore.path: certs/http.p12
```

After changing the configuration, the Elasticsearch node/cluster must be restarted.

How to do it...

To configure Apache Spark to communicate with Elasticsearch, we will perform the following steps:

1. We need to download the Elasticsearch Spark `.jar` file, as follows:

```
wget -c https://artifacts.elastic.co/downloads/
elasticsearch-hadoop/elasticsearch-hadoop-8.1.0.zip
unzip elasticsearch-hadoop-8.1.0.zip
```

Alternatively, if the file is missing in some releases, you can download the JAR directly from Maven via the following:

```
wget https://repo1.maven.org/maven2/org/elasticsearch/
elasticsearch-spark-30_2.12/8.1.0/elasticsearch-spark-
30_2.12-8.1.0.jar
```

2. A quick way to access the Spark shell in Elasticsearch is to copy the Elasticsearch Hadoop file that's required in Spark's `jar` directory. The file that must be copied is `elasticsearch-spark-30_2.12-8.1.0.jar`.

> **Note**
>
> The versions of Scala that are used by both Apache Spark and Elasticsearch Spark (3x and 2.12 in this case) must match!

To store data in Elasticsearch using Apache Spark, we will perform the following steps:

1. In Spark's root directory, start the Spark shell to apply the Elasticsearch configuration by running the following command:

```
./bin/spark-shell \
    --conf spark.es.index.auto.create=true \
    --conf spark.es.net.http.auth.user=$ES_USER \
    --conf spark.es.net.http.auth.pass=$ES_PASSWORD
```

`ES_USER` and `ES_PASSWORD` are environment variables that hold your credentials to the Elasticsearch cluster.

2. Before using Elasticsearch's special **Resilient Distributed Dataset (RDD)**, we will import the Elasticsearch Spark implicits, as follows:

```
import org.elasticsearch.spark._
```

3. We will create two documents to be indexed, as follows:

```scala
val numbers = Map("one" -> 1, "two" -> 2, "three" -> 3)
val airports = Map("arrival" -> "Otopeni", "SFO" -> "San Fran")
```

4. Now, we can create an RDD and save the document in Elasticsearch, as follows:

```scala
sc.makeRDD(Seq(numbers, airports)).saveToEs("spark")
```

How it works...

Storing documents in Elasticsearch via Spark is quite simple. After having started a Spark shell in the shell context, the `sc` variable is available, which contains a SparkContext. If we need to pass values to the underlying Elasticsearch configuration, we need to set them in the Spark shell command line.

There are several configurations that can be set (if passed by the command line, add a `spark.` prefix); the following are the most used ones:

- `es.index.auto.create`: This is used to create indices if they do not exist.
- `es.nodes`: This is used to define a list of nodes to connect with (default `localhost`).
- `es.port`: This is used to define the HTTP Elasticsearch port to connect with (default `9200`).
- `es.ingest.pipeline`: This is used to define an ingest pipeline to be used (default `none`).
- `es.mapping.id`: This is used to define a field to extract the ID value (default `none`).
- `es.mapping.parent`: This is used to define a field to extract the parent value (default `none`).

Simple documents can be defined as `Map[String, AnyRef]`, and they can be indexed via an RDD, a special Spark abstraction on a collection.

Via the implicits that are available in `org.elasticsearch.spark`, the RDD has a new method called `saveToEs` that allows you to define the pair index or document to be used for indexing:

```scala
sc.makeRDD(Seq(numbers, airports)).saveToEs("spark")
```

See also

You can refer to the following for further references related to this recipe:

- To download the latest version of Elasticsearch Hadoop, go to the official page at `https://www.elastic.co/downloads/hadoop`.

- The official documentation for installing Elasticsearch Hadoop can be found at `https://www.elastic.co/guide/en/elasticsearch/hadoop/current/install.html`. This page also provides some `border` cases.

- For a quick-start guide on using Spark, I suggest the Spark documentation at `http://spark.apache.org/docs/latest/quick-start.html`.

- For a detailed list of configuration parameters that can be set in the Spark config, look at `https://www.elastic.co/guide/en/elasticsearch/hadoop/7.x/configuration.html`.

Indexing data with meta using Apache Spark

Using a simple map for ingesting data is not good for simple jobs. The best practice in Spark is to use the **case class** so that you have fast serialization and can manage complex type checking. During indexing, providing custom IDs can be very handy. In this recipe, we will see how to cover these issues.

Getting ready

You need an up-and-running Elasticsearch installation, as we described in the *Downloading and installing Elasticsearch* recipe in *Chapter 1, Getting Started*.

You also need a working installation of Apache Spark.

How to do it...

To store data in Elasticsearch using Apache Spark, we will perform the following steps:

1. In the Spark root directory, start the Spark shell to apply the Elasticsearch configuration by running the following command:

```
./bin/spark-shell \
    --conf spark.es.index.auto.create=true \
    --conf spark.es.net.http.auth.user=$ES_USER \
    --conf spark.es.net.http.auth.pass=$ES_PASSWORD
```

2. We will import the required classes, as follows:

```
import org.elasticsearch.spark.rdd.EsSpark
```

3. We will create `case class Person`, as follows:

```
case class Person(username:String, name:String, age:Int)
```

4. We will create two documents that are to be indexed, as follows:

```
val persons = Seq(Person("bob", "Bob",19),
Person("susan","Susan",21))
```

5. Now, we can create RDD, as follows:

```
val rdd=sc.makeRDD(persons)
```

6. We can index them using `EsSpark`, as follows:

```
EsSpark.saveToEs(rdd, "spark2", Map("es.mapping.id" ->
"username"))
```

7. In Elasticsearch, the indexed data will be as follows:

```
{ … truncated …
  "hits" : {
    "total" : { "value" : 2, "relation" : "eq" },
    "max_score" : 1.0,
    "hits" : [
      { "_index" : "spark2", "_id" : "bob",
        "_score" : 1.0,
        "_source" : {
          "username" : "bob", "name" : "Bob",
          "age" : 19 } },
      { "_index" : "spark2", "_id" : "susan",
        "_score" : 1.0,
        "_source" : {
          "username" : "susan", "name" : "Susan",
          "age" : 21 } } ] } }
```

How it works...

To speed up computation in Apache Spark, the case class is used to better describe the domain object we used during job processing. It has fast serializers and deserializers that allow easy conversion of the case class to JSON and vice versa. By using `case class`, the data is strongly typed and modeled:

```
case class Person(username:String, name:String, age:Int)
```

In the preceding example, we created a `Person` class that designs a standard person. (Nested `case` classes are automatically managed.) Now that we've instantiated some `Person` objects, we need to create a Spark RDD that will be saved in Elasticsearch:

```
EsSpark.saveToEs(rdd, "spark2", Map("es.mapping.id" ->
"username"))
```

In this example, we have used a special class called `EsSpark`, which provides helpers to pass metadata that's used for indexing. In our case, we have provided information on how to extract the ID from the document using `Map("es.mapping.id" ->
"username")`.

There's more...

Often, the ID is not a field of your object – it's a complex value that's computed on the document. In this case, you are able to create an RDD with a tuple (ID or document) to be indexed:

1. For the following example, we can define the function that does the ID computation on the `Person` class:

```
import org.elasticsearch.spark._
case class Person(username:String, name:String, age:Int)
{
  def id=this.username+this.age
}
```

2. Then, we can use it to compute our new RDD, as follows:

```
val persons = Seq(Person("bob", "Bob",19),Person("susan",
"Susan",21))
val personIds=persons.map(p => p.id -> p)
val rdd=sc.makeRDD(personIds)
```

3. Now, we can index them, as follows:

```
rdd.saveToEsWithMeta("spark3-person-id")
```

In this case, the stored documents will be as follows:

```
{ ... truncated ...
    "hits" : [
      { "_index" : "spark3-person-id",
        "_id" : "susan21", "_score" : 1.0,
        "_source" : { "username" : "susan",
          "name" : "Susan", "age" : 21 }  },
      { "_index" : "spark3-person-id",
        "_id" : "bob19", "_score" : 1.0,
        "_source" : { "username" : "bob",
          "name" : "Bob", "age" : 19 } } ] } }
```

Reading data with Apache Spark

In Spark, you can read data from a lot of sources, but in general, with NoSQL data stores such as HBase, Accumulo, and Cassandra, you have a limited query subset, and you often need to scan all the data to read only what is required. Using Elasticsearch, you can retrieve a subset of documents that matches your Elasticsearch query, speeding up the data reading several-fold.

Getting ready

You need an up-and-running Elasticsearch installation, as we described in the *Downloading and installing Elasticsearch* recipe in *Chapter 1, Getting Started*.

You also need a working installation of Apache Spark and the data that we indexed in the previous example.

How to do it...

To read data in Elasticsearch via Apache Spark, we will perform the following steps:

1. In the Spark root directory, start the Spark shell to apply the Elasticsearch configuration by running the following command:

```
./bin/spark-shell \
    --conf spark.es.index.auto.create=true \
    --conf spark.es.net.http.auth.user=$ES_USER \
    --conf spark.es.net.http.auth.pass=$ES_PASSWORD
```

ES_USER and ES_PASSWORD are environment variables that hold your credential to the Elasticsearch cluster.

2. Import the required classes, as follows:

```
import org.elasticsearch.spark._
```

3. Now, we can create an RDD by reading data from Elasticsearch, as follows:

```
val rdd=sc.esRDD("spark2")
```

4. We can watch the fetched values using the following command:

```
rdd.collect.foreach(println)
```

5. The result will be as follows:

```
(bob,Map(username -> bob, name -> Bob, age -> 19))
(susan,Map(username -> susan, name -> Susan, age -> 21))
```

How it works...

The Elastic team has done a good job in allowing the use of a simple API to read data from Elasticsearch.

You only need to import the implicit that extends the standard RDD with the esRDD method to allow data retrieval from Elasticsearch.

The esRDD method accepts the following parameters:

- resource: This is generally an index or tuple type.
- query: This is a query that is used to filter the results. It's in the args query format (an optional string).

- `config`: This contains extra configurations to be provided to Elasticsearch (an optional `Map[String,String]`).

The returned value is a collection of tuples in the form of the `ID` and `Map` objects.

Reading data using Spark SQL

Spark SQL is a Spark module for structured data processing. It provides a programming abstraction called DataFrames and can also act as a distributed SQL query engine. Elasticsearch Spark integration allows us to read data using SQL queries.

Spark SQL works with structured data; in other words, all entries are expected to have the same structure (the same number of fields, of the same type and name). Using unstructured data (documents with different structures) is not supported and will cause problems.

Getting ready

You need an up-and-running Elasticsearch installation, as we described in the *Downloading and installing Elasticsearch* recipe in *Chapter 1, Getting Started*.

You also need a working installation of Apache Spark and the data that we indexed in the *Indexing data using Apache Spark* recipe of this chapter.

How to do it...

To read data in Elasticsearch using Apache Spark SQL and DataFrames, we will perform the following steps:

1. Start the Spark shell as done in previous examples.

2. We will create a DataFrame in the `org.elasticsearch.spark.sql` format and load data from `spark3`, as follows:

```
val df = spark.read.format("org.elasticsearch.spark.
sql").load("spark2")
```

3. If we want to check the schema, we can inspect it using `printSchema`, as follows:

```
df.printSchema
root
 |-- age: long (nullable = true)
 |-- name: string (nullable = true)
 |-- username: string (nullable = true)
```

4. We can watch fetched values as follows:

```
df.filter(df("age").gt(20)).collect.foreach(println)
```

To read data in Elasticsearch using Apache Spark SQL through SQL queries, we will perform the following steps:

1. Start the Spark shell as done in previous examples.

2. We will create a view for reading data from `spark2`, as follows:

```
spark.sql("CREATE TEMPORARY VIEW persons USING org.
elasticsearch.spark.sql OPTIONS (resource 'spark2',
scroll_size '2000')" )
```

3. We can now execute a SQL query against the previously created view, as follows:

```
val over20 = spark.sql("SELECT * FROM persons WHERE age
>= 20")
```

4. We can watch fetched values as follows:

```
over20.collect.foreach(println)
```

The result will be the following:

```
[21,Susan,susan]
```

How it works...

The core of data management in Spark is the **DataFrame**, which allows you to fetch values from different data stores.

You can use SQL query capabilities at the top of DataFrames, and depending on the driver used (`org.elasticsearch.spark.sql`, in our case), the query can be pushed down at the driver level (a native query in Elasticsearch). For example, in our preceding example, the query is converted into a Boolean filter with a range that is executed natively by Elasticsearch.

The Elasticsearch Spark driver can do inference, reading information from the mappings, and manage the data store as a standard SQL data store. The SQL approach is very powerful and allows you to reuse very common SQL expertise.

A good approach to using Elasticsearch with Spark is to use the Spark notebooks – interactive web-based interfaces that speed up the testing phases of application prototypes. The most famous ones are Spark Notebook, available at `http://spark-notebook.io`, Jupyter at `https://jupyter.org/`, and Apache Zeppelin, available at `https://zeppelin.apache.org`.

Indexing data with Apache Pig

Apache Pig (`https://pig.apache.org/`) is a tool that's frequently used to store and manipulate data in data stores. It can be very handy if you need to import some **Comma-Separated Values** (**CSV**) in Elasticsearch very quickly.

Getting ready

You need an up-and-running Elasticsearch installation, as we described in the *Downloading and installing Elasticsearch* recipe in *Chapter 1*, *Getting Started*.

You need a working Pig installation. Depending on your operating system, you should follow the instructions at `http://pig.apache.org/docs/r0.17.0/start.html`.

If you are using macOS X with Homebrew, you can install it with `brew install pig`; in Linux/Windows, you can install it with the following commands:

```
wget -c https://downloads.apache.org/pig/pig-0.17.0/
pig-0.17.0.tar.gz
tar xfvz pig-0.17.0.tar.gz
```

How to do it...

We want to read a CSV file and write the data in Elasticsearch. We will perform the following steps to do so:

1. We will download a CSV dataset from the GeoNames site to get all the GeoName locations in Great Britain. We can fast download them and unzip them as follows:

    ```
    wget http://download.geonames.org/export/dump/GB.zip
    unzip GB.zip
    ```

2. The Commons `httpclient` library (`http://hc.apache.org/httpclient-legacy/`) is required to execute the `pig` command. You can download it via the following command:

```
wget https://repo1.maven.org/maven2/commons-httpclient/
commons-httpclient/3.1/commons-httpclient-3.1.jar
```

3. We can write `es.pig`, which contains the `pig` commands to be executed, as follows:

```
REGISTER elasticsearch-hadoop-8.1.0/dist/elasticsearch-
hadoop-pig-8.1.0.jar;
REGISTER commons-httpclient-3.1.jar;
SET pig.noSplitCombination TRUE;
DEFINE EsStorage org.elasticsearch.hadoop.pig.EsStorage(
'es.index.auto.create=true',
    'es.nodes.wan.only=true',
    'es.net.http.auth.user=elastic',
    'es.net.http.auth.pass=XXXX');
-- launch the Map/Reduce job with 5 reducers
SET default_parallel 5;
--load the GB.txt file
geonames= LOAD 'GB.txt' using PigStorage('\t') AS
(geonameid:int,name:chararray,asciiname:chararray,
alternatenames:chararray,latitude:double,longitude:double,
feature_class:chararray,feature_code:chararray,
country_code:chararray,cc2:chararray,admin1_
code:chararray,
admin2_code:chararray,admin3_code:chararray,
admin4_code:chararray,population:int,elevation:int,
dem:chararray,timezone:chararray,modification_
date:chararray);
STORE geonames INTO 'geoname' USING EsStorage();
```

4. Now, execute the `pig` command, as follows:

```
pig -x local es.pig
```

The output will be similar to the following:

```
… truncated …
2021-05-02 14:13:04,581 [LocalJobRunner Map Task Executor
#0] INFO  org.elasticsearch.hadoop.mr.EsOutputFormat -
Writing to [geoname]
2021-05-02 14:13:10,503 [communication thread] INFO  org.
apache.hadoop.mapred.LocalJobRunner - map > map
2021-05-02 14:13:10,599 [main] INFO  org.apache.
pig.backend.hadoop.executionengine.mapReduceLayer.
MapReduceLauncher - 15% complete
2021-05-02 14:13:10,599 [main] INFO  org.apache.
pig.backend.hadoop.executionengine.mapReduceLayer.
MapReduceLauncher - Running jobs are [job_
local1576646194_0001]
2021-05-02 14:13:13,508 [communication thread] INFO  org.
apache.hadoop.mapred.LocalJobRunner - map > map
… truncated …
Input(s):
Successfully read 64556 records from: "file:///Users/
alberto/Projects/elasticsearch-8.x-cookbook/ch17/GB.txt"
Output(s):
Successfully stored 64556 records in: "geoname"
… truncated …
```

After a few seconds, all the CSV data is indexed in Elasticsearch.

How it works...

Apache Pig is a very handy tool. With a small number of code lines, it's able to read, transform, and store data in different data stores. It has a shell, but it's very common to write a Pig script with all the commands to be executed.

To use Elasticsearch in Apache Pig, you need to register the library that contains `EsStorage`. This is done using the `REGISTER` script; the `jar` position depends on your installation, as follows:

```
REGISTER elasticsearch-hadoop-8.1.0/dist/elasticsearch-hadoop-
pig-8.1.0.jar;
```

By default, Pig splits the data into blocks and then combines them before sending the data to Elasticsearch. To maintain maximum parallelism, you need to disable this behavior using SET pig.noSplitCombination TRUE.

To prevent typing the full path for EsStorage, we must define the following shortcut:

```
DEFINE EsStorage org.elasticsearch.hadoop.pig.EsStorage(
'es.index.auto.create=true',
    'es.nodes.wan.only=true',
    'es.net.http.auth.user=elastic',
    'es.net.http.auth.pass=XXXX');
```

By default, the Pig parallelism is set to 1. If we want to speed up the process, we need to increase this value, as follows:

```
-- launch the Map/Reduce job with 5 reducers
SET default_parallel 5;
```

Reading a CSV in Pig is very simple – we define a file, PigStorage, with the field separator and the format of the fields, as follows:

```
--load the GB.txt file
geonames= LOAD 'GB.txt' using PigStorage('\t') AS
(geonameid:int,name:chararray,asciiname:chararray,
alternatenames:chararray,latitude:double,longitude:double,
feature_class:chararray,feature_code:chararray,
country_code:chararray,cc2:chararray,admin1_code:chararray,
admin2_code:chararray,admin3_code:chararray,
admin4_code:chararray,population:int,elevation:int,
dem:chararray,timezone:chararray,modification_date:chararray);
```

After reading the CSV file, the lines are indexed as objects in Elasticsearch, as follows:

```
STORE geonames INTO 'geoname' USING EsStorage();
```

As you can see, all the complexity in using Pig comes from managing the format of input and output. The key advantage of Apache Pig is the ability to load different datasets, join them, and store them in a few lines of code.

Using Elasticsearch with Alpakka

The Alpakka project (`https://doc.akka.io/docs/alpakka/current/index.html`) is a reactive enterprise integration library for Java and Scala, based on Reactive Streams and Akka (`https://akka.io/`).

Reactive Streams is based on components – the most important ones are **Source** (which is used to read data from different sources) and **Sink** (which is used to write data in storage).

Alpakka supports Source and Sink for many data stores, Elasticsearch being one of them.

In this recipe, we will go through a common scenario – reading a CSV file and ingesting it in Elasticsearch.

Getting ready

You need an up-and-running Elasticsearch installation, as we described in the *Downloading and installing Elasticsearch* recipe in *Chapter 1*, *Getting Started*.

An IDE that supports Scala programming, such as IntelliJ IDEA with the Scala plugin, should be installed globally.

The code for this recipe can be found in the `ch17/alpakka` directory, and the class that is referred to is `CSVToES`.

How to do it...

We are going to create a simple pipeline that reads a CSV and write the record in Elasticsearch. To do so, we will perform the following steps:

1. Add the `alpakka` dependencies to `build.sbt`:

    ```
    "com.github.pathikrit" %% "better-files" % "3.9.1",
    "com.lightbend.akka" %% "akka-stream-alpakka-csv" %
    "3.0.4",
    "com.lightbend.akka" %% "akka-stream-alpakka-
    elasticsearch" % "3.0.4",
    ```

2. Then, initialize the Akka system:

    ```
    implicit val actorSystem = ActorSystem()
    implicit val actorMaterializer = ActorMaterializer()
    implicit val executor = actorSystem.dispatcher
    ```

3. Now, we define `ElasticsearchConnectionSettings`, which provides
    settings to connect to our Elasticsearch server:

```
val connectionSettings =
    ElasticsearchConnectionSettings("http://
localhost:9200")
    .withCredentials(
        sys.env.getOrElse("ES_USER", "elastic"),
        sys.env.getOrElse("ES_PASSWORD", "password")
    )
```

4. The data will be stored in the `Iris` class, and we will also create a Scala `implicit`
    parameter to allow encoding and decoding in JSON. All of this is done using a few
    lines of code:

```
final case class Iris(label: String, f1: Double, f2:
Double, f3: Double, f4: Double)
import spray.json._
import DefaultJsonProtocol._
implicit val format: JsonFormat[Iris] = jsonFormat5(Iris)
```

5. Before initializing `sink`, we define some policies to manage back pressure and
    retry logic:

```
val sinkSettings =
  ElasticsearchWriteSettings()
    .withBufferSize(1000)
    .withVersionType("internal")
    .withRetryLogic(RetryAtFixedRate(maxRetries = 5,
retryInterval = 1.second))
```

6. Now, we can create a pipeline, and the source will be read by a CSV, for every line.
    We will create an `Iris` index message and ingest `iris-alpakka` in an index
    using the Elasticsearch Sink:

```
val graph = Source.single(ByteString(Resource.
getAsString("com/packtpub/iris.csv")))
    .via(CsvParsing.lineScanner())
    .drop(1)
    .map(values => WriteMessage.createIndexMessage[Iris](
        Iris(
```

```
            values(4).utf8String,
            values.head.utf8String.toDouble,
            values(1).utf8String.toDouble, values(2).
  utf8String.toDouble,
            values(3).utf8String.toDouble)))
      .runWith(
        ElasticsearchSink.create[Iris](
          ElasticsearchParams.V7("iris-alpakka"),
          settings = sinkSettings
        )
      )
```

7. The preceding pipeline is executed asynchronously; we need to wait for it to finish processing the items, and then we need to close all the used resources:

```
val finish = Await.result(graph, Duration.Inf)
actorSystem.terminate()
Await.result(actorSystem.whenTerminated, Duration.Inf)
```

How it works...

Alpakka is one of the most used tools to build modern **Extract, Transform, and Load** (**ETL**) workflows with a lot of convenient features:

- **Back pressure**: If a data store is under high load, it automatically reduces the throughput.

- **Modular approach**: A user can replace Source and Sink to read and write to other data stores without massive code refactoring.

- **Numerous operators**: A plethora of operators (map, `flatMap`, `filter`, `groupBy`, `mapAsync`, and so on) can be used to build complex pipelines.

- **Low-memory footprint**: It doesn't load all the data in memory like Apache Spark, but it streams data from Source to Sink. It can be easily Dockerized and deployed on a Kubernetes cluster for large scaling of your ETL.

The Elasticsearch Source and Sink use an Akka-optimized streaming HTTP/S client to interact with Elasticsearch. As you can see from the following code, we need to define the parameters used for the connections:

```
val connectionSettings =
  ElasticsearchConnectionSettings("http://localhost:9200")
    .withCredentials(
      sys.env.getOrElse("ES_USER", "elastic"),
      sys.env.getOrElse("ES_PASSWORD", "password")
    )
```

The client variable must be implicit so that it is automatically passed to the Elasticsearch Source and Sink or flow constructors.

By default, it supports serializing and deserializing case classes in JSON using spray.json, using the jsonFormatN Scala macro (with *N* being the number of fields):

```
implicit val format: JsonFormat[Iris] = jsonFormat5(Iris)
```

Also, in this case, the variable should be implicit so that it can be automatically passed to the required methods.

We can customize the parameters used to write in Elasticsearch using ElasticsearchWriteSettings:

```
ElasticsearchWriteSettings()
  .withBufferSize(1000)
  .withVersionType("internal")
  .withRetryLogic(RetryAtFixedRate(maxRetries = 5,
retryInterval = 1.second))
```

The most common methods of this class are the following:

- withBufferSize(size:Int): The number of items to be used for a single bulk.

- withVersionType(vType:String): This sets the type of record versioning in Elasticsearch.

- `withRetryLogic(logic:RetryLogic)`: This sets the retry policies. `RetryLogic` is a Scala trait class that can be extended to provide different implementations. You can implement your `RetryLogic` policy by extending the trait class; the following ones are already available by default:

- `RetryNever`: To never retry.

- `RetryAtFixedRate(maxRetries: Int, retryInterval: scala.concurrent.duration.FiniteDuration)`: This allows `maxRetries` at a fixed interval (`retryInterval`).

An Elasticsearch Sink or flow accepts only `WriteMessage[T,PT]`, where *T* is the type of the message and *PT* is a possible `PassThrough` type (used, for example, if you want to pass a Kafka offset and commit it after the Elasticsearch `write` response).

`WriteMessage` has helpers to create the most commonly used messages, which are as follows:

- `createIndexMessage[T](source: T)`: This is used to create an `index` action.

- `createIndexMessage[T](id: String, source: T)`: This is used to create an `index` action with the provided ID.

- `createCreateMessage[T](id: String, source: T)`: This is used to build a `create` action.

- `createUpdateMessage[T](id: String, source: T)`: This is used to create an `update` action.

- `createUpsertMessage[T](id: String, source: T)`: This is used to create an `upsert` action (it tries to update the document; if the document doesn't exist, it creates a new one).

- `createDeleteMessage[T](id: String)`: This is used to create a `delete` action.

To create these messages, the most common practice is to use a `map` function to do the transformation from a value to the required `WriteMessage` type.

After having created `WriteMessage`, we can create a `sink` function that will write the records in Elasticsearch. The required parameters for this `sink` function are as follows:

- `indexName:String`: The index to be used.

- `typeName:String`: The mapping name (usually `_doc` in Elasticsearch 7.x). This will probably be removed in future releases of Alpakka Elasticsearch.

- settings: ElasticsearchWriteSettings (optional): The setting parameters for write, which we have discussed previously.

In this short introduction, we have only scratched the surface of Akka and Alpakka, but it is easy to understand how powerful this system is for orchestrating simple and complex ingestion jobs.

See also

You can refer to the following URLs for further information related to this recipe:

- The official website of Akka at https://akka.io/ and Alpakka at https://doc.akka.io/docs/alpakka/current/index.html
- Alpakka documentation about Elasticsearch at https://doc.akka.io/docs/alpakka/current/elasticsearch.html, which provides more examples of using Source and Sink

Using Elasticsearch with MongoDB

MongoDB (https://www.mongodb.com/) is one of the most popular documented data stores, due to its simple installation and the large community that is using it.

It's very common to use Elasticsearch as a search or query layer and MongoDB as a more secure data stage in many architectures. In this recipe, we'll see how simple it is to write in MongoDB, reading from an Elasticsearch query stream using Alpakka.

Getting ready

You need an up-and-running Elasticsearch installation, as we described in the *Downloading and installing Elasticsearch* recipe in *Chapter 1, Getting Started*.

An IDE that supports Scala programming, such as IntelliJ IDEA with the Scala plugin, should be installed globally.

A local installation of MongoDB is required to run the example. You can install it faster with Docker via the following command:

```
docker run -d -p 27017:27017 --name example-mongo mongo:latest
```

The code for this recipe can be found in the ch17/alpakka directory, and the referred class is ESToMongoDB. We will read the index created in the previous recipe.

How to do it...

We are going to create a simple pipeline that reads the previous CSV data in Elasticsearch and writes the records in MongoDB. To do so, we will perform the following steps:

1. We need to add the `alpakka-mongodb` dependencies to `build.sbt`:

```
"org.mongodb.scala" %% "mongo-scala-bson" % "4.5.0",
"com.lightbend.akka" %% "akka-stream-alpakka-mongodb" %
"3.0.4",
```

2. The first step is to initialize the Akka system:

```
implicit val actorSystem = ActorSystem()
implicit val actorMaterializer = ActorMaterializer()
implicit val executor = actorSystem.dispatcher
```

3. Now, we define `ElasticsearchConnectionSettings`, which provides settings to connect to our Elasticsearch server:

```
val connectionSettings =
  ElasticsearchConnectionSettings("http://
localhost:9200")
    .withCredentials(
      sys.env.getOrElse("ES_USER", "elastic"),
      sys.env.getOrElse("ES_PASSWORD", "password")
  )
```

4. The data will be stored in an `Iris` class; we will also create a Scala `implicit` to allow encoding and decoding in JSON and a codec for MongoDB. All of this is done using a few lines of code:

```
import spray.json._
import DefaultJsonProtocol._
implicit val format: JsonFormat[Iris] = jsonFormat5(Iris)
val codecRegistry =
  fromRegistries(fromProviders(classOf[Iris]), DEFAULT_
CODEC_REGISTRY)
```

5. Now, we can create `irisCollection`, which will store our data in MongoDB:

```
private val mongo = MongoClients.create("mongodb://
localhost:27017")
private val db = mongo.getDatabase("es-to-mongo")
val irisCollection = db
  .getCollection("iris", classOf[Iris])
  .withCodecRegistry(codecRegistry)
```

6. Finally, we can create a pipeline; the source will be `ElasticsearchSource`, and all the records will be ingested in MongoDB using its `MongoSink`:

```
val graph =
  ElasticsearchSource
    .typed[Iris](
      ElasticsearchParams.V7("iris-alpakka"),
      query = """{"match_all": {}}""",
      settings =
ElasticsearchSourceSettings(connectionSettings).
withBufferSize(1000)
    )
    .map(_.source) // we want only the source
    .grouped(100) // bulk insert of 100
    .runWith(MongoSink.insertMany[Iris](irisCollection))
```

7. The preceding pipeline is executed asynchronously, and we need to wait for it to finish processing the items and then close all the used resources:

```
val finish = Await.result(graph, Duration.Inf)
mongo.close()
actorSystem.terminate()
Await.result(actorSystem.whenTerminated, Duration.Inf)
```

How it works...

Using MongoDB Source and Sink in a pipeline is very easy. In the preceding code, we have used `ElasticsearchSource`, which, given an index via `ElasticsearchParams`, a query **Domain Specific Language** (**DSL**), and a source configuration (`ElasticsearchSourceSettings`), is able to generate a typed stream of items:

```
ElasticsearchSource
  .typed[Iris] (
    ElasticsearchParams.V7("iris-alpakka"),
    query = """{"match_all": {}}""",
    settings = ElasticsearchSourceSettings(connectionSettings).
withBufferSize(1000)
  )
```

`ElasticsearchParams` has two helpers:

- V5, used for Elasticsearch version 5.x/6.x, which requires us to define an index and a type

- V7, used for Elasticsearch version 7.x or above, which requires us to define only the index name

`ElasticsearchSourceSettings` requires `ElasticsearchConnectionSettings` to be initialized, and the most used method is `withBufferSize`, which allows us to change the default fetch size (10 records) to a greater value to speed up fetching records.

The returned type is `Source[ReadResult[T], NotUsed]`, where *T* is our type (which is `Iris` in the example).

`ReadResult[T]` is a wrapper object that contains the following:

- `id:String`: This is the Elasticsearch ID.

- `source:T`: This is the source part of the document, converted to the T object.

- `version:Option[Long]`: This is an optional version number.

To write in MongoDB, we need to create a connection, select a database, and get a collection:

```
private val mongo = MongoClients.create("mongodb://
localhost:27017")
private val db = mongo.getDatabase("es-to-mongo")
val irisCollection = db
  .getCollection("iris", classOf[Iris])
  .withCodecRegistry(codecRegistry)
```

In this case, we have defined that irisCollection is of the Iris type, and we have provided a codec to do data marshaling (conversion).

codecRegistry is built using a Scala macro:

```
import org.bson.codecs.configuration.CodecRegistries.{
  fromProviders,
  fromRegistries
}
import org.mongodb.scala.MongoClient.DEFAULT_CODEC_REGISTRY
import org.mongodb.scala.bson.codecs.Macros._
…truncated …
val codecRegistry =
  fromRegistries(fromProviders(classOf[Iris]), DEFAULT_CODEC_
REGISTRY)
```

To speed up the writing, we have chosen to execute a bulk write in MongoDB of 100 elements, so we first convert the stream into a group of 100 elements using the following:

```
.grouped(100) // bulk insert of 100
```

The results are written in the collection using MongoSink:

```
.runWith(MongoSink.insertMany[Iris](irisCollection))
```

In the case of streaming, I suggest always writing in bulk, in both Elasticsearch and MongoDB.

In the preceding examples, we have seen how to use Elasticsearch Source and Sink and MongoDB Sink, and it's very easy to understand how you can combine different Sources or Sinks to build your own pipelines.

Because Alpakka provides a native connection with a lot of data stores (`https://doc.akka.io/docs/alpakka/current/index.html`), creating a fast, efficient, and type-safe ETL pipeline is very fast and enjoyable using this kind of technology.

See also

Check out the Alpakka documentation about MongoDB at `https://doc.akka.io/docs/alpakka/current/mongodb.html`, which provides more examples of using Sources and Sinks for MongoDB, and the entry point of all Alpakka documentation with a lot of connectors at `https://doc.akka.io/docs/alpakka/current/index.html`.

18
X-Pack

X-Pack is a set of modules/plugins that expands the standard capabilities of Elasticsearch, such as the following:

- Security
- Alerting
- Reporting
- Index management
- Monitoring
- Machine learning
- SQL support
- A common extension on mapping, analyzer, search, and ingester processors

It is a commercial component that is managed by the Elastic company, and it's installed by default in Elasticsearch. All the features can be used in trial mode for 30 days, but many of them are free of charge for life in the free tier. The complete map of capabilities offered by different **X-Pack tiers** is available at `https://www.elastic.co/subscriptions`.

In this chapter, we will see the most commonly used feature of X-Pack related to everyday usage. For furthermore insights, I suggest checking the official documentation on the Elastic website (`https://www.elastic.co`). The topics that we will cover in this chapter are as follows:

- ILM – managing the index life cycle
- ILM – automating rollover and data streams
- Using the SQL Rest API
- Using SQL via JDBC
- Using X-Pack security
- Using alerting to monitor data events

ILM – managing the index life cycle

Index Lifecycle Management (**ILM**) in Elasticsearch is a system that allows you to set up policies to manage different aspects of your index life, such as the following:

- Manage daily, weekly, and monthly indices with backup, optimization, and movement in the different serving tiers (hot and cold)
- Create and switch to new indices when some indices reach a certain size or number of documents
- Delete obsolete indices at the end of their life
- Delete stale indices to enforce retention standards (that is, **General Data Protection Regulation** (**GDPR**) and 3-month retention)

ILM policies can be defined via an API or Kibana (go to **Stack Management | Index Lifecycle Policies**):

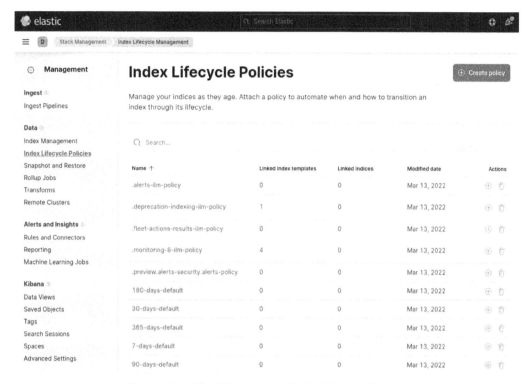

Figure 18.1 – The Kibana view of Index Lifecycle Policies

Getting ready

You need an up-and-running Elasticsearch installation, as we described in the *Downloading and installing Elasticsearch* recipe in *Chapter 1, Getting Started.*

To execute these commands, any HTTP client can be used, such as `curl` (`https://curl.haxx.se/`), Postman (`https://www.getpostman.com/`), or similar. I suggest that you use the Kibana console, as it provides code completion and better character escaping for Elasticsearch.

How to do it...

We want to create a policy that will be able to delete all the indices with data that is older than 90 days (a standard 3-month retention policy).

We will execute the following steps:

1. We will create the policies with the following command:

```
PUT _ilm/policy/90d_policy
{"policy": {
    "phases": {
        "delete": { "min_age": "90d",
            "actions": { "delete": {} } } } } }
```

2. If everything is okay, we will have the following result:

```
{ "acknowledged" : true }
```

3. Now, we will extend the index template used in *Chapter 2, Using Index Components and Templates*, adding the policy previously created:

```
PUT _index_template/order
{ "index_patterns": ["order-*"],
  "template": {
    "settings": {
        "number_of_shards": 1,
        "index.lifecycle.name": "90d_policy",
        "index.lifecycle.rollover_alias": "order" },
      "mappings": {
        "properties": { "id": { "type": "keyword"}}}},
    "priority": 200,
    "composed_of": ["timestamp-management", "order-data",
"items-data"], "version": 1,
    "_meta": {"description": "My order index template"}}
```

4. If everything is okay, we will have the following result:

```
{ "acknowledged" : true }
```

How it works...

ILM is one of the most useful capabilities of X-Pack to manage the life of your indices. It works by defining **policies** that are applied to your indices.

The best place to put these policies is in an index template so that they will be automatically provided to newly created indices.

The URL used to create a policy is the following:

```
PUT _ilm/policy/<policy_name>
```

A **policy** is a nested json compose of one or more of these optional phases:

- **Hot**: The index is actively being updated and queried. it is the most common live index.

- **Warm**: The index is no longer being updated but is still being queried. This index is generally a previous week/month one that doesn't receive any new data.

- **Cold**: The index is no longer being updated and is queried infrequently. It's searchable, but the queries are slower. The index data is often saved on rotative disks.

- **Frozen**: The index is no longer being updated and is queried rarely. In this case, the queries are extremely slow, and they are generally offline indices stored on S3 or other remote data stores.

- **Delete**: The index data is no longer needed, or it has reached the maximum retention time and can be removed.

The migration between phases is defined by the minimum age for every phase, which defaults to 0.

> **Note**
>
> To prevent issues with phase migration and ILM, your cluster must be *green*. In the case of a yellow state, the phase change could be unpredictable.

During phase execution, the state of the index is saved in index metadata. ILM is executed periodically by the Elasticsearch cluster, and the polling time is defined by the `indices.lifecycle.poll_interval` settings (a default of 10 minutes).

The supported actions are the following ones in the order of execution steps:

- **Set Priority**: This sets the priority in the executing steps.
- **Unfollow**: This sets a cross-cluster index in a standard one.
- **Rollover**: This sets up a rollover index.
- **Read-Only**: This sets the index in read-only mode. It follows a rollover command.
- **Shrink**: This reduces the number of shards of the current index.
- **Force Merge**: This optimizes the number of segments in the indices.
- **Allocate**: This changes the node that can host the index.
- **Migrate**: This changes the tier of the index (and can reduce the number of shards on request).
- **Freeze**: This freezes an index (removing it from memory).
- **Searchable Snapshot**: This mounts a snapshot as a searchable index.
- **Wait for Snapshot**: This waits for a complete snapshot of the index to ensure that there is a backup copy of the index before deletion.
- **Delete**: This removes the index.

Not all of the actions are available for the different phases; the following table summarizes the actions for the phases:

Action	Hot	Warm	Cold	Frozen	Delete
Allocate		X	X		
Delete					X
Force Merge	X	X			
Freeze			X		
Migrate		X	X		
Read-only	X	X			
Rollover	X	X			
Searchable snapshot			X	X	
Set Priority	X	X	X		
Shrink	X				
Unfollow	X	X	X		
Wait for snapshot					X

Figure 18.2 – Phase action mapping

There's more...

ILM can be easily integrated into existing indices via a simple update of their settings. To add a policy to an existing index, you must provide only the policy name:

```
PUT myindex/_settings
{ "index": {
    "lifecycle": { "name": "50gb_30d_delete_90d_policy" }}}
```

You can use wildcards in an index name to apply the changes to multiple indices.

See also

You can refer to the following URLs for further reference, which are related to this recipe:

- The official documentation of ILM at https://www.elastic.co/guide/en/elasticsearch/reference/current/index-lifecycle-management.html

- A video tutorial about this feature at https://www.youtube.com/watch?v=p-AhtrfbEjE and some more samples in this blog at https://blog.nviso.eu/2019/06/17/optimizing-elasticsearch-part-2-index-lifecycle-management/

ILM – automating rollover

ILM is often used in conjunction with rolling-over indices. These indices usually contain append-only data such as logs.

The standard management of the rollover indices is as follows:

- Creation of the index, generally managed by an index template.

- Rolling over the index when it reaches a particular size (the default is 50 GB) or a predefined number of documents.

- If required, move the indices to different tiers.

- Delete the index to be aligned with retention policies.

The rollover functionality is generally used with managed indices called **data streams**. They are **append-only** data indices that automatically roll over a new index when it meets some criteria of size and number of documents.

These indices are used mainly for the following:

- Logs
- Events
- Metrics
- Continuously generated data

Getting ready

You need an up-and-running Elasticsearch installation, as we described in the *Downloading and installing Elasticsearch* recipe in *Chapter 1, Getting Started.*

To execute these commands, any HTTP client can be used, such as `curl` (https://curl.haxx.se/), Postman (https://www.getpostman.com/), or similar. I suggest you use the Kibana console, as it provides code completion and better character escaping for Elasticsearch.

How to do it...

We want to create a policy that will be able to do the following:

- Automatically roll over an index when it reaches 50 GB or 30 days of data
- Delete all the indices with data that is older than 90 days (a standard 3-month retention policy)

We will execute the following steps:

1. We will create the policies with the following command:

```
PUT _ilm/policy/50gb_30d_delete_90d_policy
{ "policy": {
    "phases": {
        "hot": {
            "actions": {
                "rollover": {
                    "max_size": "50GB", "max_age": "30d" }}},
        "delete": {
            "min_age": "90d",
            "actions": { "delete": {} } } } } }
```

2. If everything is okay, we will have the following result:

```
{ "acknowledged" : true }
```

3. Now, we will create a data stream, adding the policy previously created:

```
PUT _index_template/timeseries_template
{ "index_patterns": ["timeseries"],
  "data_stream": { },
  "template": {
    "settings": {
      "number_of_shards": 1, "number_of_replicas": 1,
      "index.lifecycle.name": "50gb_30d_delete_90d_
policy" } } }
```

4. If everything is okay, we will have the following result:

```
{ "acknowledged" : true }
```

5. We can add a record to the index to force its creation:

```
POST timeseries/_doc
{ "message": "this is a message",
  "@timestamp": "2022-03-16T15:03:03" }
```

The result will be as follows:

```
{ "_index" : ".ds-timeseries-2022.03.15-000001",
  "_id" : "osN7cHkBtG0xN2Ak2au2", … truncated … }
```

6. Now, we can check whether the policy is applied to the
    `.ds-timeseries-2022.03.15-000001` index via the following API call:

```
GET .ds-timeseries-*/_ilm/explain
```

7. The result will be something similar:

```
{ "indices" : {
    ".ds-timeseries-2022.03.15-000001" : {
      "index" : ".ds-timeseries-2022.03.15-000001",
      "managed" : true,
      "policy" : "50gb_30d_delete_90d_policy",
      "lifecycle_date_millis" : 1621089728349,
      "age" : "3.3m", "phase" : "hot",
```

```
        "phase_time_millis" : 1621089728462,
        "action" : "rollover",
        "action_time_millis" : 1621089728533,
        "step" : "check-rollover-ready",
        "step_time_millis" : 1621089728533,
        "phase_execution" : {
          "policy" : "50gb_30d_delete_90d_policy",
          "phase_definition" : {
            "min_age" : "0ms",
            "actions" : {
              "rollover" : {
                "max_size":"50gb","max_age" : "30d"}}},
          "version" : 2,
          "modified_date_in_millis" : 1621089704907 }}}}
```

How it works...

If you are working with append-only data such as logs and events, ILM with rolling policies will be one of the easiest ways to manage this kind of data.

In the previous example, we used the following rolling action:

```
"rollover": { "max_size": "50GB", "max_age": "30d" }
```

It uses two optional parameters (at least one of them should be provided):

- `max_size`, which defines the maximum size of the index
- `max_age`, which defines the maximum age of the index

I will automatically create a new index when one of the preceding constraints is matched.

There's more...

ILM can be easily integrated into existing indices via a simple update of their settings.

To add a policy to an existing index, you must provide only the policy name:

```
PUT myindex/_settings
{ "index": {
    "lifecycle": { "name": "50gb_30d_delete_90d_policy" }}}
```

You can use wildcards in the index name to apply the changes to multiple indices.

As we see in the previous recipe, it's important to define the ILM policy in the index template to automatically apply it to every newly created index.

It is done via the definition of `index.lifecycle.name` in the `template/settings` section:

```
"template": {
  "settings": {
    "number_of_shards": 1, "number_of_replicas": 1,
    "index.lifecycle.name": "50gb_30d_delete_90d_policy" }}
```

We can check the ILM applied rule for every index via the REST API:

```
GET <index_name>/_ilm/explain
```

The result contains a lot of information, but the most important fields are as follows:

- `managed` (Boolean): If the index is ILM-managed.
- `policy`: The name of the applied policy. It must be checked mainly for debug purposes.
- `age`: The time of the index.
- `phase`: The actual phase of the index.
- `action`: The last applied action.
- `step`: The step of the action. Some actions can have one or more steps.
- `phase_execution`: The description of the actual phase execution.

Using this Elasticsearch capability provided by X-Pack is very easy, and it also simplifies all the index management for streaming data.

See also

You can refer to the following URLs for further reference, which are related to this recipe:

- The official documentation of ILM for a rolling index at `https://www.`
  `elastic.co/guide/en/elasticsearch/reference/current/`
  `getting-started-index-lifecycle-management.html#ilm-gs-`
  `apply-policy`

- The official documentation about a rollover index at `https://www.elastic.`
  `co/guide/en/elasticsearch/reference/master/indices-`
  `rollover-index.html`

Using the SQL Rest API

X-Pack allows you to bring the power of SQL, the standard language for query data, to Elasticsearch to simplify both the usage of data users and the integration of external applications.

Getting ready

You need an up-and-running Elasticsearch installation, as we described in the *Downloading and installing Elasticsearch* recipe in *Chapter 1*, *Getting Started*.

To execute these commands, any HTTP client can be used, such as `curl` (`https://` `curl.haxx.se/`), Postman (`https://www.getpostman.com/`), or similar. I suggest you use the Kibana console, as it provides code completion and better character escaping for Elasticsearch. We will use the datasets that were populated in *Chapter 4*, *Exploring Search Capabilities* and *Chapter 7*, *Aggregations*.

How to do it...

We will execute the following steps:

1. We want to return the first five books ordered by quantity; we will use the following API call:

```
GET _sql?format=txt
{ "query": "SELECT * FROM mybooks ORDER BY quantity DESC
LIMIT 5" }
```

2. If everything is okay, we will have the following result:

```
           date           |          description
|   position   |     price     |   quantity    |
title     |     uuid
-----------------------+-------------------------------
----+--------------+---------------+-------------+----
-----------+---------------
2015-10-22T00:00:00.000Z|Joe Testere nice guy
|1             |4.3            |50                      |Joe
Tester    |11111
2016-06-12T00:00:00.000Z|Bill Testere nice guy
|2             |5.0            |34                      |Bill
Baloney   |22222
2017-09-21T00:00:00.000Z|Bill is not
                        nice guy|3              |6.0
|33                      |Bill Klingon   |33333
```

3. We want to return records paginated by 5 elements ordered by descending price in json format; we will execute the following query:

```
GET _sql?format=json
{ "query": """SELECT in_stock, date, name, age, price
FROM "index-agg" ORDER BY price DESC""",
  "fetch_size": 5 }
```

4. If everything is okay, we will have the following result:

```
{ "columns" : [
    {"name" : "in_stock", "type" : "boolean"},
    {"name" : "date", "type" : "datetime"},
    {"name" : "name", "type" : "text"},
    {"name" : "age", "type" : "long"},
    {"name" : "price", "type" : "float"}
  ],
  "rows" : [
    [false, "2013-01-26T16:46:02.828Z", "Bova", 48,
99.983025],
    [true, "2013-09-11T16:46:01.837Z", "Achilles", 87,
99.91804],
    [false, "2015-08-05T16:46:02.666Z", "Silver Samurai",
```

```
44, 99.86804],
    [false, "2012-10-10T16:46:02.964Z", "Dweller-in-
Darkness", 94, 99.78011],
    [true, "2018-12-08T16:46:02.849Z", "Sunfire", 6,
99.62338 ]
 ],
 "cursor" :
"x8qyAwFaAXN4RkdsdVkyeDFaR1ZmWTI5dWRHVjRkRjkxZFdsa0RYRj
FaWEo1UVc1a1JtVjBZMmdCRmw4MlRuTTRRMloyVVZSaGMzWXlNR1ZRZV
VSdFNIY0FBQUFBQUFBRmhSWTBVMnh3Tmppkb1ZGUmZiVXBpUkcxZmIwcG5
iRTlS/////
w8FAWYIaW5fc3RvY2sBB2Jvb2xlYW4AAAFmBGRhdGUBCGRhdGV0aW1l
AAABZgRuYW1lAQR0ZXh0AAABZgNhZ2UBBGxvbmcAAAFmBXByaWNlAQ
VmbG9hdAAAAAR8=" }
```

How it works...

X-Pack provides support for using SQL _sql endpoints, which allow us to convert standard SQL queries into Elasticsearch queries and run them on the required indices.

The API REST has the following format:

```
POST _sql
```

To customize the returned response, it's possible to pass the format parameter as query arguments. It accepts these possible values – csv, json, tsv, txt, yaml, cbor, and smile.

Inside the POST body, the JSON objects can contain the following parameters:

- query: This is the mandatory field that contains the SQL to be executed.
- params: An optional array of values to be passed to replace the "value placeholders" (?) inside the query.
- fetch_size: The maximum of rows to return in a single response (a default of 1000).
- filter: An optional Query DSL to be used to filter out values.
- runtime_mappings: A dictionary of keys and scripts used to compute additional fields at runtime.

fetch_size allows us to use pagination, returning a cursor until all the records are fetched.

To access the following chunk of results, the _sql endpoint must be called using the returned cursor value, similar to the scroll API used in *Chapter 4, Exploring Search Capabilities*:

```
POST /_sql?format=json
{ "cursor": "x8qyAwFaAXN4RkdsdVkyeDFaR1ZmWTI5dWRHVjRkRjkxZFdsa0
RYRjFaWEo1UVc1a1JtVjBBZMmdCRmw4M1RuTTRRM1oyVVZSaGMzWX1NR1ZRZVVSd
FNIY0FBQUFBQUFBBRmhSWTBBVMnh3Tmppbk1ZGUmZiVXBpUkcxZmIwcG5iRTlS///
//w8FAWYIaW5fc3RvY2sBB2Jvb2xlYW4AAAFmBGRhdGUBCGRhdGV0aW1lAAABZg
RuYW11AQR0ZXh0AAAABZgNhZ2UBBGxvbmcAAAFmBXByaWNlAQVmbG9hdAAAAR8="
}
```

The cursor keeps in memory locked values, so it must be cleared at the end of the fetching or to finish the scrolling via a _sql/close call, similar to the following:

```
POST /_sql/close
{ "cursor": "x8qyAwFaAXN4Rkds…. }
```

There's more...

In a more complex scenario, the Elasticsearch SQL layer allows you to compute complex return values that can be expressed via painless scripting.

For example, if we want to return the day of the week and the price with a 20% discount for every item, we can define them in runtime_mappings in the following way:

```
GET _sql?format=csv
{ "runtime_mappings": {
    "day_of_week": { "type": "keyword",
      "script": """
        emit(doc['date'].value.dayOfWeekEnum.toString())
      """ },
    "final_price": { "type": "double",
      "script": """
        emit((doc['price'].value * 0.8).toString())
      """ } },
  "query": """SELECT in_stock, date, name, age, price FROM
"index-agg" ORDER BY price DESC""",
  "fetch_size": 5
}
```

> **Note**
>
> When you are defining a script in `runtime_mappings` for returning the value of a painless script, the value must be passed to the `emit` method.

Another important feature of the SQL layer is the ability to translate the SQL SELECT statement to a standard Elasticsearch query via the `_sql/translate` endpoint.

We can translate the preceding query via the following code:

```
GET _sql/translate
{ "query": """SELECT in_stock, date, name, age, price FROM
"index-agg" ORDER BY price DESC""" }
```

The result is a valid Elasticsearch query:

```
{ "size" : 1000, "_source" : false,
  "fields" : [
    {"field" : "in_stock"},
    {"field" : "date", "format" : "strict_date_optional_time_
nanos"},
    {"field":"name"},{"field":"age"},{"field" : "price"}],
  "sort" : [
    { "price" : { "order" : "desc", "missing" : "_first",
        "unmapped_type" : "float" } } ] }
```

See also

You can refer to the following URLs for further reference, which are related to this recipe:

- The official documentation of SQL Access at `https://www.elastic.co/guide/en/elasticsearch/reference/master/xpack-sql.html`

- SQL limitations at `https://www.elastic.co/guide/en/elasticsearch/reference/master/sql-limitations.html`

Using SQL via JDBC

The SQL access layer for Elasticsearch is able to provide data access via the two most used **Database Management System (DBMS)** protocols, **Java Database Connectivity (JDBC)** and **Open Database Connectivity (ODBC)**. These drivers can be installed in the client application to easily integrate Elasticsearch with the most popular tools, such as the following:

- DBeaver
- Microsoft Excel
- Microsoft Power BI
- Qlik Sense Desktop
- Tableau

Getting ready

You need an up-and-running Elasticsearch installation, as we described in the *Downloading and installing Elasticsearch* recipe in *Chapter 1, Getting Started*.

We will use the datasets that were populated in *Chapter 4, Exploring Search Capabilities* and *Chapter 7, Aggregations*. In this recipe, we will configure **DBeaver** (https://dbeaver.io/) with Elasticsearch; you need to download (https://dbeaver.io/download/) and install it.

A trial license must be active to use a SQL JDBC connection to Elasticsearch; it can be easily set up in Kibana under **Stack Management | License Management**. This can also be done by executing the following API call:

```
POST /_license/start_trial?acknowledge=true&pretty
```

How to do it...

We will query our data via DBeaver using the following steps:

1. We need to configure a new database connection. Select **Elasticsearch** under the full-text section of DBMS:

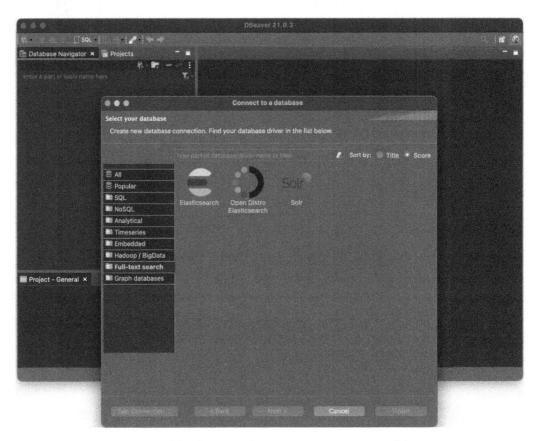

Figure 18.3 – The Elasticsearch connection type options in DBeaver

2. In the setup dialog, we will configure the **Host** address, **Port**, and credentials
    if needed:

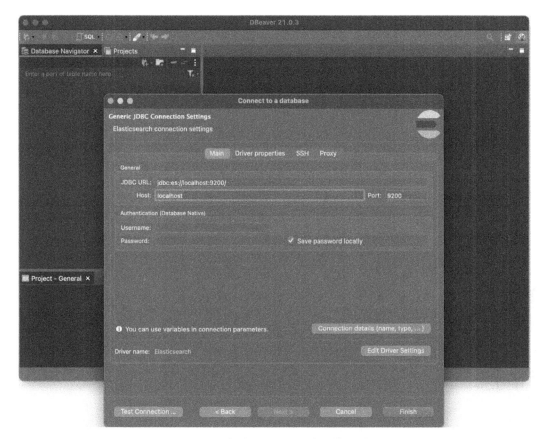

Figure 18.4 – Setting the host, port, and credentials in DBeaver

3. If you try to test the connection and some drivers are required, they can be easily downloaded by pressing the **Download** button:

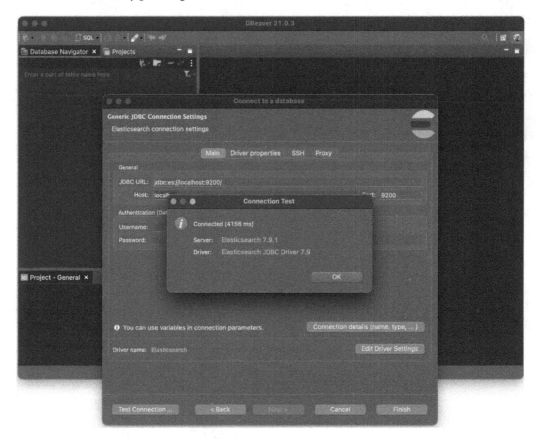

Figure 18.5 – A connection check in DBeaver

4. Now, you can navigate and query the data in Elasticsearch using SQL:

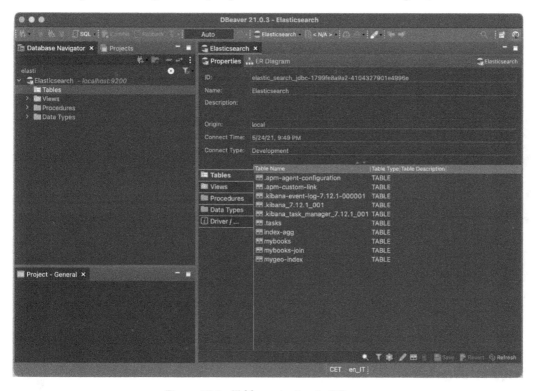

Figure 18.6 – Table navigation in DBeaver

5.   Selecting a table/index, you can see the data inside it:

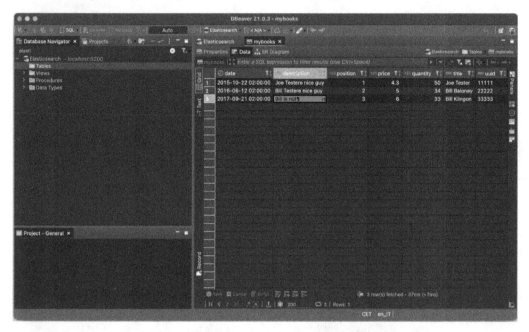

Figure 18.7 – Data exploration in DBeaver

How it works...

The SQL JDBC driver for Elasticsearch uses the standard Elasticsearch port 9200, and it wraps all connections used from different applications in a standard JDBC query and ResultSet, allowing seamless integration with Elasticsearch.

> **Note**
>
> Pay attention to the driver compatibility; the JDBC Elasticsearch driver version 7.x is not compatible with 8.x. You should use Elasticsearch drivers that match your installed Elasticsearch version.

After having created a connection in DBeaver, Elasticsearch can be used as the other most common DBMS (along with MySQL, PostgreSQL, Oracle, and Microsoft SQL Server), and you can simply use the standard SQL SELECT to execute queries. The results are returned as a standard SQL ResultSet, as you can see in the following screenshot:

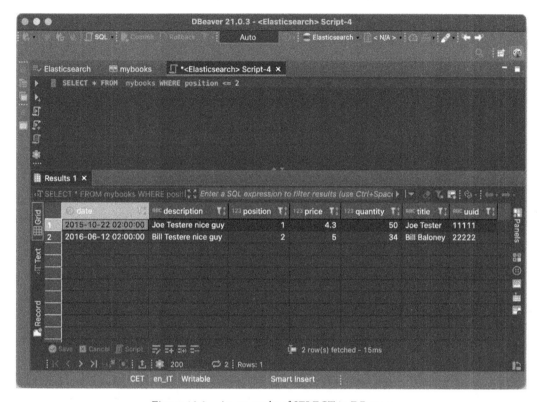

Figure 18.8 – An example of SELECT in DBeaver

See also

You can refer to the following URLs for further reference, which are related to this recipe:

- The official documentation of SQL access at https://www.elastic.co/guide/en/elasticsearch/reference/master/xpack-sql.html

- JDBC-related official documentation for more settings and examples of integration in Java applications at https://www.elastic.co/guide/en/elasticsearch/reference/master/sql-jdbc.html

- Official documentation about client integration via the SQL access layer that contains more examples of integration between Elasticsearch and other applications at `https://www.elastic.co/guide/en/elasticsearch/reference/master/sql-client-apps.html`

Using X-Pack Security

X-Pack Security allows users to secure their production cluster using the security best practices for Elasticsearch.

It covers the following:

- The use of credentials to access an Elasticsearch cluster. Credentials can be a standard user/password stored internally in Elasticsearch or a more complex solution can be used, such as Active Directory and **Lightweight Directory Access Protocol (LDAP)**.

- The use of a **Transport Layer Security (TLS)**-encrypted connection between the Elasticsearch nodes to prevent traffic sniffing and decoding using custom generated keys.

- The use of SSL endpoints for HTTP calls to secure communication with the cluster.

Getting ready

You need the Docker desktop (`https://www.docker.com/`) installed on your computer. To be able to execute the Docker commands, you should set up Docker for a production environment; see the Elasticsearch reference: `https://www.elastic.co/guide/en/elasticsearch/reference/current/docker.html#docker-prod-prerequisites`. All the resources for executing the previous commands are available in the `ch18/elasticsearch-docker-security` repository folder.

How to do it...

Our target is to build a cluster with three nodes and a Kibana that use SSL for inter-node communications and HTTP connections.

To be able to set up Elasticsearch security, we need to execute the following steps:

1. We need to create a `.env` file that contains settings to be passed to our `docker-compose.yml` file; we will use the following `.env` file:

```
# Password for the 'elastic' user (at least 6 characters)
ELASTIC_PASSWORD=myESPassword
```

```
# Password for the 'kibana_system' user (at least 6
characters)
KIBANA_PASSWORD=myKibiPassword
# Version of Elastic products
STACK_VERSION=8.1.0
# Set the cluster name
CLUSTER_NAME=docker-cluster
# Set to 'basic' or 'trial' to automatically start the
30-day trial
LICENSE=trial
# Port to expose Elasticsearch HTTP API to the host
ES_PORT=9200
# Port to expose Kibana to the host
KIBANA_PORT=5601
# Increase or decrease based on the available host memory
(in bytes)
MEM_LIMIT=1073741824
COMPOSE_PROJECT_NAME=es-book
```

2. We will set up a mini-cluster with three instances of Elasticsearch and Kibana using docker-compose.yml. The following is the main part to set up certificates for the first Elasticsearch server:

```
version: "2.2"
services:
  setup:
    image: docker.elastic.co/elasticsearch/
elasticsearch:${STACK_VERSION}
    volumes:
      - certs:/usr/share/elasticsearch/config/certs
    user: "0"
    command: >
      bash -c '
        if [ x${ELASTIC_PASSWORD} == x ]; then
          echo "Set the ELASTIC_PASSWORD environment
variable in the .env file";
          exit 1;
        elif [ x${KIBANA_PASSWORD} == x ]; then
```

```
            echo "Set the KIBANA_PASSWORD environment
variable in the .env file";
            exit 1;
        fi;
        if [ ! -f certs/ca.zip ]; then
            echo "Creating CA";
            bin/elasticsearch-certutil ca --silent --pem
-out config/certs/ca.zip;
            unzip config/certs/ca.zip -d config/certs;
        fi;
… truncated code that setup certs and set password…
        echo "All done!"; '
    healthcheck:
      test: ["CMD-SHELL", "[ -f config/certs/es01/es01.
crt ]"]
      interval: 1s
      timeout: 5s
      retries: 120
  es01:
    depends_on:
      setup:
        condition: service_healthy
    image: docker.elastic.co/elasticsearch/
elasticsearch:${STACK_VERSION}
    volumes:
      - certs:/usr/share/elasticsearch/config/certs
      - esdata01:/usr/share/elasticsearch/data
    ports:
      - ${ES_PORT}:9200
    environment:
      - node.name=es01
      - cluster.name=${CLUSTER_NAME}
      - cluster.initial_master_nodes=es01,es02,es03
      - discovery.seed_hosts=es02,es03
      - ELASTIC_PASSWORD=${ELASTIC_PASSWORD}
      - bootstrap.memory_lock=true
      - xpack.security.enabled=true
```

```
        - xpack.security.http.ssl.enabled=true
        - xpack.security.http.ssl.key=certs/es01/es01.key
        - xpack.security.http.ssl.certificate=certs/es01/
es01.crt
        - xpack.security.http.ssl.certificate_
authorities=certs/ca/ca.crt
        - xpack.security.http.ssl.verification_
mode=certificate
        - xpack.security.transport.ssl.enabled=true
        - xpack.security.transport.ssl.key=certs/es01/es01.
key
        - xpack.security.transport.ssl.certificate=certs/
es01/es01.crt
        - xpack.security.transport.ssl.certificate_
authorities=certs/ca/ca.crt
        - xpack.security.transport.ssl.verification_
mode=certificate
        - xpack.license.self_generated.type=${LICENSE}
    mem_limit: ${MEM_LIMIT}
    ulimits:
      memlock:
        soft: -1
        hard: -1
    healthcheck:
      test:
        [
          "CMD-SHELL",
          "curl -s --cacert config/certs/ca/ca.crt
https://localhost:9200 | grep -q 'missing authentication
credentials'",
        ]
      interval: 10s
      timeout: 10s
      retries: 120
```

We will start all the instances via a Docker `compose` command:

```
docker compose -f docker-compose.yml up -d
```

or

```
docker-compose up -d
```

3. We can access Kibana at `https://0.0.0.0:5601` (because the SSL certificate is self-signed, you need to accept the security warning provided by the browser). If everything is meticulously done, you should see the following Kibana login page, which you can access with one of the previously generated credentials:

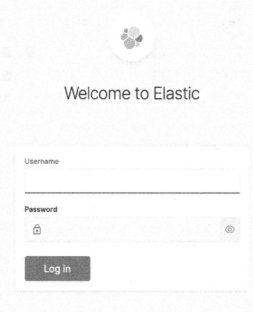

Figure 18.9 – The Kibana secure login page

> **Note**
>
> To be able to use the advanced features in Kibana, I added the following entry to the default environment variables of it to enable the use of alerting and reporting:

```
- XPACK_ENCRYPTEDSAVEDOBJECTS_ENCRYPTIONKEY=<random bunch of
numbers/letters>
```

How it works...

Elasticsearch allows us to have a different layer of security depending on customer needs. The following diagram summarizes the different security scenarios (each one built on the feature of the previous one) that can be implemented:

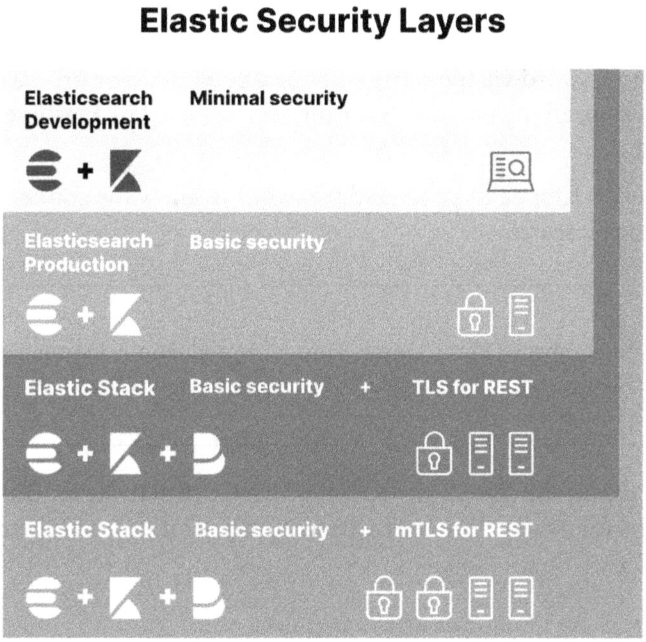

Figure 18.10 – Security scenarios

The scenarios are as follows:

- Elasticsearch development, in which you generally set up user/password access with built-in users

- Elasticsearch production, which extends development by adding a TLS to prevent data sniffing between nodes

- **Elasticsearch, Logstash, and Kibana (ELK)** production with TLS for REST (SSL) endpoints to secure encrypted communication between client applications and Elasticsearch

> **Note**
> Using self-signed certificates can have issues with third-party application integration; for this reason, in the case of TLS for REST, the best practice is to have them trusted by a **Certificate Authority (CA)**.

In this recipe, to be able to apply Elasticsearch top-level security, we have executed the following steps:

1. We created the certificate used for both internal TLS communication between the nodes and for REST communication.

2. We started the cluster to set up an initial secured working cluster between Elasticsearch nodes.

3. We set up the password via the `bin/elasticsearch-setup-passwords` script. In this step, all the passwords for the services used by Elasticsearch are generated.

4. We propagated to Kibana its generated password to allow the connection between Kibana and Elasticsearch.

> **Note**
>
> The minimum number of Elasticsearch nodes in a production environment is three. For this reason, in the actual Docker setup, we have fired three Elasticsearch instances.

To use more of the Elasticsearch security feature, we have to activate an Elasticsearch trial license that lasts for 30 days. We have automatically set it up on Docker via the environment variable:

```
- xpack.license.self_generated.type=trial
```

The Elasticsearch security layer has a large number of options to be able to cover a large number of production scenarios, due to the different systems, tools, and technologies used in today's architecture, such as Active Directory, LDAP, **Identity Provider (IdP)**, and **Single Sign On (SSO)**.

The most commonly used settings are the following:

- `xpack.security.enabled` (default – `false`), which is used to activate the security layer

- `xpack.security.http.ssl.enabled` (default – `false`), which is used to activate the SSL layer for REST communication between clients and Elasticsearch nodes on port 9200.

- `xpack.security.http.ssl.*`, which groups all the SSL options such as certificates and key management

- `xpack.security.transport.ssl.enabled` (default – `false`), which is used to activate TLS communication between Elasticsearch nodes on port `9300`.

- `xpack.security. transport.ssl.*`, which groups all the TLS options such as certificates and key management for node communication

At the Kibana level, we have used different environment variables to set up the security layer, such as the following:

- `ELASTICSEARCH_URL` and `ELASTICSEARCH_HOSTS`, which are pointing to HTTPS Elasticsearch endpoints

- `ELASTICSEARCH_USERNAME` (best practice – `kibana_system`) and `ELASTICSEARCH_PASSWORD`

- `ELASTICSEARCH_SSL_CERTIFICATEAUTHORITIES`, which points to the CA used by Elasticsearch nodes

- `SERVER_SSL_ENABLED`, which enables SSL for Kibana

- `SERVER_SSL_KEY` and `SERVER_SSL_CERTIFICATE`, which set up the SSL certificate in Kibana

- `XPACK_ENCRYPTEDSAVEDOBJECTS_ENCRYPTIONKEY` (at least a 32-character string), which is used to encrypt all the data in Kibana to prevent spoofing of values. It is not set up; it will be regenerated at every Kibana restart.

See also

Because security is a very wide matter that covers a lot of aspects, we suggest that you further investigate the following links:

- The official documentation about the secure Elasticsearch stack at `https://www.elastic.co/guide/en/elasticsearch/reference/master/secure-cluster.html`

- The startup differences between the development and production clusters at `https://www.elastic.co/guide/en/elasticsearch/reference/master/bootstrap-checks.html#dev-vs-prod-mode`

- A blog post regarding the different security concepts and idioms at `https://opensource.com/article/19/11/internet-security-tls-ssl-certificate-authority`

Using alerting to monitor data events

Alerting is one of the most used X-Pack components because it allows us to fire *alert* events on data that is processed in the cluster.

The main concepts behind Elasticsearch alerting are as follows:

- **Conditions**: These define what needs to be detected.
- **Schedule**: These define the frequency of how the checks run.
- **Actions**: These define how to respond to an alert.

Elasticsearch is able to cover the following:

- Infrastructural alerting such as issues about load on the server, disk space, and node being down
- ETL flow alerting such as the reduction of ingested records in some indices
- Business alerting with rules defined by a business user on data quality or features on their data
- Predictive alerting using the **Machine Learning** (**ML**) X-Pack component, which is able to detect an anomaly in ingested data

Getting ready

Alerting only works on a full setup environment with security enabled; we will use the one that we have prepared for the previous recipe.

How to do it...

An alert can be mainly configured using the Kibana interface, using the following steps:

1. We need to set up a connector that is called when our alert is fired. We can access them from the Kibana interface under **Stack Management | Alerts and Insights | Rules and Connectors |** the **Connectors** tab| **Create connector**, as shown in the following screenshot:

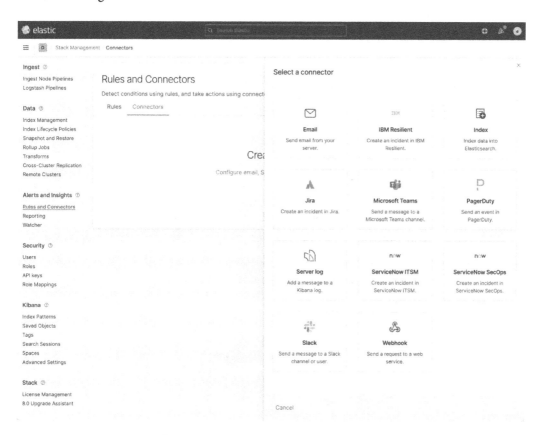

Figure 18.11 – Connector creation in Kibana

2. We will set up an index connector that will write the alerts in an index called `my-alert-index`:

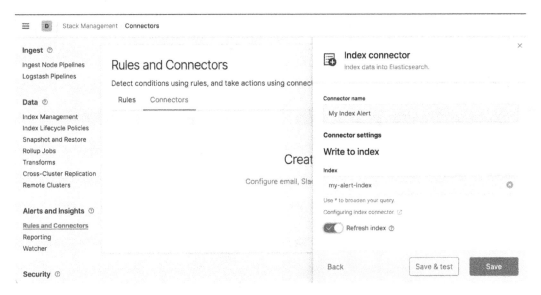

Figure 18.12 – Setting up the index connector

To better manage the index, we will create an index with the following mappings:

```
PUT my-alert-index
{ "settings": { "number_of_shards": 1 },
  "mappings": {
    "properties": {
      "rule_id": { "type": "text" },
      "rule_name": { "type": "text" },
      "alert_id": { "type": "text" },
      "context_message": { "type": "text" } } } }
```

3.   Now, we can create a rule defining its name, frequency, query, and alert thresholds:

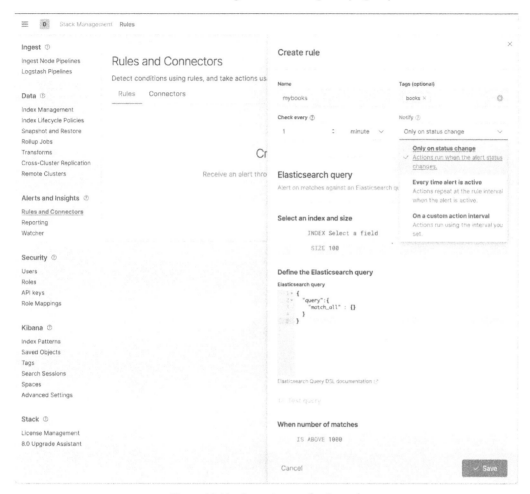

Figure 18.13 – Inserting an alerting rule

After setting the rule parameters, we need to set up an action. We will select a connector type, **Index**; then, we will choose the previous setup connection, **My Index Alert**, and put the JSON to be indexed:

Figure 18.14 – Setting up an alert action

4. To test the rule, we can execute the `alert_fire.sh` script present in `ch18/elasticsearch-docker-security`, which will ingest data to fire our alert.

5. We can check that the rule is fired by executing a simple search on the following:

```
GET my-alert-index/_search
```

The result will be something similar to the following:

```
{ ... truncated ...
    "hits" : [
        { "_index" : "my-alert-index",
            "_id" : "_5J7vXkBdJi1a6eRO5KZ","_score" : 1.0,
            "_source" : {
                "rule_id" : "",
                "context_message" : """"alert 'mybooks' is
active:

- Value: 1000
- Conditions Met: Number of matching documents is greater
than 10 over 5m
- Timestamp: 2021-05-30T13:33:57.745Z""",
                "rule_name" : "",
                "alert_id" : "f8df5ba0-c0a0-11eb-9407-
25f0a6fc2466" } ] } }
```

How it works...

X-Pack allows us to set up alerts in a different part of the ELK stack, such as in MLanomaly detection, metrics monitoring, and other custom parts.

The core of an alerting system is to be able to integrate with different tools/applications. X-Pack alerting out of the box supports the following:

- **Email**: This sends an email to users or groups. This is the most commonly used way to notify exceptions or send warnings to IT users.

- **Index**: This stores alerts in Elasticsearch and then external systems connect to it via the Elasticsearch REST API or JDBC/ODBC integration.

- **Webhooks**: This sends the notification to another REST service. It allows a microservice approach to an alerting system.

- **Slack/Microsoft Teams**: These send a notification to group channels to activate their needs.

- **Jira/ServiceNow and others**: These allow tickets to be opened in ticketing systems.

The custom parts provide the ability to easily express user custom business needs. The most important parts of defining an alert are as follows:

- The rule, which expresses a condition that must be matched to fire an alert. It can be defined with a query (remember that a query can be also a painless script, so every type of business requirement can be expressed).

- The frequency, which is used internally in Kibana to execute the query on the indices to check whether the alert is fired. It automatically manages the repeating/non-overlapping of the alerting using the timestamp used to select the query.

- One or more actions that notify the user about the firing alert. The output of an action is a document/text result of a mustache template rendering, allowing simple customization of the output based on the alert.

Rules and actions can be disabled to be able to manage rules without deleting them (which is useful for maintenance).

See also

X-Pack alerting is an ELK component that is continuously developing new integrations; alert types and features are added in every release, so we suggest that you further investigate the following links:

- A blog about an alerting announcement at `https://www.elastic.co/blog/elastic-stack-alerting-now-generally-available`

- The official documentation about alerting at `https://www.elastic.co/guide/en/kibana/master/alerting-getting-started.html`

- Common troubleshooting of alert issues at `https://www.elastic.co/guide/en/kibana/master/alerting-troubleshooting.html`

Index

W

X

Other Books You May Enjoy

If you enjoyed this book, you may be interested in these other books by Packt:

Getting Started with Elastic Stack 8.0

Asjad Athick

ISBN: 9781800569492

- Configure Elasticsearch clusters with different node types for various architecture patterns

- Ingest different data sources into Elasticsearch using Logstash, Beats, and Elastic Agent

- Build use cases on Kibana including data visualizations, dashboards, machine learning jobs, and alerts

- Design powerful search experiences on top of your data using the Elastic Stack

- Secure your organization and learn how the Elastic SIEM and Endpoint Security capabilities can help

- Explore common architectural considerations for accommodating more complex requirements

Machine Learning with the Elastic Stack - Second Edition

Rich Collier, Camilla Montonen, Bahaaldine Azarmi

ISBN: 9781801070034

- Find out how to enable the ML commercial feature in the Elastic Stack
- Understand how Elastic machine learning is used to detect different types of anomalies and make predictions
- Apply effective anomaly detection to IT operations, security analytics, and other use cases
- Utilize the results of Elastic ML in custom views, dashboards, and proactive alerting
- Train and deploy supervised machine learning models for real-time inference
- Discover various tips and tricks to get the most out of Elastic machine learning

Packt is searching for authors like you

If you're interested in becoming an author for Packt, please visit `authors.packtpub.com` and apply today. We have worked with thousands of developers and tech professionals, just like you, to help them share their insight with the global tech community. You can make a general application, apply for a specific hot topic that we are recruiting an author for, or submit your own idea.

Share Your Thoughts

Now you've finished *Elasticsearch 8.x Cookbook*, we'd love to hear your thoughts! Scan the QR code below to go straight to the Amazon review page for this book and share your feedback or leave a review on the site that you purchased it from.

`https://packt.link/r/1-801-07981-1`

Your review is important to us and the tech community and will help us make sure we're delivering excellent quality content.